LIFE, SPACE, AND TIME

LIFE, SPACE, AND TIME

A Course in Environmental Biology

Barry Fell

*Museum of Comparative Zoology,
Harvard University*

Harper & Row, Publishers
NEW YORK, EVANSTON, SAN FRANCISCO, LONDON

Sponsoring Editor: Joe Ingram
Project Editor: David Nickol
Designer: Rita Naughton
Production Supervisor: Will C. Jomarrón

Library of Congress Cataloging in Publication Data
Fell, Howard Barraclough, 1917–
Life, space, and time.

1. Ecology. 2. Natural history. I. Title.
QH541.F38 574.5 74–495
ISBN 0-06-042033-2

Contents

Preface

A text dealing with the environments of the world requires about 1000 color photographs of representative scenes, plants, and animals to impart that dimension which words alone cannot give. But the cost of producing a work so illustrated would in effect deny it to readers to whom this book is addressed. Therefore, instead, the color illustrations are being made available by the publisher in sets of approximately 100 35-mm color slides for each biome, each set prepared by an authority on the region concerned and accompanied by technical information on the subjects illustrated. In this way it is hoped that teachers may be able to present the visual materials in whatever way best suits the needs of particular courses, while the student members of any course need not be deprived of the essential experience. At the time of this writing sets have been issued, or are in production, for the marine, freshwater, and Antarctic and Arctic environments; others are in preparation and will be published as opportunity offers.

A book covering so broad a field must inevitably draw largely from the writings or experiences of others. As to the biomes that I have personally studied, the sections on tundra and montane forests are based largely on travel in Lapland and on many mountain climbs in various countries. Boreal forests and faunas were studied in Sweden and Canada, summergreen forests in Europe and the United States. Tropical forests are known to me mainly from visits to examples in New Britain, Polynesia, and Panama. Desert environments were observed in Arabia and Egypt, chaparral biomes in Australia, New Zealand, Italy, and California; and the various austral forests of New Zealand were visited on numerous occasions. Lawrence Palkovic and David Barrington contributed information on the American biomes generally; Charlene Long collected material in South America and Florida; all three also assisted in teaching a course based on this book. John Dearborn of the University of Maine contributed much information on Arctic and Antarctic environments, and Julian Fell also made observations in southern Chile and west Antarctica. Many useful facts on a range of biomes were added by Paul S. Martin, who, in addition, gave generously of his time and knowledge in final revisions of the text itself. The chronological treatment of the biblical materials in Chapter 18 owes much to the advice of Norman Totten. The historical data in Chapter 8 derives in part from my own researches and decipherments of early Polynesian tablets and Kawi steles in Java; an extended treatment of these is planned for separate publication. I am also indebted to Joseph Germano, whose photographic skill has given renewed life to the many sketches and engravings by older naturalists, here reproduced as line blocks. Drawings by Desmond FitzGerald, Irene Fell, James Clark, and David Moynahan are acknowledged in the captions. Haris Lessios also photographed chaparral biomes in Greece, and Peter Garfall recorded the New England scene. These, like many other photographs used in preparing the text, were treated as sources of data, though not actually reproduced here. I am grateful to all these colleagues and friends.

A feature of the black-and-white illustrations is that they have been chosen for the most part to represent man's concept of the world from ancient times. The artists, many of whose names are now unknown, recorded the world of nature as they saw it in Stone Age cave paintings, in Bronze Age tomb inscriptions, on ancient coins and in rare books and manuscripts, as well as in the enlightened diaries of travel by Victorian naturalists. These illustrations, as also those of the Victorian anatomists, may be traced to their original sources by means of the list that follows, which is also a formal acknowledgment.

Alvin Abbott and his colleagues at Harper & Row made numerous practical suggestions during the preparation of the manuscript, and latterly Joe Ingram helped in bringing various threads together. I wish to thank these friends for their help.

Museum of Comparative Zoology

H. B. F.

SOURCES OF OTHER ILLUSTRATIONS

Illustrations other than those prepared especially for this book have been sought among the extensive archives of sketches made by naturalists since the close of the last ice age. The sources are indicated by serial numbers, placed in parentheses, in the descriptive captions under the figures, in accordance with the practice followed by the Geological Society of America. Following is the key to the numbers:

(1) *Ca.* 10,000 B.C. to 8,000 B.C. Anonymous Magdalenian artists of Europe; paintings and engravings on cave walls, stone, antlers and bone, collected by Piette, Breuil *et al.* in S. Reinach, *Repertoire de l'art quaternaire* (Paris, 1913).
(2) 2,700 B.C. Anonymous Egyptian artist, Tomb of Ti, fifth dynasty; (2a) 1,800 B.C. Anonymous artist of the twelfth dynasty.
(3) Seventh century B.C. Anonymous Assyrian artist of relief at Kuyundjik, near Nineveh, Babylonia.
(4) 530 B.C. (Attributed to) Pythagoras son of Mnesarchos, engraver of coin dies of Croton, South Italy, staters of Metapontum.
(5) 482 B.C. Anonymous Greek engraver, drachmae of Acragas, Italy.
(6) Fourth century B.C. (Attributed to) Theodotos of Klazomenae, near Ephesus.
(7) A.D. 512. Anonymous artist of the Codex Vindebonensis (a manuscript of Dioscorides the botanist), in the National Library of Vienna, Austria.
(8) 1864. Anonymous traveler. View of Harrison Lake and Cascade Mountains. *Illustrated London News* (45, 616).
(9) 1864. Anonymous artist. Högtorp, Sweden. *Illustrated London News* (45, 424); Petropolis, Brazil (45, 576).
(10) 1864. Anonymous artist. Lochnagar, Scotland. *Illustrated London News* (45, 257).
(11) 1470. Sandro Botticelli (detail).
(12) 1864. F. Boyle, scenes in Borneo. *Illustrated London News* (45, 473).
(13) 1864. British Association. Report on meeting at Bath. Ibid. (45, 305–8).
(14) 1903. Calman, Sars, Giesbrecht, Alcock, Cunningham; copies of drawings incorporated in *Treatise on Zoology, Crustacea,* ed. R. A. Lankester and C. Black (London).
(15) 1885. C. Claus and A. Sedgwick. *Text-book of Zoology,* Sonnenschein, London.
(16) 1898. J. M. Curran. *Geology of Sydney* (Angus and Robertson, Sydney).
(17) 1884. W. S. Dallas. Orthoptera, *Cassell's Natural History,* 6 (Cassell, London). Insectivora, ibid., 1; Rodentia, ibid., 3.
(18) 1884. W. B. Dawkins. Proboscidea, ibid., 2; Ungulata Perissodactyla, ibid., 2.

(19) 1888. J. W. Dawson. *Geological History of Plants* (Kegan, Paul, London).
(20) 1864. W. Duffield, (detail). *Illustrated London News* (45, 53).
(21) 1884. P. M. Duncan. Apes and Monkeys, *Cassell's Natural History*, 1 (Cassell, London). Edentata and Marsupialia, ibid., 3.
(22) 1884. P. M. Duncan and J. Murie. Lemurs, ibid., 1.
(23) H. B. Fell. Illustrations from publications on oceanography and echinoderms, 1950 onward, and other illustrations not previously published.
(24) 1903. W. Furneaux. *The Sea Shore* (Longmans, Green, London).
(25) 1904. W. Furneaux. *Life in Ponds and Streams* (Longmans, Green, London).
(26) 1884. A. H. Garrod. Ungulata Ruminantia, *Cassell's Natural History*, 3 (Cassell, London).
(27) 1864. A. Gilbert. *Illustrated London News* (45, 404).
(28) 1875. F. B. Goodrich. *History of the Sea* (Lyon, New York).
(29) 1910. S. F. Harmer et al. Guide to the Crustacea, etc., in the Department of Zoology in the British Museum, London.
(30) 1864. J. D. Herbert. Elephant Kraal in Ceylon. *Illustrated London News* (45, 169).
(31) 1854. J. D. Hooker. *Himalayan Journals* (Ward, Lock, London).
(32) 1864. H. Jutsum. The Woods in Autumn (details), *Illustrated London News* (44, 225).
(33) 1917–22. Kidston and Lang, *Transactions of the Royal Society of Edinburgh*, 51–52.
(34) 1884. W. F. Kirby. Lepidoptera, *Cassell's Natural History*, 6 (Cassell, London).
(35) 1885. W. F. Kirby. *Text-book of Entomology* (Swan, Sonnenschein, London).
(36) 1902. E. R. Lankester et al. Guide to the Mammalia of the British Museum, London.
(37) 1452–1541. Leonardo da Vinci (details). Notebooks.
(38) 1912. J. Murray and J. Hjort. *The Depths of the Ocean* (Macmillan, London).
(39) 1877. H. A. Nicholson. *Ancient Life-History of the Earth* (Blackwood, Edinburgh).
(40) 1884. W. K. Parker. The Land Carnivora, *Cassell's Natural History*, 2 (Cassell, London).
(41) 1896. Strasburger, Noll, Schenk, and Schimper. *Text-book of Botany* (Macmillan, London).
(42) 1869. A. R. Wallace. *The Malay Archipelago* (Macmillan, London).
(43) 1906. K. Zittel, in E. Haeckel. *The Evolution of Man* (Watts, London).

1

LIFE, SPACE, AND TIME

On at least four occasions since our earliest human ancestors appeared on earth great ice sheets have spread across the northern continents and as many times retreated again. Thus eight or more drastic reversals of climates and environmental conditions have taken place during man's sojourn on his planet, and each time the stress of change was felt by every living creature. In still earlier periods in the earth's history many other changes have affected its environments and their denizens.

THE INSTABILITY OF LIFE ZONES. The life zones of our planet and the plants and animals that inhabit them are therefore not stable entities that man alone disturbs. On the contrary, one lesson of prehistory is that communities of living organisms are subject to continuing change, varying not only in response to the natural processes of evolution by genetic means, but also in direct response to the endless northward and southward swing of the warm and cold belts of the earth and the changes in precipitation and in the character of the winds and ocean currents generated by the ever-fluctuating climates of the planet. If the environment of a particular region changes its character, the denizens of that environment may either become extinct, adapt to the change, or (as is more usually the case) simply migrate elsewhere to occupy an amenable territory; later, if the conditions reverse, the descendants of the original emigrant communities may return to territories formerly occupied. So temporal and spatial changes in dispersion of living organisms are normal features of the earth.

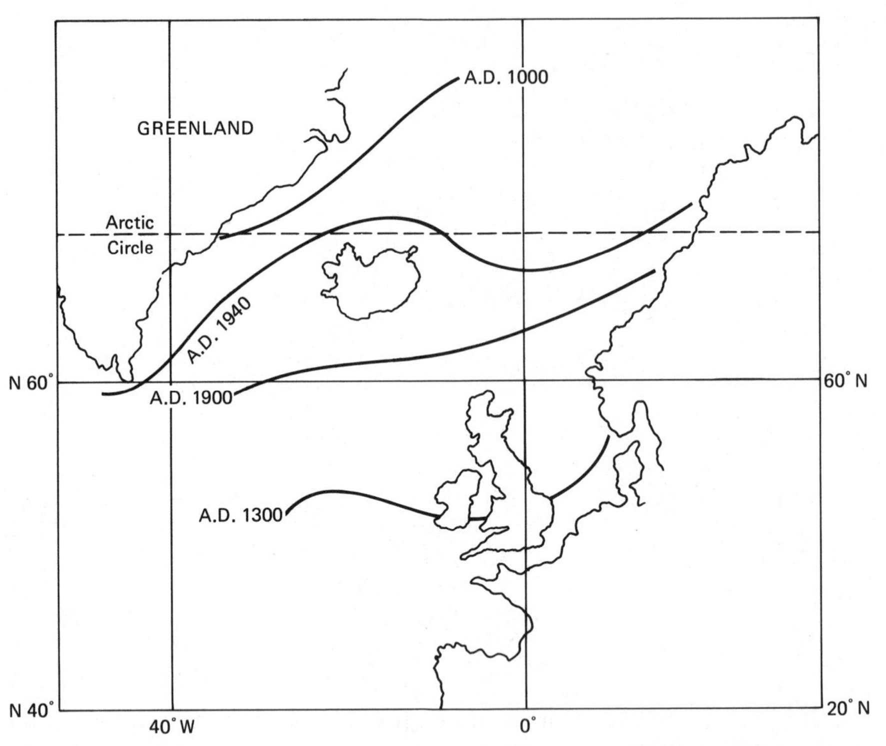

Figure 1. Changes in the summer range of herring shoals in the north Atlantic during the past 1000 years. In the viking period, and in the twentieth century, herring and codfish spawned off Greenland and Iceland, whereas during the cold medieval period the schools, and hence the fishery, were located in the North Sea. Climatic change is thought to account for the facts (23).

Many disciplines, including some of the humanities, contribute to man's understanding or contemplation of his environment. Stone Age men observed nature and recorded what they saw in cave paintings; and ever since their descendants have watched or kept records. Initially man, the savage, appeared coincidentally in his own records as just one of the living species depicted. Then, with the growth of civilization, his influence on the environment became disproportionately significant, culminating in our own age when, seemingly, no part of the earth or moon is now immune to his visitations.

This book is devoted to an account of the main environments of the earth today, with some background study of how the animals and plants came to occupy them. The inquiring naturalist may be led to ponder how man might better assume the role of protector rather than destroyer, that all may become the beneficiaries of wiser policies when once we can agree on what such policies should be. The events of time past color whatever picture we may conceive of the future and certainly explain much of what we can perceive of the present.

The convulsions of the ice ages effected far-reaching changes in the populations of plants and animals all over the world, and in the more northern and far southern lands the very landscape itself was greatly altered; forests were destroyed and later re-created, and the grasslands sometimes spread toward the poles, sometimes toward the equator. The vastness of these changes within the past 20,000 years alone teaches us that world environments are resilient entities, perhaps capable of repair even after the unprecedented destruction effected by modern man.

ECOLOGY IN ROMAN TIMES. History, too, has some encouraging lessons. For example, the reign of the emperor Trajan brought to the Western world an era of transient prosperity previously unknown. By A.D. 100 Roman engineers had mastered the technology of delivering copious fresh water to barren regions, and the whole coast of north Africa and the ravaged states of the Levant became flourishing farmlands, graced by noble cities. Now, after centuries of gross neglect, nearly all these lands have been overwhelmed by desert sands. Elsewhere in this book these events are traced in some detail, but here it is sufficient to note that human endeavor brought prosperity to Trajan's domains. Was it an accident of history that this occurred during his administration? Perhaps, but I do not think so. I suspect that the naturalist-minded Plinian family had a decided influence in directing the emperor's attention to ecology.

The few writings we now possess from the hand of Trajan disclose a magnanimous mind responsive to wise counsels and sensitive to the aspirations of his period. It so happens that among the most touching documents that have come down to us from ancient times is a sheaf of letters exchanged by the emperor and his personal friend Pliny the Younger. They were written over the period of Pliny's service as governor of Judea, and in one of them Trajan commends Pliny for his moderate handling of a dissident minority known as the Christians. A deliberate repressive policy, though in accordance with existing Roman law, would be, writes Trajan, "contrary to the spirit of our age." Herein lies a clue to the character of his administration. Although it is possible that his liberal views stemmed from his unorthodox background, for his boyhood in the highlands of his native Spain could scarcely have held even the remotest prospect that one day the Senate would offer him the imperial robe, it would seem more probable that he was influenced by his friends, among whom the members of the Plinian family were certainly the closest. The elder Pliny had been the leading naturalist of his day. He lost his life while observing the eruption of Vesuvius that entombed Pompeii, but his influence lived on in his writings, which, even today, tell us much of how the educated Roman viewed his environment. It seems scarcely possible that a mind as sensitive as Trajan's could have failed to absorb something of the environmental insights of his companions or to have failed to turn them to the advantage of his people.

It is given to few men to share Trajan's capacity for perceiving the spirit of an age, but when the history of our own times comes to be written, surely a prevailing theme will be mankind's awakening to the threat posed by the unbridled exploitation of natural environments, and by the unrestricted reproduction of his own species. As most thinking people are now aware, an urgent solution must be found to the problem that faces us. A solution to any problem demands a clear recognition of the nature of the problem, and in this case in turn presupposes an understanding of the original environment before man began to inflict such grievous injuries on it. To such matters we may now direct our attention.

The lively cave paintings executed by our Stone Age ancestors show that man has been interested in his environment, and in his fellow denizens, for thousands of years. But, for most of time, this interest has been focused on the immediate environment. For a brief period in classical times the planet at large opened up in broad vistas of strange tropical lands and vast unsailed seas, as Greek and Roman sailors made contact with Chinese explorers, and the magic attraction of the Golden Chersonese and the Silklands penetrated to the Mediterranean peoples. Then the misery of the Dark Ages supervened, nearly all learning being confined to a few monks and the occasional curious pilgrim. As the medieval era dawned books of travel were written or dictated by pilgrims to the Holy Land, by crusaders and their retainers, and by mountebanks such as had the gift of storytelling, and the image of benighted *straunge londes* took on the aspect of the fabulous. Cotton became the wool of a lamb tree of Tartary, dragons were reported from many quarters, and the peoples of the Far East were found to have mouths and eyes on the thorax, or the heads of intelligent dogs. The age of faith was, it would seem, an age of total credulity. Columbus and his successors replaced all of this by a matter-of-fact account of the real world and soon learned Jesuit priests were presenting the first scientific accounts of the American wilderness. So dawned the modern concept of the biosphere.

THE BIOSPHERE. The rocky portion of the earth's crust, including the emergent continents and the submerged seafloor, is called the *lithosphere.* Beneath the lithosphere lies a molten or semimolten material called *magma;* continents have an average thickness of about 30 km (about 20 miles) and they seem to float like rafts on the magma. Oceans cover 71 percent of the lithosphere; they are collectively termed the *hydrosphere.* The average depth of the oceans is 3.8 km (about 2½ miles); the ocean floor has a thickness of between 5 and 10 km, below which lies molten magma. Surrounding the lithosphere and hydrosphere is the *atmosphere.* Living

Figure 2. Such medieval books as the *Travels of Sir John Mandeville* depicted far distant lands as inhabited by dragons and monsters. After 1492 scientific accounts of the American plants and animals set a new pattern for man's view of the world at large (28).

organisms occupy the lowest part of the atmosphere, the hydrosphere, and the uppermost part of the lithosphere; this habitable shell of the planet, about 20 miles deep, is collectively termed the *biosphere.*

ZONATION OF THE BIOSPHERE. The kinds of animals and plants that inhabit particular regions of the world are not distributed uniformly. Instead we find that some species constitute the flora and fauna of tropical rain forests, others occur only in arid desert regions or in the temperate zones, still others occur only on high mountains and in the polar regions, and so on. These facts constitute the central body of data the study of which may be termed *environmental biology*. Paleolithic man developed special hunting techniques adapted to the environments he encountered, and when he occupied an *ecotone*—that is, an overlap region between two well-marked kinds of environment—he acquired techniques of hunting adapted to each. The Australian aboriginals were still in a Paleolithic phase of development when Cook and Banks described them in 1770, yet they were acquainted with simple bark canoes, in which they fished the offshore waters of New South Wales, and with spears and boomerangs, by means

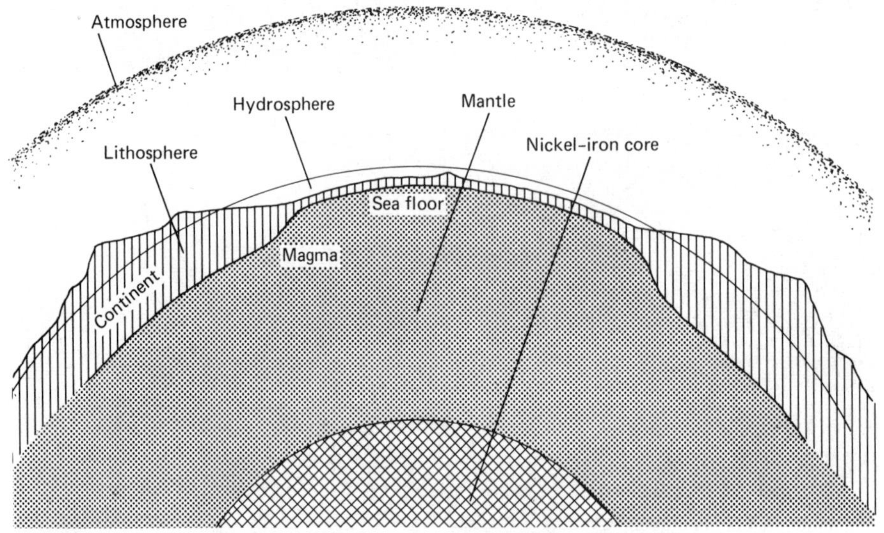

Figure 3. The biosphere and adjacent portions of the earth (23).

of which they hunted terrestrial birds and mammals. So evidently man's concept of environmental biology must be very ancient. In its modern guise we might characterize environmental biology as an interdisciplinary study, drawing materials from several branches of science, notably the following:

1. Planetary astronomy, in particular the branch of astronomy that deals with the evolution of the earth-moon system. It may seem surprising, but it is becoming clear that minute features of the growth patterns of living organisms are sensitive to and respond to changes in the distance of the moon from the earth, or variations in the rate at which the earth turns on its axis. More obvious is the inference that a change in the positions of the North and South poles would neces-

sarily affect the distribution patterns of plants and animals. Changes of this kind are relevant to our present study and will be discussed.

2. Historical geology, the study of the successive changes of the earth's lands and seas and of the succession of floras and faunas that have inhabited the earth. For we cannot hope to understand present patterns of environments unless we take into account those that previously existed.
3. Evolutionary biology, the study of the natural relationship between the various kinds of plants and animals, their classification in accordance with such relationships, their geographical distribution (biogeography), and their special interrelations with and adaptions to given environments (ecology), and so on.

EARLY ESTIMATES OF THE EARTH'S SIZE. When man had emerged from the prehistoric phases of development, and had learned to write inscriptions that have a meaning for us today, we find that he had already become a navigator and explorer. Homer's Bronze Age warriors were told by their bards that to the south lay Ethiopia with dark-skinned peoples. Herodotus some centuries nearer our own time could add numerous details about the animals and plants he saw in Egypt and also tell of peoples and animals living in a far, frigid north. Several centuries later again the geographers Ptolemy and Strabo wrote that the earth is a sphere, banded by zones of climates parallel to the line they called the equator, and with special kinds of living things inhabiting the zones. Strabo further predicted that it would be possible to sail west from Spain across the Atlantic to reach India.

These matters were well known to learned men in the medieval period; in England, Alfred (A.D. 849–899) could find time from the task of ruling the unruly to write a geography for the West Saxons, in the course of which he compares the earth to the yolk of an egg, suspended in space, which he compares to the white of the egg. Columbus had access to Ptolemy's geography, brought to Portugal by the learned Hebrews whom Henry the Navigator enticed to his court, with their precious Arabic translations of the Greek texts that once lay in the library of Alexandria. So the notion that the earth is a sphere banded by latitudinal zones is a very old one; it was faintly nurtured during the Dark Ages, revived early in the Middle Ages, and brought to a triumphant demonstration by the great navigations of the fifteenth and sixteenth centuries.

Although this is not the place to trace the growth of the idea in detail, I do think that one outstanding investigator deserves mention. He is Eratosthenes, the librarian at Alexandria in the third century B.C., who not only cared for his charges, but found time to read them, too. He read in one book that on one day of the year, near the summer solstice, a man drawing water at a well in Syene (Aswan) could see the midday sun shining back at him from the water at the bottom of the well. Realizing that this meant the sun must stand overhead at Syene at that time, Eratosthenes took the trouble of measuring the angular elevation of the sun at midday at Alexandria on the summer solstice; he found the sun stood one forty-eighth part of a circle (7.5°) from the zenith. The only logical explanation of the difference in altitude of the sun must be that the line joining Alexandria to Syene must in fact be a curved arc, subtending an angle of 7.5° at the center of the earth. To find the circumference of the earth, therefore, he had only to multiply the measured distance between the two cities by 48. He obtained the result 250,000 stadia. The stadium at Olympus measures 607 feet long, whence Eratosthenes' measure of the circumference of the earth becomes 28,000 miles. The difference from the true result (24,000 miles) is partly due to the fact that Syene does not lie due south of Alexandria, as Eratosthenes had mistakenly supposed, so he had omitted one correction. Ptolemy, 300 years later, studied the logs of Greek navigators and deduced that Eratosthenes had overestimated the size of the earth. His corrected figure, unfortunately, was worse than Eratosthenes' and made the earth too small. Columbus, who used Ptolemy's estimate, therefore expected to encounter land earlier than otherwise would have been the case. Had he used Eratosthenes' figure, his voyage would have cost him less sleep.

ZONATION OF THE EARTH'S SURFACE. Examples in condensed form are given in Table 1 of terrestrial east-west zones and corresponding zones of the oceans and lower atmosphere. When opportunity offers, glance at a globe of the earth with this table before you. Notice that the Southern Hemisphere is approximately, but not exactly, a mirror image of the Northern Hemisphere, the equator defining the axis of symmetry on a Mercator's map, or the plane of symmetry in reality, as on a globe.

Some of the entries in Table 1 may puzzle you. They are explained in later chapters but are inserted at this juncture because they illustrate the matching symmetry of the climatic zones of the earth's Northern and Southern hemispheres.

HOW OLD IS THE EARTH? Recognizing that ecosystems and their contained biota have changed in the course of time, the environmental biologist is naturally led to inquire how long a time span has occurred since the planetary ecosystem first began to evolve. Obviously no ecosystem can antedate the formation of the earth itself, so for an upper limit for the antiquity of life on earth most biologists accept the best available estimate of the age of the planet provided by sister sciences. Astrophysics presently incorporates a general theory of the evolution of stars from which it is inferred that the earth, and probably also the other planets of the solar system, arose as a consequence of a disruptive explosion of some large star in the transient phase called a *supernova,* as a result of which the internal heavy elements of the star were discharged in space, subsequently condensing as planets and being captured by our local star, the sun. Study of the radioactive decay of elements in the earth's crust, supposed to have originated in definite proportions in the interior of a

Table 1. Zonation of the surface of the earth.

LATITUDE	NATURE OF LAND SURFACE	CONDITION OF SEA SURFACE	ATMOSPHERIC FEATURES	MEAN ANNUAL SURFACE TEMPERATURE
90°–70° North	Arctic ice desert	Floating sea ice	Arctic high pressure	Below zero
70°–50° North	Arctic tundra			0° C
50°–40° North	Boreal needle-leaf forest	Currents flow from west	Sea winds blow from west	6° C
40°–30° North	Summergreen forest and grasslands			12° C
35° North	Savannas, deserts	Seasonally reversing currents Northern limit of coral reefs	Seasonally reversing sea winds	18° C
30° North–0°	Tropical forests	Currents flow from east	Sea winds blow from northeast (trade winds)	26° C
0°–30° South	Tropical forests	Currents flow from east Southern limit of coral reefs	Sea winds blow from southeast (trade winds)	26° C
35° South	Savannas, deserts	Seasonally reversing currents	Seasonally reversing sea winds	18° C
30°–40° South	Temperate evergreen forests and grasslands			12° C
40°–50° South	Austral beech forests	Currents flow from west	Sea winds blow from west	6° C
50°–60° South	Island tundras			3° C
60°–90° South	Antarctic ice desert	(No ocean)	Antarctic high pressure	Below zero

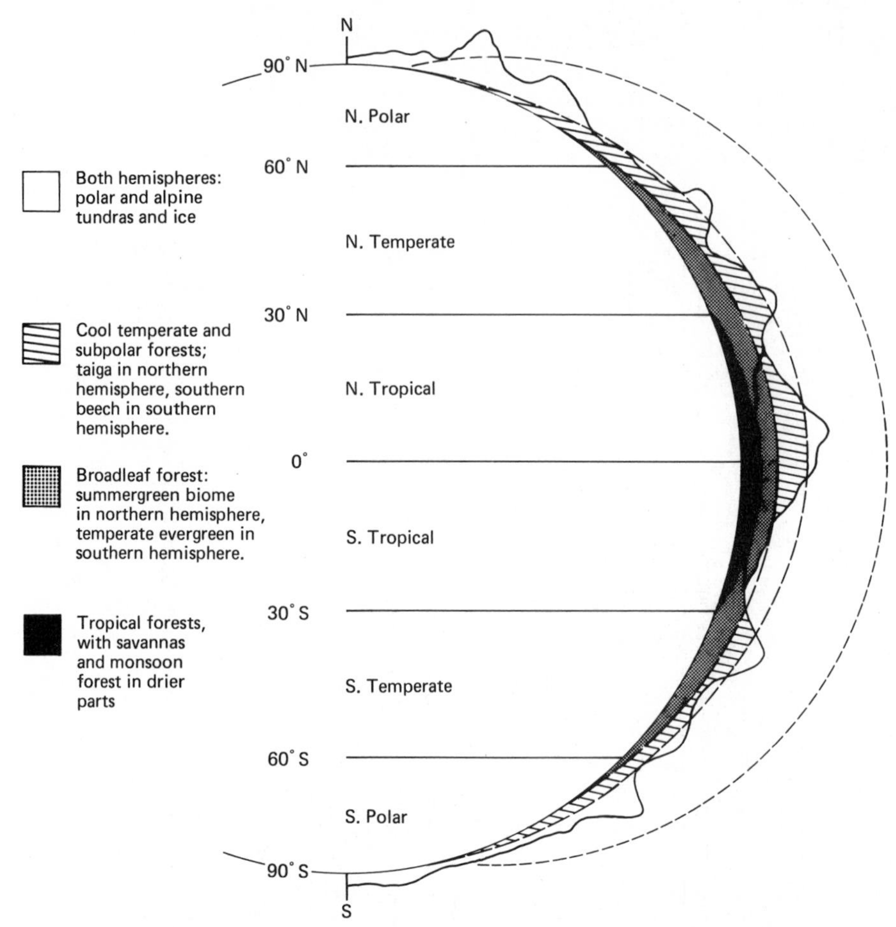

Figure 4. The main life zones (or biomes) of the earth are related both to latitude and to altitude or depth. The diagram shows zonation on the lithosphere; a similar diagram can be drawn for the hydrosphere, where increasing depth corresponds to increasing altitude on land. The topic is discussed further in Chapter 4.

supernova, yields a calculated age for the earth of 4.5 billion years. The oldest known fossil remains of recognizable organisms on the earth date from about 3 billion years ago. So it would appear that the planetary ecosystem began to evolve sometime between 3 and 4.5 billion years ago.

Study of the phenomena of volcanic eruptions shows that when magma reaches the earth's surface the loss of pressure causes it to differentiate into steam and a residual that solidifies as rock (lava). In 1951 W. W. Rubey, an American geologist, noticed that the annual production of water vapor and lava by the world's active volcanoes seems to be in the correct quantity and proportions to account for the creation of the oceans and the continents over a time span of about 4 billion years, that is, the same order of time as that postulated by the theory of radioactivity. Theory implies, therefore, that the continents and the oceans began to form as soon as the earth itself was formed and that magma has been continuously rising to the surface and differentiating into oceans and continents ever since. It is a matter of no small interest that the oldest rocks so far found on the moon also indicate an age of about 4.5 billion years, a fact that suggests that the earth and the moon are sister planets formed by some common event.

ECOLOGY AND THE ECOSYSTEM. The organisms that inhabit any particular region or environment are collectively termed the *biota.* Biota comprise plants and animals. Most plants can capture the energy radiated as light and heat by the sun (electromagnetic energy) and can use it to convert inorganic constituents of the environment into organic materials. Organisms with this power are termed *autotrophs.* Animals, on the other hand, need preexisting organic materials for their nutrition; in this role animals are collectively termed *heterotrophs.* Animals that feed on plants are termed *primary heterotrophs* and those that feed on other animals *secondary heterotrophs.* The interlocking nexus of autotrophs, primary heterotrophs, and secondary heterotrophs is termed a *food web,* made up of many distinct *food chains.* The ultimate products of any food web are feces, carrion, gases, and other lifeless materials of organic origin. These are converted by bacteria into simpler inorganic substances, and thus returned to the environment to become available again to the autotrophs. Any such self-sustaining food web, together with the input of solar energy, and the environment containing the food web, constitutes an *ecosystem.* Ecosystems may be large or small; the entire ocean conforms to the definition just given and therefore may be viewed as a single ecosystem, but so also does a forest pond or a desert oasis. The study of organisms in relation to their environment is named *ecology.* A primary task of the ecologist is to identify the various plants and animals in an ecosystem and to determine the roles they play in it. The identification of organisms is the task of *systematic biology,* which also seeks to determine the evolutionary relationships of organisms. Organisms have evolved through geological time, whence it follows that ecosystems must have undergone change, for their constituent biota cannot always have comprised the same species of animals or plants. Thus temporal as well as spatial elements enter into the study of ecosystems; the forest ecosystem of New Hampshire today, for example, bears little similarity to that of the same region 10,000 years ago or 10 million years ago. Most ecosystems can be subdivided into parts, some of which may themselves qualify as independent ecosystems; usually ecosystems are, at least in part, interdependent. The entire biosphere may be viewed as comprehending a single *planetary ecosystem,* itself made up of numerous subsidiary ecosystems.

ANCIENT ENVIRONMENTS. In the course of this book, as each environment is reviewed, particular elements of the biota will be isolated as carrying out some particular role in an ecosystem. One question that then arises is: How old is that ecosystem? Sometimes we can supply an answer, if we happen to know how far back in time fossil remains can be found of the organism in question. For example, in Chapter 2 the overwhelming importance of diatoms in the existing oceanic ecosystem is noted, for the food chain of virtually every marine animal is found to begin with diatoms. But then, if we examine the fossil record of diatoms, we discover the disconcerting fact that fossil diatoms are common as fossils only as far back as the Cretaceous period, some 120 million years ago. Thus we are obliged to seek an explanation of how the marine ecosystem operated before Cretaceous times (a matter discussed in the chapter in question). Frequent reference will be made to the various periods of the past into which geologists divide the history of our planet, so for convenience in following such discussion the most commonly employed divisions of the geological time chart should now be mastered. They are set out in Table 2.

THE INTERDISCIPLINARY APPROACH. As must already have become apparent, environmental biology is by no means a subsidiary branch of biology alone. It involves the study of living organisms in relation to their planetary setting, and this inevitably demands that we call on relevant data supplied by the sister sciences of astronomy and geology. We cannot disregard the fact that environments change with time, as illus-

Table 2. The geological chronology.

ERA	PERIODS AND EPOCHS		MILLIONS OF YEARS BEFORE PRESENT	ENVIRONMENTS AND LIFE
CENOZOIC	QUATERNARY PERIOD	Recent (Holocene) epoch		Modern climates; first agriculture
		Pleistocene epoch	2	Ice-age tundras; modern man
	TERTIARY PERIOD	Pliocene epoch	10	Apemen appear
		Miocene epoch	25	First extensive grasslands
		Oligocene epoch	40	Mammals become dominant
		Eocene epoch	60	Mammals become conspicuous
		Paleocene epoch	70	Extinction of dinosaurs
MESOZOIC	Cretaceous period		135	First flowering plants
	Jurassic period		180	First podocarp forests
	Triassic period		225	First araucarian forests
PALEOZOIC	Permian period		270	Conifers become forest trees
	Pennsylvanian period		310	First reptiles
	Mississippian period		350	First swamp forests
	Devonian period		400	Lungfishes first appear
	Silurian period			First bog plants
			440	First land animals (scorpions)
	Ordovician period		500	Marine invertebrates conspicuous
	Cambrian period		600	Marine invertebrates conspicuous
	(Precambrian time)			Marine invertebrates; probably lichens on land
	Creation of the earth		4500	

Figure 5. This restored swamp ecosystem of Sydney, Australia, in the Triassic period (about 200 million years ago) depicts a tropical scene. The plants and animals are identified in Chapter 11. The elements of this ecosystem contradict geomagnetic hypotheses that Sydney lay close to the Triassic geomagnetic and geographic South Poles. Similarly tropical fossil remains from Antarctica imply that continent cannot always have been the site of the South Pole (16).

trated in the first paragraph of this introduction, so we must rely on the conclusions of paleontologists (whose field is the study of fossils) and of glaciologists (who include in their inquiries the study of ice ages). Again, man has set down in writing his varying concepts of the world around him in a continuing tradition of literature since the closing part of the Bronze Age; such writings, often poetical, and exemplified by the Homeric epics and by many of the books of the Bible, prove to be valuable sources of information on former states of the Mediterranean flora and fauna. Still earlier, primitive man painted or carved extremely accurate representations of the animals he hunted, so archeology yields data on prehistoric faunas, as also on such matters as the level of the sea at different times in the past (for we can be sure that submerged temples were not originally constructed beneath the sea). Again, matters of Roman agricultural policy, to be gleaned from a study of ancient literature, can have a bearing on our interpretation of environmental change, as we noted briefly above in the case of Trajan. These are all aspects that are discussed at greater length in appropriate parts of this book.

This interdisciplinary approach is all-pervading in environmental studies, for it applies also at a lower level, within biology itself. I am by training and experience a marine biologist, yet it is impossible for me to appreciate the changing environments within the sea without paying due attention to what my botanist colleagues have to say about contemporary forests, for example. Often a suspected former climatic situation, inferred on the basis of fossil seashells, can be tested and proved or disproved by checking what the contemporary fossil pollen deposits reveal. By combining the evidence from several disciplines a whole may be derived that is greater than the sum of its parts, for several different viewpoints yield a degree of depth to our perception; this way of combining disciplines to improve our understanding of a problem is sometimes referred to as a *gestalt* approach, since German philosophers drew attention to its value and employed that word for it. Studies of this kind have a special fascination not only for biologists, but for very many people in other walks of life. For they show us a new way of looking at the world around us and greatly enrich our appreciation of the complexity of the earth's history and of the many ways in which men, animals, and plants all interact with one another and with their surroundings. They also throw light on the problems of pollution, for pollution after all is but the symptom of man's failure hitherto to appreciate the nature of his own role in world ecosystems.

MAJOR PLANETARY ENVIRONMENTS. Quite different techniques have to be used in the study of different ecosystems. The methods of investigating life on the seabed, for example, are necessarily unlike those used in forest ecology. Therefore it is a matter of simple practical convenience to separate the major planetary environments into (1) oceanic, (2) freshwater, and (3) terrestrial ecosystems. The sequence given is that in which it is believed they were occupied by living organisms during the history of the earth. They are studied in the same order in this book.

THE CONTINENTAL DRIFT THEORY. The study of environmental biology in the context of geological time leads to significant inferences as to the former climates of the earth. For example, the luxuriant flora with associated crocodile-sized amphibians, such as that illustrated in Figure 5, tells us that such an ecosystem would require at least a subtropical climate, with illumination by the sun throughout the year. The figure illustrates the Triassic environment of Sydney, Australia, and thus related to an epoch some 200 million years ago. However, on the basis of geophysical theories, especially the evidence of fossil rock-magnetism of Australian Triassic deposits, some geophysicists have suggested that Sydney lay at about 80° south latitude in Triassic times, that is to say, so near the then South Pole as to suffer a long Antarctic night every winter. Such an inference is clearly unacceptable to the student of fossil ecosystems.

Again, on the basis of fossil rock magnetism of Jurassic and Cretaceous strata of India and of Antarctica, some geophysicists infer that these two continents were united with Australia and that the whole supercontinent (called Gondwanaland) lay over the southern region adjacent to present Antarctica. India is claimed to yield rock magnetism whose angle of dip implies that India lay some 60° of latitude south of its present position. On the other hand, some Indian geologists have recently held that India has always formed part of Asia. Now, during the Jurassic and Cretaceous periods (from 150 to 100 million years ago) coral reefs formed on the margins of peninsular India, and their fossil remains are preserved so well that we can identify the genera present in the reef populations at that time. They prove to be similar to those of other fossil coral reefs of lands that were coral-girt at those epochs and that no one doubts formed in the tropics. Hence of the two theories offered by geologists, the student of fossil ecosystems must of course favor that which views India as having remained joined to Asia, lying therefore in the same tropical belt as it still occupies today.

POLAR WANDERING. But the matter is complicated. Antarctica itself yields fossil marine organisms, and some of them are also tropical in character; others are temperate. Was the entire world tropical at one time, then? This is unlikely, and besides, even if Antarctica were warm, if it lay over the South Pole as it does today there would still have been a long winter night. And coral reefs require regular daily illumination by strong sunlight in order to generate the oxygen in excess (formed by algae imbedded in the reef and in the tissues of corals), for the dense population of a reef would suffocate at night but for the excess oxygen produced in the daylight hours. The most promising explanation of puzzles of this kind seems to lie not so much in a theory of drifting continents as rather in an alternative theory that the position of the earth's poles may change with time. This and related problems are discussed from the biological viewpoint in later chapters.

2

THE OCEAN AS AN ABODE OF LIFE

Biological oceanography is the study of the ocean in its role as an abode of life. Marine biology is the study of living organisms that inhabit the ocean. The two disciplines therefore have essentially the same subject matter but differ somewhat in approach and stress.

Topics to be discussed in this chapter are the physical characteristics of the oceans that determine the habitats of living organisms, the physical structure of marine ecosystems, marine productivity, the primary producers of living tissue in the oceans, the consumer organisms that feed on the other organisms, and the ecological relationships between living organisms and the marine environment.

Physical Characteristics of the Oceans

The term *parameter* is strictly part of the language of mathematics, where it is applied to any variable quantity, each possible value of which determines the form of some expression. However, in the sister sciences the word has been widely adopted in a somewhat metaphorical sense to signify any variable factor that influences the conditions under study. In the present context parameter may be used to include any variable such as *temperature, pressure, salinity,* or *illumination,* and so on; for any one of these measurable states helps to determine in some way the overall conditions of life in the sea. Following are illustrations of how some of these parameters operate.

TEMPERATURE. The *temperature* of the sea at any place varies according to the *season,* the *latitude,* and the *depth* at which the measurement is made. The warmest sea water is found at the surface near the equator, where the average temperature of the water over the whole year is about 26.6°C. The coldest seas are those of the polar regions, with a minimum possible value of −1.9°C (when sea water freezes). The total range of marine water temperatures is thus about 30°C, a modest figure as compared with the fluctuations observed in terrestrial environments. So in marine animals we do not find any pronounced adaptations for protection against temperature changes or extremes, such as are observed in polar terrestrial organisms or desert organisms. Because many marine animals are tolerant of only a minor range of temperatures, we find coral reefs developed only in or near the tropics, where the water temperatures do not drop below about 18°C, and some marine plants are intolerant of equatorial warmth, such as the large brown kelps, found only in cool latitudes, and killed if accidentally swept by currents into tropical seas. Cold water is denser than warmer water over most of the temperature range encountered, so in any latitude the temperature decreases with increasing depth. This has one interesting effect on distribution, for it is possible for polar species to exist in deep water in quite low latitudes. Some species occur in surface waters of both the Arctic and the Antarctic seas and have a continuous distribution, occurring in progressively deeper water nearer the tropics and crossing the equator in very deep water. Obviously temperature is an important parameter in the cases of such organisms, for their distribution shows that they are obliged to conform to appropriate isotherms, without regard to depth.

The cases just cited also illustrate another feature of marine organisms, namely, their general insensitivity to pressure. Water is almost incompressible, so all organisms whose bodies are permeated by fluids suffer very little inconvenience if pressures change. Most invertebrates fall in this category. On the other hand, animals such as fishes, many of which have a gas gland in the abdomen (used for adjusting the specific gravity of the body, to permit rapid movement), and air-breathing marine animals all experience serious physiological difficulty if transferred out of their normal depth; such animals are, of course, distributed in such a way as to conform to their depth requirements, or *bathymetric parameters.*

All organisms require a source of energy, and most obtain it by oxidizing substances such as sugars or fats in their body tissues. So *oxygen* in solution is an important environmental parameter in seawater. In some enclosed seas, such as the Mediterranean and parts of the Caribbean, the deeper levels are deficient in oxygen, with the result that stagnant lifeless zones exist. The pollution of the sea by man has exacerbated this condition, and it seems likely that seas such as the Mediterranean will become almost totally lifeless unless sewage pollution is brought under control.

SALINITY. The *salinity* of the sea is its best-known characteristic and also one of the most interesting and important environmental parameters. Average sea water contains about 33 parts per thousand (33 ‰) of dissolved salts, mostly sodium chloride. Enclosed seas such as the Baltic, where the influx of rivers is relatively high, have reduced salinities (Baltic, 8 ‰, Aral Sea, 10 ‰, Caspian, 12 ‰, Black Sea, 18 ‰). A few seas have lost their connection with the world ocean and if they occupy regions subject to evaporation greater than their input of rain or streams their salinity undergoes a constant increase as their water volume dwindles; the Dead Sea of Israel and the Great Salt Lake of Utah are examples. But these are exceptional. The oceans at large do not vary so widely, the lowest salinities being usually about 27 ‰ where equatorial rains are heavy and rivers such as the Congo and Amazon discharge, and the highest salinities occur under the opposite circumstances, where evaporation is high and rainfall low, as in the Red Sea, with salinities reaching about 40 ‰. These varying salinities are important parameters restricting the life of marine animals. For example, sea urchins and starfishes will not tolerate lowered salinities and avoid such regions.

The origin of the sea salt is to be found in the minute quantities of dissolved salts carried down to the seas by rivers. The amount of dissolved salt delivered each year by the world's rivers can be estimated. When the total salt of the sea is compared with the annual delivery rate, it appears that it would take 100 million years to produce the observed oceanic salinity. When this figure was first calculated it was taken to be a measure of the age of the oceans (and hence of the earth). It is now realized that 100 million years is the average time that a salt molecule spends in solution before its constituent atoms are extracted by natural processes and converted into other minerals and deposited, given the observed conditions. Evidently, therefore, the oceans acquired their present salinity soon after the ocean began to form 4.5 billion years ago, and they must have retained the same salinity, more or less, ever since. It is often stated, with little logic, that the amount of salt in the blood of land animals is a perpetuated salinity of the oceans at the time when land vertebrates left the sea some 300 million years ago; but such ideas are clearly incorrect, because, as noted above, the salinity of the world ocean stabilized within about 100 million years, and so the salt content of the ocean in Paleozoic times must have been the same as it is today. And further, different kinds of fishes have differing blood salt content, depending on their kidney physiology and also on the stage of the life history.

Some fishes — for example, sharks — freely range from the sea into rivers and even enter freshwater lakes, such as communicate with rivers. Some of the worst shark attacks in Australia have occurred in rivers or even on river banks, hundreds of miles inland. The ability to tolerate a wide range of salinities depends on the water-balance physiology of the fish concerned, mediated by the kidneys; puffer fishes, flounders, and eels are other examples, and the migrations of salmon from the sea to the rivers, and of young salmon from rivers to the sea, are other well-known cases of such tolerance. Most marine fishes do not tolerate such changes in salinity. The fishes of the Baltic and of the Caspian and other seas of low salinity are essentially freshwater species plus a few tolerant marine species.

WATER MASSES. The oceans can be shown to be stratified into various layers, or *water masses,* having varying salinities and also varying temperatures, such that any particular water mass can be defined as having a stated *temperature-salinity* (TS) range. These defined water masses are believed to originate from steady flow processes in the sea, by means of which sea water of, for example, Antarctic origin is conveyed to particular parts of the Atlantic and the Pacific. As a result, particular species of fishes that favor particular TS values are found in particular water masses. The relationship is so pronounced that an ichthyologist studying the contents of a trawled sample of fishes from a given depth can often predict correctly what the TS values of the sea water must be. Facts such as these illustrate the importance of physical parameters of the oceans in determining the characters of local marine faunas. Plants are even more susceptible to limitations set by salinity, so much so that very few plants can grow in both fresh and salt water. So it is a commonplace observation that seaweed does not occur in lakes, and lake algae do not grow in the sea.

The amount of suspended insoluble matter in the oceans depends on the distance from the effluence of rivers, which deliver suspended particles carried in their swift currents; it also depends on the degree of disturbance of the sea itself, for disturbed waters sweep up silt from the bottom. The relative *turbidity* of the sea affects the penetration of light and hence sets a limit to the depth at which light-demanding plants can grow. Some animals, too, such as crinoids (relatives of starfishes), require clear water, for suspended silt seems to clog their respiratory systems. Turbidity can be measured by determining the visibility of standardized sets of spaced lines, or by similar devices, suspended in the water and observed from measured distances through the water.

There are also some purely mechanical limiting parameters in the sea. The boundary between the surface and the overlying air, called the *air-sea interface,* is an obvious frontier between aquatic life and aerial life. A few animals such as flying fishes and oceanic birds cross this boundary, but not for far or for long, and most organisms cannot cross it. The other interface, that between the bottom water and the seafloor, is analogous; a number of burrowing animals cross it, but not for far, and most animals do not cross it. The general term for the material of the seabed immediately in contact with the water is *substrate.* We will examine its characters later on (Chapter 3).

The World Ocean

Since all the oceans are interconnected, geographers speak of a world ocean or, more simply, the ocean, as a collective entity of all the saline waters of the hydrosphere. Its total area amounts to 360 million km^2, or 70.8 percent of the earth's surface. If we plot the entire ocean as a circle representing an area of 360 million km^2, then the areas of the separate oceans are yielded in proportion by sectors subtending angles of 82° each for the north Pacific and the Indian oceans, 66° for the south Pacific Ocean, 47° for the north Atlantic Ocean, 37° for the south Atlantic Ocean, 32° for the Antarctic Ocean, and 14° for the Arctic Ocean. The average depth of the entire ocean is 3.8 km. The deepest ocean is the Indian, with an average depth of 4.3 km; the shallowest is the Arctic, with 1.2 km.

The continents are bordered by a zone of relatively shallow seabed called the *continental shelf,* which slopes gradually downward until it lies at about a depth of 200 m, beyond which the slopes become steeper. These steeper parts of the seabed are called the *continental slope,* and they continue to a depth of about 4 km, at which level the seafloor flattens to form the *abyssal plain* (in fact the abyss is undulating rather than a plain). In a few regions (defined in Chapter 5) there are deep *trenches,* down to 11 km.

The waters and the associated biota of the continental shelf region are called *neritic;* thus one speaks of neritic faunas, neritic habitats, neritic seas. The corresponding terms used for the other regions of the ocean are as follows: for the open sea, beyond the shelf, *pelagic;* for the continental slope, *bathyal* or *archibenthic* (both terms are in current use); for the abyss, *abyssal;* and for trenches, *hadal.* Biota and habitats on the margin of the sea and the land are called *littoral.* Biota and habitats between the upper and lower limits of the tides are called *intertidal.*

These features are illustrated diagrammatically in Figure 6.

Biota that are restricted to a limited range of temperatures are said to be *stenothermal;* biota that tolerate wide temperature ranges are *eurythermal.* Corresponding terms with respect to depth tolerance are *stenobathic* and *eurybathic;* and for geographic range, *stenotopic* and *eurytopic.*

The Physical Structure of the Marine Ecosystem

As has already been stated in Chapter 1, any ecosystem is a self-sustaining community of living organisms comprising autotrophs, which produce organic substances from inorganic environmental materials in the presence of sunlight, together with heterotrophs, which consume the autotrophs or one another and yield lifeless organic products in the forms of carrion, feces and carbon dioxide, and other excreted substances; the latter substances are degraded by bacteria, converted back to

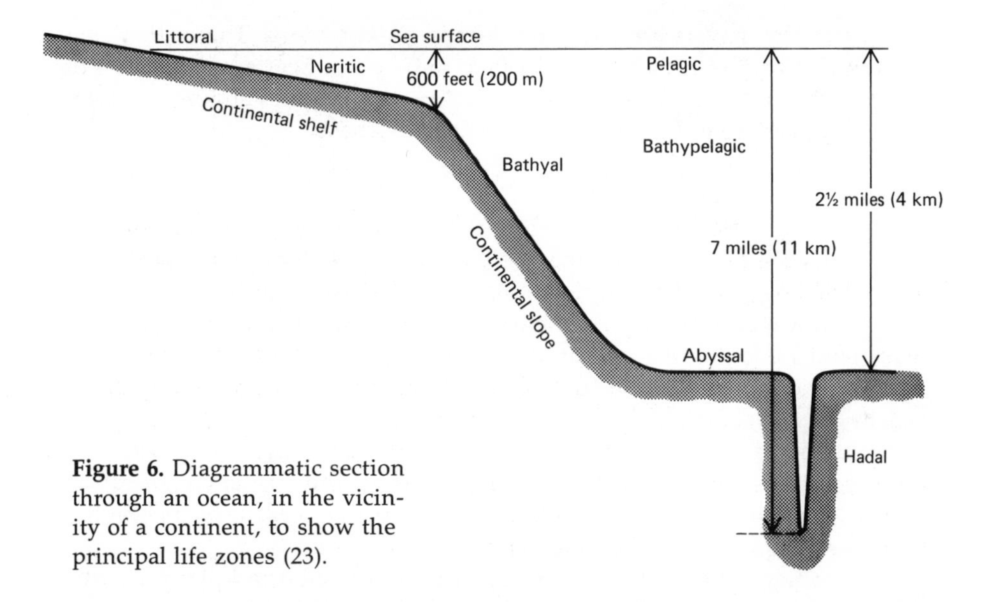

Figure 6. Diagrammatic section through an ocean, in the vicinity of a continent, to show the principal life zones (23).

inorganic constituents of the habitat, and so recycled to the renewed use of the autotrophs. So long as the solar energy continues to be supplied the ecosystem is permanent.

A few more definitions may be introduced at this juncture. One useful term is *biomass,* of which a crude definition is the wet weight of living tissue, including unassimilated organic materials ingested by animals but not yet digested or defecated. In recent years the world's oceanographers have devoted much time to measuring the amounts of biomass in various parts of the sea. The amount of biomass existing at any given instant under a stated area of the sea surface is called the *standing crop.* It can be measured in the first instance as so many pounds per square yard, or more usually in scientific contexts the amount is given as kilograms per square meter (kg m^{-2}). The wet weight varies with the water content, which is not constant for various kinds of organisms; so a more useful measure is the dry weight, or still better the ash-free dry weight, which is the difference between the dry weight and the residual inorganic salts after total combustion of the dry biomass. Chemists can measure the energy stored in the biomass by a calorimeter and so give its calorific value as calories or kilocalories per square meter. If the area under investigation is studied at fixed intervals of time, the increase in the standing crop per day can be determined. This value is termed the *productivity,* which can be reported as kilocalories per square meter per day (Cal m^{-2} day^{-1}). If the biomass be broken down into that of the autotrophs alone, we have a measure of what is called *primary productivity.* Such measures are obviously of great importance in determining the relative richness of communities in different regions of the sea and their ability to yield crops of fish for the use of mankind. The measurements have shown that the most productive areas of the sea lie in the cool northern oceans and also where upwelling of deep water takes place, bringing dissolved mineral nutrients to the surface. The warm tropical oceans are relatively low in productivity. This rather surprising fact can be explained if we take into account certain properties of the minute floating plants of the sea called the phytoplankton, a matter more appropriate to Chapter 3, where it is discussed.

THE AUTOTROPHS. The primary producers, or autotrophs, of the oceans are of course the marine plants, nearly all of which are called *algae.* The largest and best-known algae are the seaweeds, common on all coasts. However, the most important in their contribution to overall biomass (and hence to primary productivity) are the microscopic one-celled algae called *diatoms.* These tiny plants float in the surface waters of all oceans, unnoticed by the casual observer, though in actual fact they are the most abundant living things on the earth at the present epoch. The term *alga* is no longer (as it once was) a part of the formal classification system of plants, but it is so widely used, probably because of its brevity, that its meaning should be made clear at this point.

In ordinary scientific contexts the terms autotrophs and plants may be taken as synonymous. These organisms are grouped together in one large assemblage called in formal classification the kingdom Plantae. The assemblage is further subdivided by botanists into two sections, as follows:

1. subkingdom Thallophyta: plants without leaves, roots, or vascular tissue, in which the sex cells are released when they are formed
2. subkingdom Embryophyta: plants with leaves (or leaflike organs of nutrition), in which the female sex cells are retained in the sex organ until after fertilization and the embryo has begun to develop

The first of these groups, the Thallophyta, comprises a number of subsidiary assemblages, called phyla, each phylum with some distinctive character. Some of the phyla include plants that possess chlorophyll, the pigment that produces the green color of plants and that has a complex molecule capable of causing sugars to be synthesized in the presence of sunlight. All Thallophyta that possess chlorophyll are termed *algae.* Some supposedly degenerate Thallophyta lack chlorophyll; if multicellular, they are called *fungi;* if unicellular, *bacteria.* So the fungi are organisms that have the structure of plants but behave like animals (heterotrophs) in the matter of nutrition, obtaining their energy from carrion or dead plant materials. The bacteria, as already noted, break down lifeless organic residues into inorganic components, thereby obtaining energy; some bacteria obtain energy from inorganic sources.

A summary of the main phyla of algae is given in Table 3.

With respect to the subkingdom Embryophyta, these are almost entirely plants of terrestrial and freshwater environments. One group, the phylum Tracheophyta, or vascular plants with roots and leaves, includes a few marine members, of which the best-known are the eelgrasses, which grow in shallow seas on soft bottom; and the mangroves, which are low tropical trees found in estuarine environments and along muddy shores subject to tidal waters.

Table 3. The phyla of marine algae.

PHYLUM	CHARACTERISTICS	EXAMPLES
Cyanophyta	Minute floating one-celled plants that lack a distinct cell nucleus (the genetic material being dispersed in the cell); often colored blue-green on account of the presence of the pigment phycocyanin	Blue-green algae
Chrysophyta	Minute floating or attached one-celled plants, with a distinct nucleus, and having the cell body enclosed in a 2-valved secreted silica capsule; usually brown or yellow pigments present, plus chlorophyll	Diatoms
Pyrrophyta	Extremely minute floating and swimming one-celled plants (requiring a magnification of at least 500 X to reveal detail); having 1 or 2 threadlike flagella used as swimming organs, and usually with several cellulose plates forming a jointed capsule	Dinoflagellates
Phaeophyta	Conspicuous brown seaweeds attached to rocks on coasts and to hard substrates of the continental shelf; some have hollow flotation bladders enabling them to float if their attachment is lost; contain brown pigments plus chlorophyll	Kelp Sargasso weed
Chlorophyta	Green seaweeds, some minute and one-celled or of a few cells, others larger and resembling moss or lettuce, the larger kinds attached to the seabed; chlorophyll the main or only pigment	Sea lettuce
Rhodophyta	Red or purple seaweeds, usually a few inches long, never massive, containing red pigments as well as chlorophyll; sometimes also containing lime (in which case they resemble encrusting coral or paint spills on rocks); usually attached to substrate	Dulse Corallines

THE HETEROTROPHS. The consumers, or heterotrophs, of the oceans are the animals. Unlike the plants, the marine animals are very varied in complexity of structure and are accordingly classified in nearly all the various phyla into which the animal kingdom is divided. Many of them have distinct preferences for particular marine habitats, and these matters are discussed in the chapters that follow, where the classification is also more explicitly set out. It is a matter of practical convenience to use the term *Invertebrata* to cover all the phyla of animals that lack a backbone; they are mostly rather small animals. The Vertebrata are the larger and better-known animals, all members of the one phylum Chordata and (except for some transitional forms called protochordates) distinguished by having a vertebral column.

Table 4 summarizes the characters of the more conspicuous phyla of marine invertebrates. Although this is not an appropriate context for discussing the classification of organisms, since space forbids it, a few comments on the characters may be helpful here. Botanists do not recognize any particular importance in the matter of whether a plant comprises a

Table 4. The most conspicuous phyla of marine invertebrates.

PHYLUM	CHARACTERISTICS	EXAMPLES
Protozoa	Minute one-celled animals, marine forms often with a limy or silica capsule; differing from minute algae by lacking chlorophyll; floating or benthic	Forams Radiolarians
Porifera	Multicellular anchored forms with no mouth or central digestive cavity; grow on seabed	Sponges
Coelenterata	Multicellular anchored or floating forms with a radially symmetrical body and a mouth leading into a central digestive chamber, but no coelom; solitary or colonial	Jellyfishes Sea anemones Corals
Echinodermata	Radially symmetrical coelomates, mostly free roaming on the seabed, sometimes attached or floating; surface of body commonly spiny; solitary	Sea urchins Starfishes
Platyhelminthes	Bilaterally symmetrical flattened wormlike forms, with mouth and digestive cavity present in free living forms, but lost in some parasitic forms	Turbellarians Fish flukes Tapeworms
Aschelminthes	Body cylindrical, wormlike, tapering at tips, with mouth and digestive cavity; coelom imperfectly developed, no obvious segmentation; often parasitic	Roundworms
Annelida	Body elongated, wormlike, flattened or cylindrical, with well-developed segmentation, mouth, and digestive tract, and a segmented coelom; mainly benthic	Bristle worms
Arthropoda	Body bilaterally symmetrical, covered by an external chitinous skeleton, and having well-developed jointed paired limbs, mouth with jaws, and complex gut	Shrimps Crabs Lobsters
Mollusca	Body bilaterally symmetrical (though sometimes coiled in a helix), usually secreting a protective limy shell of 1, 2, or several valves; mouth and gut well developed, but coelom secondarily reduced	Sea snails Clams Octopuses

single cell or several or many cells and refer both one-celled and many-celled plants to various phyla. Zoologists, on the other hand, draw a sharp line between one-celled animals (all of which are referred to a single phylum called Protozoa), and many-celled animals, which are referred to other phyla. So any one-celled animal is immediately recognizable as a member of the phylum Protozoa. Animals may lack obvious symmetry, or have a radially repeated symmetry such as starfishes or jellyfishes, or have a bilateral symmetry in which the left side is the mirror image of the right, as in lobsters or bristleworms or man. These contrasted patterns of symmetry provide valuable diagnostic characters in recognizing the phyla, as the table shows. Sponges (phylum Porifera) are peculiar in having no single mouth opening and no single internal digestive chamber. Of the phyla that have a mouth, some have only a single internal digestive chamber; others have a general body cavity (coelom), within which the digestive chamber (enteron) is separately placed; these latter are called coelomates.

Table 5. The classes of marine vertebrates (phylum Chordata).

<table>
<tr><th>CLASS</th><th>VERNACULAR NAMES</th><th>SKELETON</th><th>BODY CHARACTERS</th><th>REMARKS</th></tr>
<tr><td>Cyclostomata</td><td>Lampreys and hagfishes</td><td>Cartilage only</td><td>Body eellike, no paired fins, mouth suctorial, without jaws</td><td rowspan="3">Aquatic animals swimming by tail motion, fins acting as stabilizers; and breathing by means of gills in water</td></tr>
<tr><td>Chondrichthyes</td><td>Sharks and rays</td><td>Cartilage only</td><td rowspan="2">True fishes with paired fins and paired jaws (one or both pairs of paired fins occasionally lacking)</td></tr>
<tr><td>Osteichthyes</td><td>Bone fishes, codfishes, flounders, herrings, eels, perches, and so on</td><td>Bone and cartilage</td></tr>
<tr><td>Reptilia</td><td>Marine crocodiles, marine turtles, and sea snakes</td><td>Bone and cartilage</td><td>Cold-blooded egg-laying, scaled tetrapods</td><td rowspan="3">Originally terrestrial tetrapods, readapted to life in the sea; using limbs as paddles, and breathing by means of lungs in air</td></tr>
<tr><td>Aves</td><td>Oceanic and shore birds, penguins, gulls, terns, and so on</td><td>Bone and cartilage</td><td>Warm-blooded egg-laying, feathered tetrapods</td></tr>
<tr><td>Mammalia</td><td>Dolphins and whales, seals, manatees, and so on</td><td>Bone and cartilage</td><td>Warm-blooded viviparous, haired, milk-secreting tetrapods</td></tr>
</table>

The skeleton may be lacking, as in flatworms (phylum Platyhelminthes), or take the form of an external limy shell (phylum Mollusca), an exoskeleton investing the whole animal (phylum Arthropoda), or an internal bone or cartilaginous endoskeleton (phylum Chordata). Table 4 shows how various combinations of these contrasted diagnostic characters serve to define the different phyla.

The phylum Chordata, of which the best-known members are the vertebrates, is so well represented in terrestrial environments that nearly everyone already knows the principal subdivisions by their everyday vernacular names. The phylum is subidvided into a number of classes, corresponding for the most part to groupings known by vernacular names. The marine representatives, with an outline of the diagnostic characters, are set out in Table 5, which is self-explanatory. Most of the classes shown in Table 5 are referred to in Chapter 4. You should, however, begin now to memorize the technical names of the phyla and classes, for they will frequently be cited without further definition in the pages that follow.

THE DISPERSION OF MARINE BIOTA. Marine organisms differ conspicuously according to their means of dispersion. Organisms that float in the sea, or that swim too feebly to be able to overcome motions imparted by ocean currents, are termed the *plankton*, the corresponding adjective being *planktonic*. Those which live on the bottom, either as fixed *(sessile)* forms or creeping about, are spoken of as the *benthos* and are described as *benthic*. Those which swim actively, mainly fishes, and can overcome the effects of ocean currents, are termed *nekton* and described as *nektonic*. The differing habitats and degrees of mobility of these three categories influence the nature of the distributions achieved by marine organisms; they also each require special collecting apparatus and techniques. Thus in the study of marine biology, plankton, benthos, and nekton are important concepts. It is generally believed that the earliest forms of life on earth were planktonic organisms, so we begin a review of oceanic life with the planktonic ecosystem, discussed in the next chapter.

3

PLANKTON

Plankton is a collective noun applied to floating organisms, also to a sample collected for laboratory study. Marine plankton occurs at all depths in all oceans. It comprises *phytoplankton* (floating plants) and *zooplankton* (floating animals). The phytoplankton is mainly concentrated near the surface of the sea, within the first hundred fathoms, where the sunlight is strongest: it comprises various kinds of unicellular minute plants, especially those called diatoms, together with some larger seaweeds (algae) equipped with flotation organs. The animals of the plankton, on the other hand, occupy all depths and, instead of depending on flotation organs alone, are commonly provided with muscular swimming organs; the planktonic animals are very varied in structure and relationships, and most of them are unfamiliar to the average reader.

Plankton is usually collected by towing a fine-meshed net through the sea, the size of the mesh being chosen to match the type of organism sought (Table 6). Zooplankton commonly occupies the surface waters during the night, sinking to greater depths during the daylight hours; so the richest

Table 6. Relationship of mesh size to organisms trapped in plankton nets.

CODE NUMBER OF NYLON NETTING	NUMBER OF MESHES PER CENTIMETER	APERTURE OF EACH MESH (MM)	SMALLEST ORGANISMS TRAPPED
000	9	1.1	Copepods (page 37)
0	15	0.6	Invertebrate larvae (page 40), larger forams (page 40)
2	20	0.4	Forams (page 24), chain diatoms (page 30)
25	80	0.07	Solitary diatoms (page 31), aggregated dinoflagellates (page 31)

hauls are made in darkness if the net is operated at the surface. During the day rich hauls can generally be made if the net is weighted so as to operate at a depth of 35 to 50 m (20 to 30 fathoms), or some greater depth. The plankton net is usually towed at speeds of about 1 knot (at speeds of over 1.5 knots the water tends to spill out instead of passing through the mesh).

Plankton becomes progressively scarcer with increasing depth, so much larger plankton nets must be used in order to filter larger volumes of water. A large plankton net such as may be employed by an ocean-going research vessel may measure up to 5 m or so in diameter. The techniques of sampling deep-water plankton were developed in the closing years of the nineteenth century by the German *Valdivia* expedition, which brought to light many kinds of planktonic animals previously unknown. One of the outstanding discoveries which followed the development of these techniques was that made by a Danish expedition of 1908–1909, led by Johannes Schmidt: namely, the disclosure that the common freshwater eels descend into the depths of the ocean in their breeding season, for the youngest larval stages of eels were found only in such locations. To make a deep-water plankton sample requires about two hours' towing, plus several hours extra for the lowering and raising of the net.

Certain animals habitually feed on plankton, so examination of their stomach contents often yields interesting results, though we cannot of course tell at what depth the sample originated. Oceanic birds such as petrels feed at night on deep-water plankton which, at those hours, rises to the surface. Pelagic fish of the smaller kinds, such as herring, and some larger species, such as mackerel, often feed on plankton, near or at the surface. Deeper down, fishes such as *Alepisaurus* feed on deep-water plankton, and their stomach contents may be exceedingly varied and informative. When planktonic animals die, their bodies begin to sink. On the way down they are eaten by deeper-water denizens, but their skeletons and shells eventually reach the seabed. Thus samples of seafloor sediments always contain some remains of the planktonic organisms that once occupied the water column above. The shells of unicellular planktonic organisms sometimes constitute a substantial proportion of the material of deep-water sediments.

All marine food chains are based ultimately on the primary producers (autotrophs), nearly all of which form a part of the phytoplankton. Even in the deepest parts of the ocean the bottom-dwelling organisms depend indirectly on the plankton, for their only source of energy lies in the input of materials derived from the sunlit layers far above. The small particles that sink to the sea floor are utilized by bottom-dwelling or bottom-visiting animals. However, some food chains are located exclusively in the upper layers of the open ocean, and these chains involve plankton elements in all but the final one or two stages.

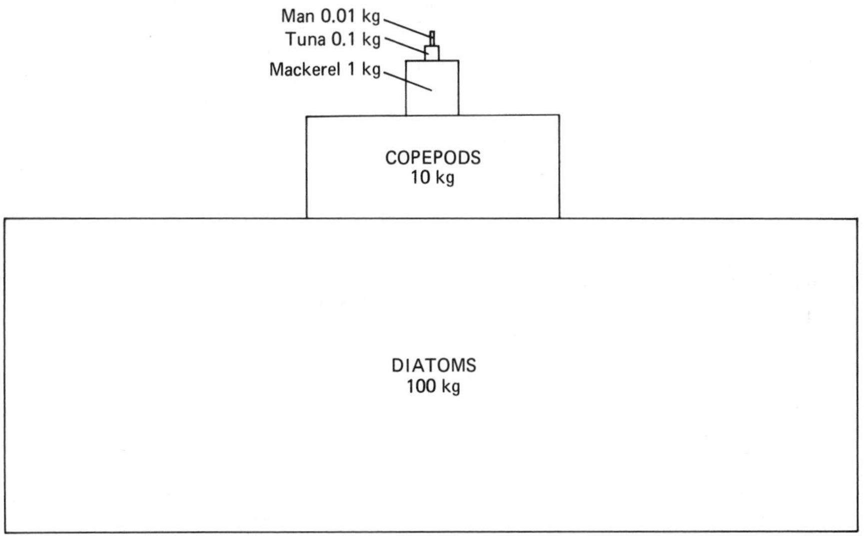

Figure 7. Trophic pyramid for a dominant predator such as tuna. Since it takes about 100 kg of diatoms to produce 1/10 of 1 kg of tuna flesh, and the same amount of diatoms produces 1 kg of mackerel, a more productive harvest would be that concentrating on mackerel. Mercury and other toxins unexcreted by the kidneys become concentrated in the higher tiers of the pyramid, so tuna flesh can be dangerous for man to consume (23).

FOOD CHAINS. Diatoms and other minute floating plants are eaten by somewhat larger floating organisms called copepods (remotely related to shrimps), and these in turn are eaten by mackerel, and that fish species is commonly eaten by tuna, which in turn may be fished and eaten by man. Any such sequence is termed a *food chain.* Measurements show that it commonly takes about ten times the weight of mackerel to produce equivalent weight of tuna flesh, and this ratio holds for other parts of the food chain. So all the combined weight of a food chain makes a pyramid, in which the broad base is made up of the large amount of initial phytoplankton, and the apex is the relatively small weight of resultant tuna flesh (or human flesh). Pollutant substances such as DDT or mercury are retained in the tissues of organisms. Thus each level in the food chain, or *trophic pyramid,* has an increased concentration of toxic pollutants, the increase factor being about ten times for each stage. This is why tuna may

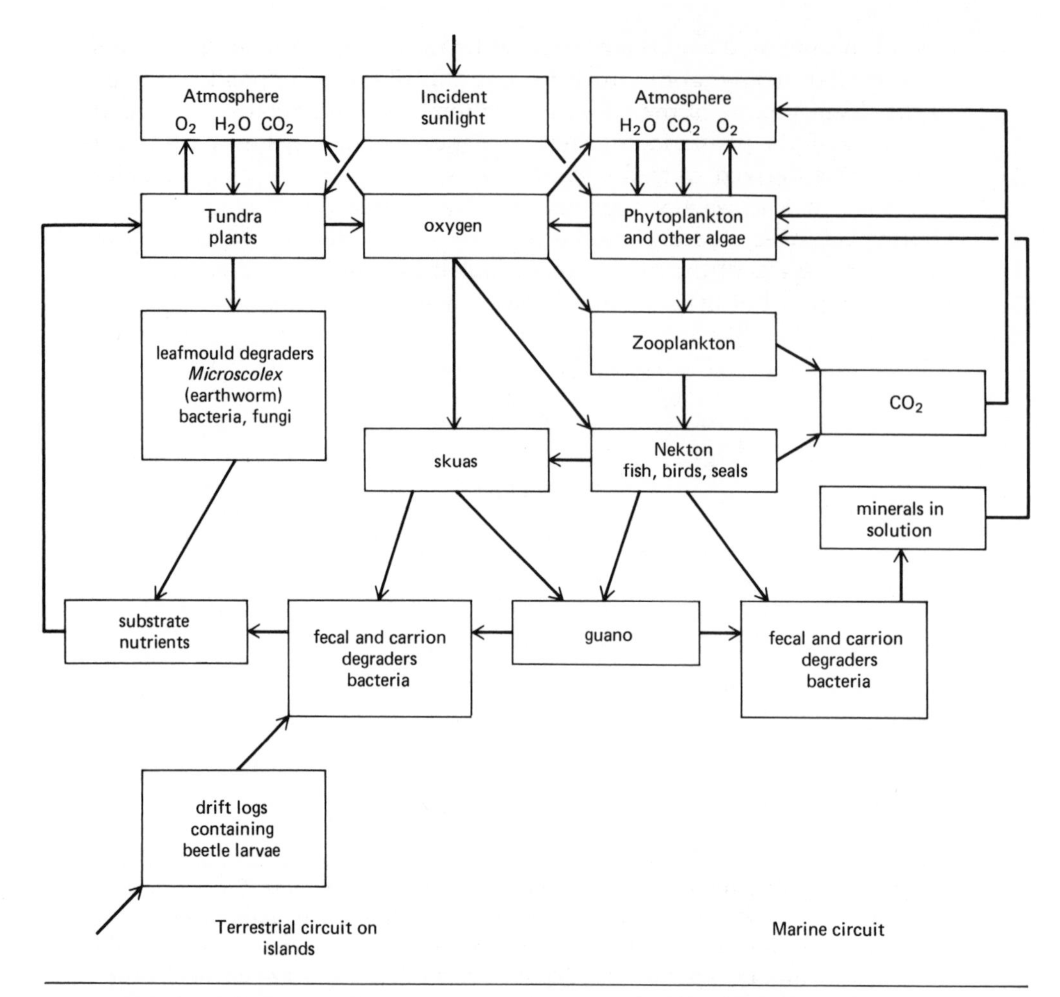

Figure 8. Example of a food chain in the Antarctic Ocean. Plankton forms the basis of the marine cycle shown to the right of the diagram, and tundra plants on the oceanic islands (such as Macquarie Island) are the main base of the terrestrial cycle, shown on the left. Grazing animals are lacking on the land. Guano, derived ultimately from the plankton, contributes to land fertility (23).

be dangerous for human consumption in the present polluted state of the oceans. Table 7 sets out the classification of phytoplankton.

Phytoplankton

Diatoms are single-celled plants that utilize a small part of the incident solar radiation energy, producing carbohydrates and oils. The synthetic process is mediated by chlorophyll, a pigment contained in small bodies in the diatom called plastids, which also contain a yellow pigment—xanthophyll—and carotenes. Characteristic is the two-valved siliceous capsule that a diatom secretes about its cell body. Many diatoms form

Table 7. Key to the phyla of marine phytoplankton.

DIAGNOSTIC CHARACTERS	PHYLUM
Minute floating 1-celled plants showing detail only under a microscope	
Nonmotile cells, lacking flagella, each with a 2-valved siliceous capsule, showing detail under 50X	Chrysophyta
Motile cells, with 1 or 2 flagella, not aggregating in chains, usually with cellulose epithecal plates, showing detail under 500X	Pyrrophyta
Macroscopic seaweeds showing detail to the naked eye	
Floating brown seaweed buoyed by air-filled bladders; or attached to floating logs	Phaeophyta
Green seaweed epiphytic on floating brown seaweed or attached to floating logs	Chlorophyta
Red or purple seaweed epiphytic on floating brown seaweed or attached to floating logs	Rhodophyta

chains. Marine diatoms are the most abundant life on earth, exceeding in mass all other living organisms in the sea and on land. During the summer bloom, when diatoms are most abundant, their concentration may reach 0.5 gm per liter of sea water, or 0.5 percent, imparting a pale golden-brown tint to the surface. Diatoms are classified as a phylum of algae named Chrysophyta, defined by the characters noted above. They form part of the plant subkingdom Thallophyta, defined as plants without roots, leaves, or vascular tissue. The term algae is used to comprehend those phyla of Thallophyta that possess the pigment chlorophyll. *Coccoliths* are a second group of the phylum Chrysophyta found in large numbers at the surface of the sea. They are symmetrical one-celled algae, with lime platelets in the cell wall, the platelets usually cemented together so they remain intact after the plant dies and sinks to the seafloor. The skeleton is remarkably symmetrical, though the whole organism seldom exceeds a few thousandths of a millimeter in diameter. They are found in association with *Globigerina,* and on the seafloor about one-third of the so-called globigerina ooze is actually made up of the remains of coccoliths.

Dinoflagellates are one-celled plants also, but they differ from diatoms in several respects. They are motile, swimming by means of one or two contractile threads called flagella. They lack the siliceous capsule, having instead one or more cellulose plates on the outer surface of the cell body, some kinds having a naked cell body. They are much smaller than diatoms and can only be seen distinctly under the higher powers of the microscope; they have some brown pigments accompanying the chlorophyll and sometimes also a red pigment. Enclosed seas where dinoflagellates bloom periodically include the Red Sea and the Vermilion Sea (California), both taking their names from the red dinoflagellate blooms. Dinoflagellates are probably much more ancient than diatoms. Diatoms did not become abundant in the earth's oceans until the Cretaceous period, about 100 million years ago. Probably, therefore, earlier food chains may have depended

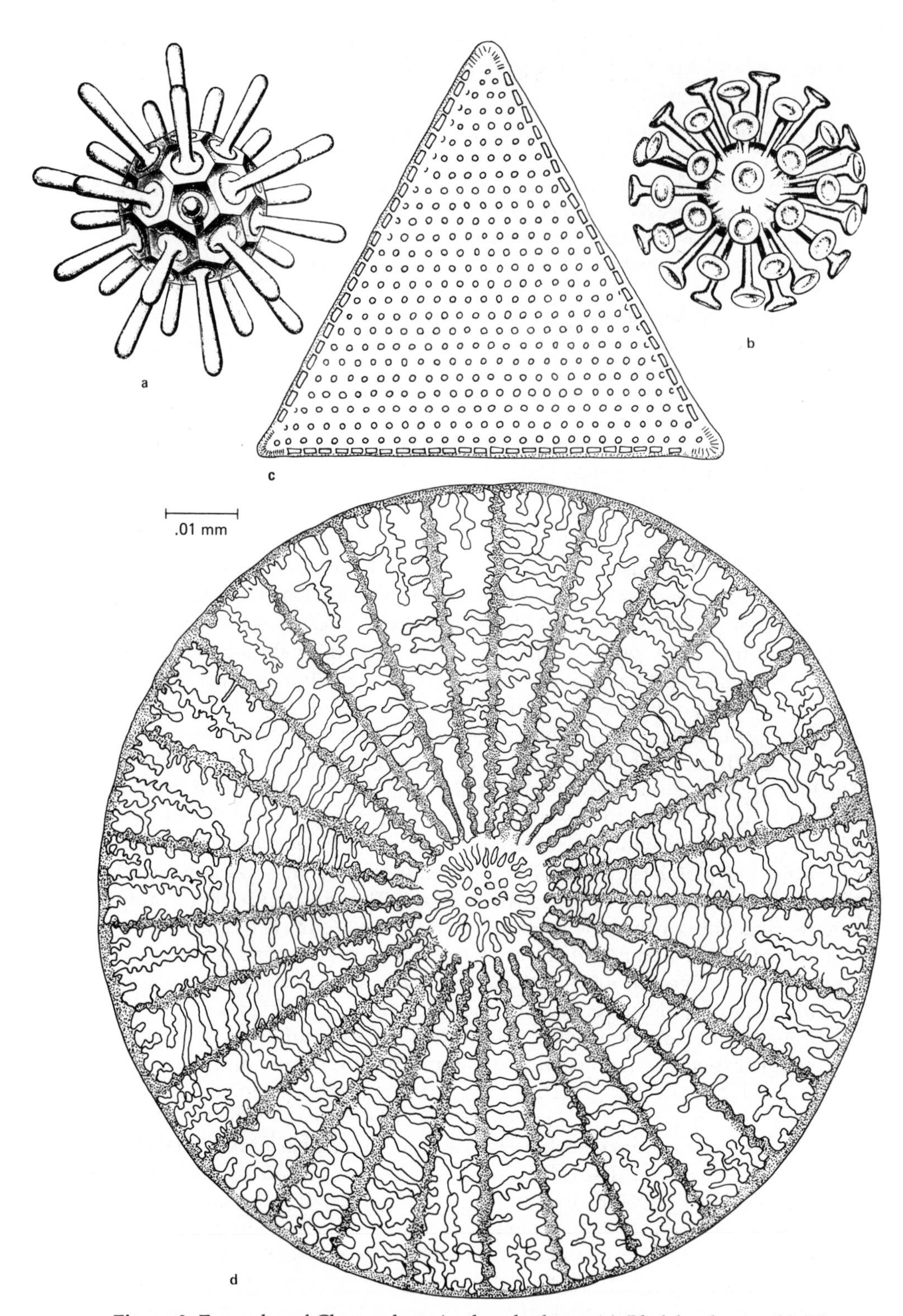

Figure 9. Examples of Chrysophyta in the plankton. (a) *Rhabdosphaera*; (b) *Discosphaera*, X 500; both coccoliths (38); (c) *Triceratium*, X 100; (d) *Arachnodiscus*, X 50; both diatoms (23).

on dinoflagellates as primary autotrophs. Dinoflagellates are sometimes very toxic to fishes, and mass mortality of fishes is a cyclic and little-understood phenomenon connected with dinoflagellate ecology. Dinoflagellates are regarded as constituting a phylum, Pyrrophyta, of the subkingdom Thallophyta.

Some of the massive marine algae, or *seaweeds,* participate in the phytoplankton. In the tropical Atlantic, a bottom-growing brown seaweed named Sargassum breaks loose from the seabed and floats with the Gulf Stream, as the so-called sargasso weed, discovered by Columbus in the mid-Atlantic in 1492. In the southern Pacific, and to a lesser extent in the northern Pacific, large bottom-growing brown kelps break loose in storms and float away from the coast by means of their air bladders, borne on the current. They remain alive for long periods and drift thousands of miles on the open sea. They often carry a whole miniaturized ecosystem of on-board passengers, underswum by a cloud of kelp fishes. In all oceans objects such as logs drift to sea and float for great distances with the ocean currents. Such logs may carry growing kelp and other seaweeds. The seaweeds in turn may carry a variety of epiphytes, both plant and animal. Benthic animals, such as sea urchins and starfishes have been taken in situations pointing very strongly to the probability that they have crossed ocean gaps as passengers on board such rafts.

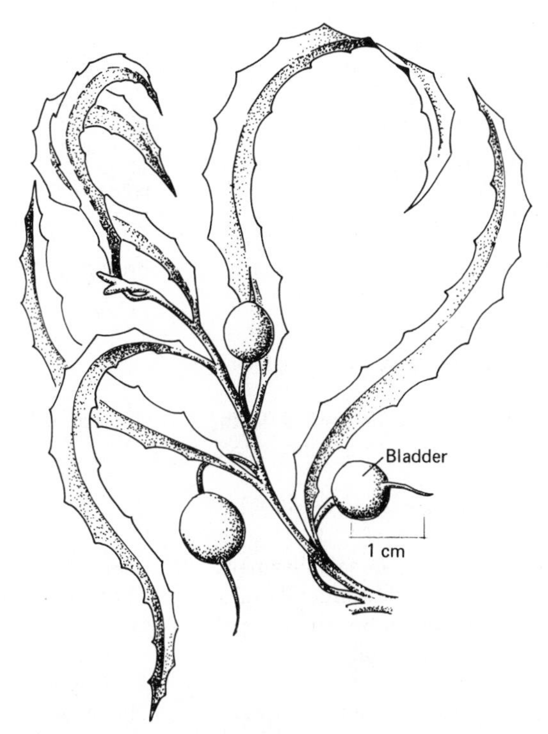

Figure 10. Piece of sargasso weed (*Sargassum*), natural size (23).

Zooplankton

Zooplankton is much more varied than phytoplankton, for almost every phylum of animals is represented in the marine plankton. Table 8 gives the classification of the main groups of zooplankton.

FORAMINIFERA. In vernacular scientific English members of this group of Protozoa are often called *forams*. Forams are characterized by their external calcareous chambered shell. The one-celled animal that secretes the shell resembles an amoeba with very much more numerous and more slender pseudopodia (called filopodia). The filopodia are extended through small holes (foramina) in the walls of the chambers, and they serve to capture diatoms as food. Most Foraminifera live on the bottom of the sea, and thus belong to the benthos; however, about 2 percent of forams are planktonic. The planktonic forams belong to two families: (1) Globigerinidae, in which the capsule looks like a collection of small spheres of various sizes stuck together; *Globigerina* is the best-known genus and also the commonest planktonic genus of forams; (2) Globorataliidae, with the type genus *Globorotalia*, in which the chambers of the shell are arranged in an expanding flat spiral. When surface-dwelling forams die they sink to the ocean floor, where their lime skeletons accumulate as the so-called globigerina ooze of the deep sea.

Table 8. Key to the phyla of zooplankton.

DIAGNOSTIC CHARACTERS	PHYLUM
Microscopic animals, usually transparent, lacking chlorophyll	
Body one-celled, with an external capsule of lime or silica	Protozoa
Body multicellular, with no capsule (larval forms, see p. 40)	
Animals at least 3 mm long, often much larger, easily visible to naked eye	
Body circular, saucer-shaped or cup-shaped or conical, with a ring of tentacles around the margin, or with a hollow float from which hang long stinging tentacles	Coelenterata
Body globular or bell-shaped with 8 meridian-like bands	Ctenophora
Body elongate, slender, wormlike	
Body clearly segmented, with fleshy fins on each segment	Annelida
Body unsegmented, with paired horizontal finlike flanges	Chaetopoda
Body sluglike, often with a lime shell, sometimes with a pair of fleshy winglike fins, often brightly colored	Mollusca
Body shrimplike, often red, with paired jointed limbs	Arthropoda
Body transparent, barrel-shaped and open widely at either end, or torpedo-shaped (sometimes adhering in chains); or resembling a flannel ice cream cone; or tadpolelike	Chordata

RADIOLARIA. Like forams, radiolarians extrude long slender filopodia. With these they capture copepods, diatoms, and other minute organisms. The capsule is siliceous, with highly developed three-dimensional symmetry; the filopodia are extruded through the foramina in the siliceous mesh. In parts of the ocean where radiolarians are abundant, the underlying seafloor may be covered by a sediment called radiolarian ooze, made up of a mixture of empty capsules, meteoritic dust, and other sedimentary matter; such sediments accumulate very slowly, so that drill cores in the seafloor can recover successive layers of sediment spanning 150 million years of continuous deposition. Some few radiolarians produce skeletons not of silica but of strontium sulfate.

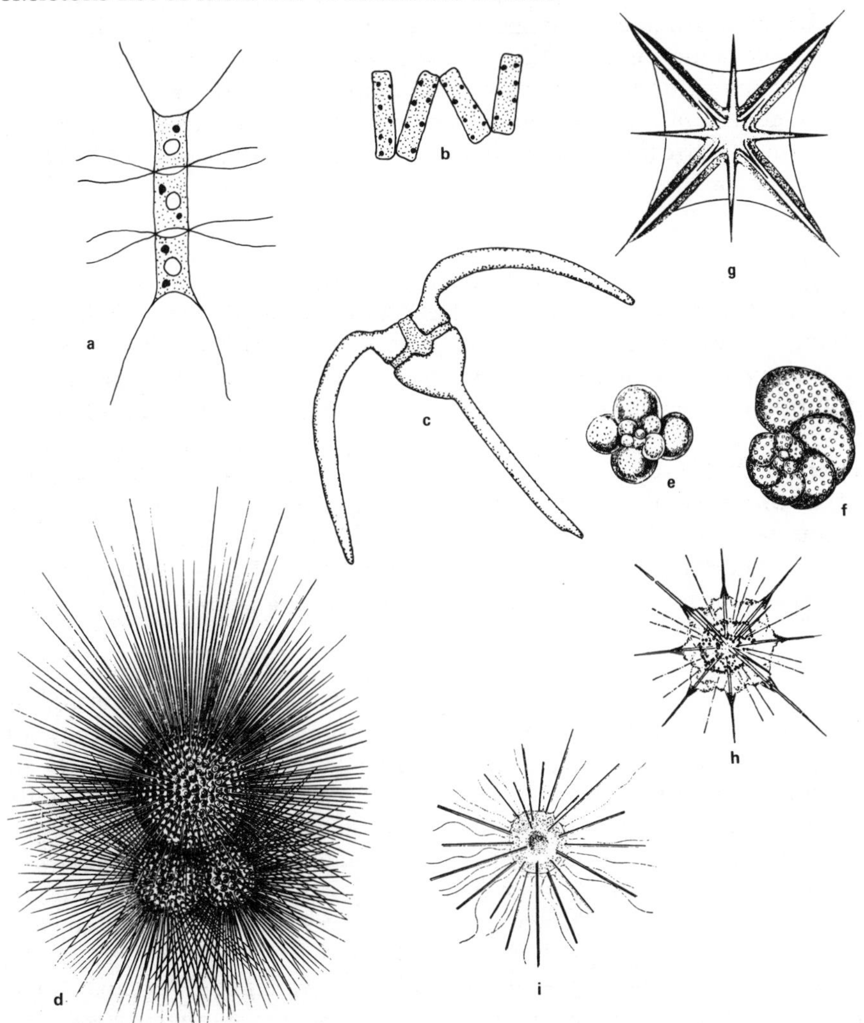

Figure 11. (a) *Chaetoceras*, X 200; (b) *Thalassiothrix*, X 400; both diatoms of the open sea (38); (c) *Ceratium*, X 250, a dinoflagellate (23); (d) *Globigerina*, with filopodia extended , X 10 (38); (e) *Globigerina* and (f) *Globorotalia*, shells (39); X 10 all forams of the open sea; (g) *Acanthostaurus*; (h) *Acanthometron*; (i) *Acanthonidium*; all radiolarians of the open sea, X 200 (38).

SIPHONOPHORIDA. Siphonophores are common in warm seas. An example is the well-known Portuguese man-o'-war *(Physalia physalis),* so called because the early European voyagers first encountered them as they crossed the edge of the continental shelf off Portugal. The common species is blue, bleaching to white in preservation. Each specimen is actually a colony, the float representing one member, and the dangling portions several hundred other members of two or more types; one type (dactylozoid) consists mainly of a stinging tentacle, for the capture of fishes and other prey; another type (gastrozoid) consists mainly of a mouth and digestive system, serving as the feeding elements for the whole colony. *Physalia* captures prey with the aid of a paralytic toxin secreted by cells in the tentacles, and some species are deadly to man. On account of the buoyancy of the gas-filled float, *Physalia* lives at the interface between the sea and the air and is subject to wind dispersion. *Velella* and *Diphyes* are other genera of siphonophores. Siphonophores are classified with sea anemones and corals and jellyfishes to form the phylum known as the Coelenterata.

SCYPHOZOA [JELLYFISHES]. *Nausithoe* is a flattened saucer-shaped jellyfish with stinging tentacles around the margin. Scyphozoans swim by means of slow rhythmical muscular contractions; the swimming movements are not strong enough to enable the animal to swim against a current, so scyphozoans are treated as part of the floating plankton. Scyphozoans feed on other planktonic animals, engulfed through a mouth opening on the lower side of the umbrella. Jellyfishes form the main part of the diet of ocean sunfish and of some other fishes. *Periphylla* is another genus of scyphozoans, related to *Nausithoe.* It is distinguished by its high conical body and by the purple pigment that lines the umbrella. Some species occur in deep water; others range into the polar seas.

CTENOPHORA [COMB JELLIES]. Ctenophores are related to coelenterates, but on account of certain structural differences, they are usually treated as forming a distinct phylum. All are marine, and all are planktonic. Distinctive are the eight bands of ciliated swimming combs arranged like meridians across the body. Some species are nearly spherical, and in these the comb bands do indeed occupy meridians. *Beroe* is a genus in which the body has a flattened form with a bell-shaped outline, the mouth being at the broad end. Ctenophores are carnivorous, feeding on planktonic animals and also on fishes. They are sometimes brightly colored in life. Some species live in the surface waters of the oceans, others range down to 1000 fathoms or more. Ctenophores are themselves eaten by fishes.

POLYCHAETA [BRISTLE WORMS]. Nearly all polychaetes live on the seafloor, but a few, such as *Alciopa,* occur in the plankton. Many are

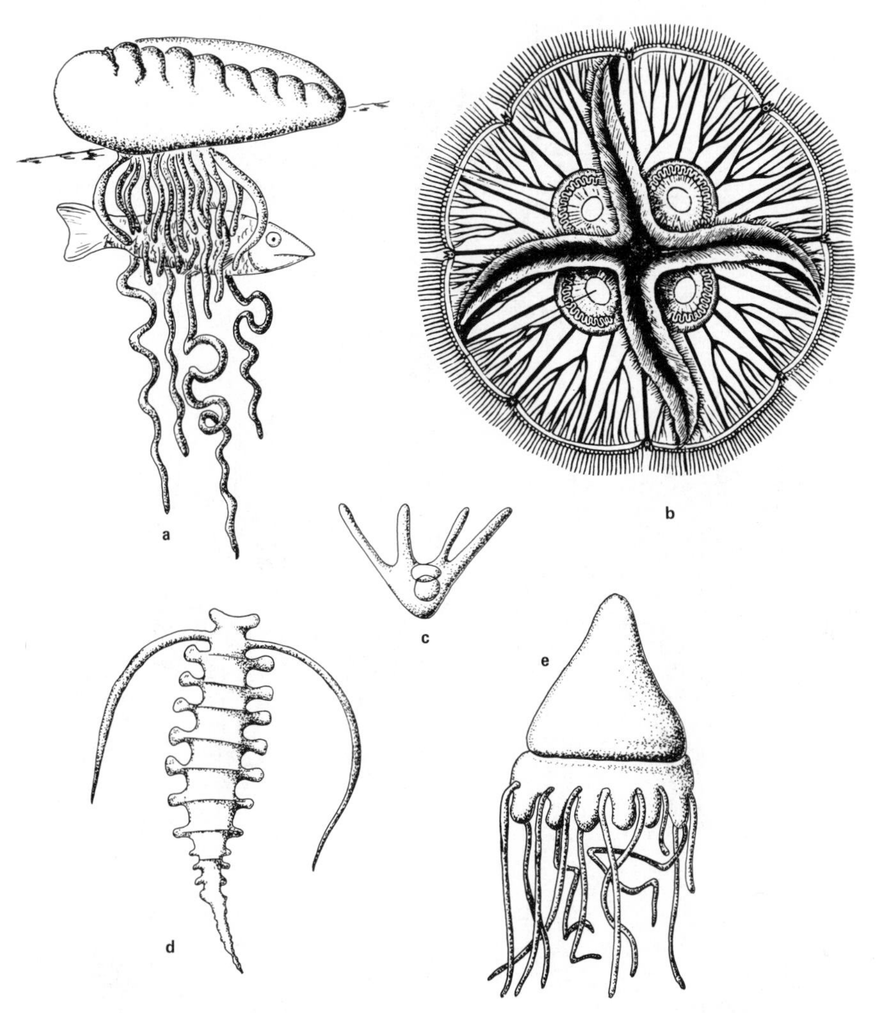

Figure 12. (a) *Physalia*, with captured fish, X 0.25, Siphonophorida (23); (b) *Aurelia*, X 0.25, Scyphozoa (15); (c) larval state (ophiopluteus) of a brittle star, X 50, Echinodermata; (d) *Tomopteris*, X 10, pelagic bristle worm, Annelida; (e) *Periphylla*, X 2, Scyphozoa (23).

carnivorous, seizing other animals by means of jaws at the anterior end of the body. They have segmented bodies, as do earthworms and leeches, to which they are related and with which they are collectively classified in a phylum called the Annelida.

COPEPODA. Copepods are small crustaceans related therefore to shrimps and crabs. Copepods are to other zooplankton what diatoms are to phytoplankton, namely, by far the commonest planktonic animals, constituting about 95 percent of it in fact. They occur in all seas in a wide range of depths, the commonest genus being *Calanus*. Species from deeper water are usually much larger than those that live near the surface. The

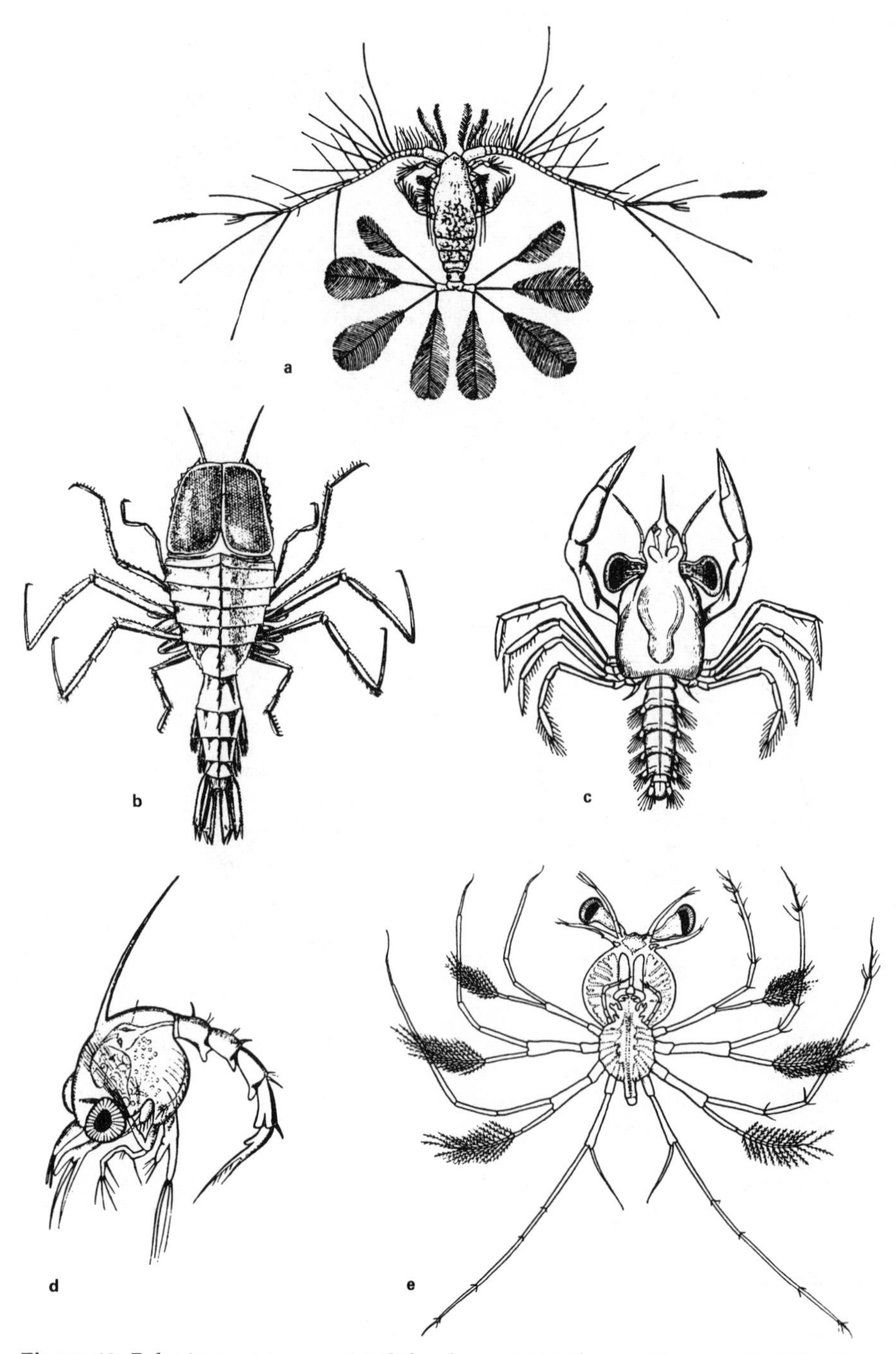

Figure 13. Pelagic crustaceans. (a) *Calocalanus*, X 10, from 500 m, north Atlantic, Copepoda (29); (b) *Cystosoma*, X 1, north Atlantic, Hyperiidea (38); (c) zoaea stage of crab *Inachus*, X 10, and (d) megalopa stage of swimming crab *Portunus*, X 10, north Atlantic (38); (e) phyllosoma stage of spiny lobster *Palinurus*, X 1, north Atlantic (29); c–e, Decapoda.

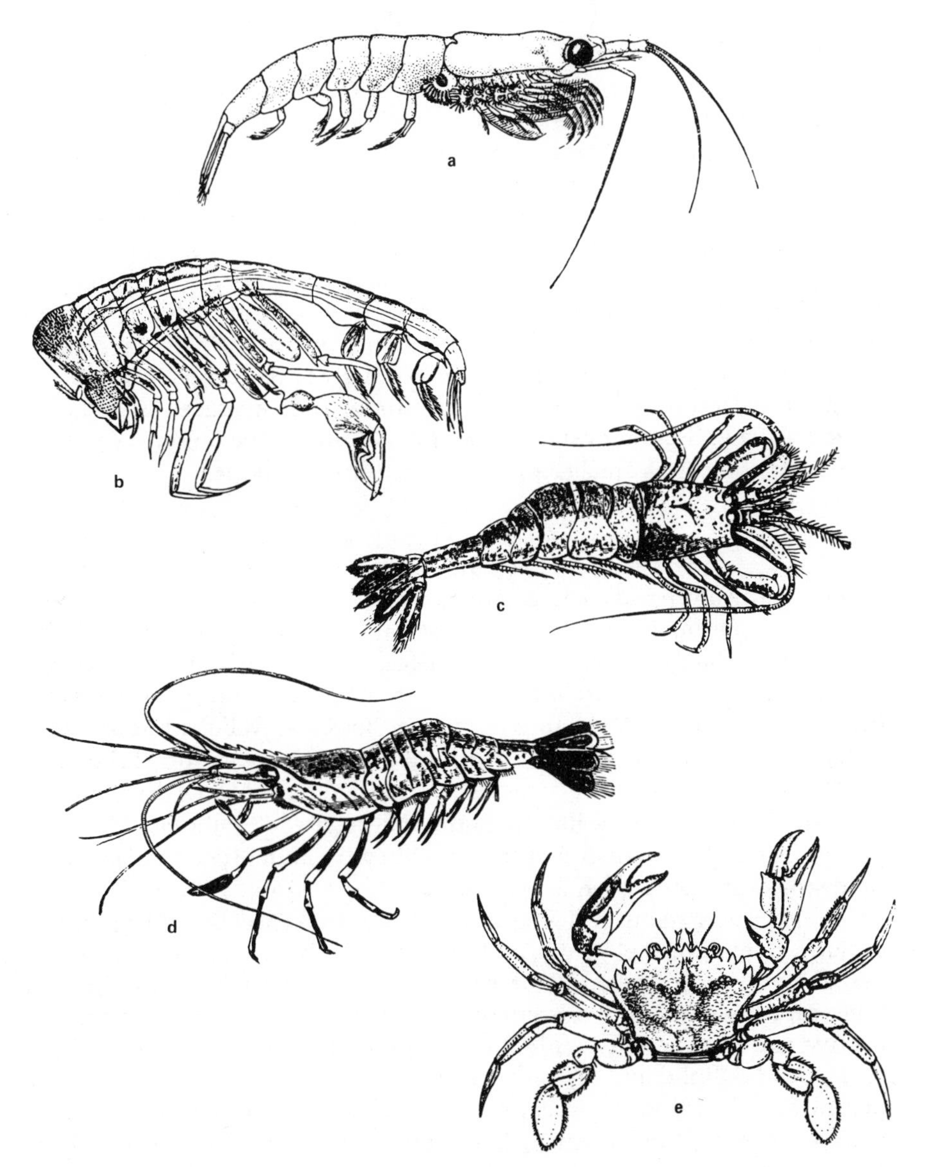

Figure 14. Pelagic crustaceans. (a) *Meganyctiphanes*, X 1, north Atlantic, Euphausiacea (29); (b) *Phronima*, X 2, north Atlantic, Hyperiidea (15); (c) *Crangon*, X 0.5, neritic; (d) *Palaemon*, X 0.5, neritic, both (24); (e) *Portunus* (swimming crab), X 0.5, mainly shallow water (38); c–e, Decapoda.

latter usually measure about 3 mm long, whereas in deeper water the copepods are likely to measure about 1 cm long. Copepods feed on diatoms in the upper waters, and in lower waters they feed on smaller copepods (larval stages) and other young stages. Copepods are themselves eaten by many other marine animals, including mackerel. The class Crustacea together with other animals with analogous structure (such as insects and spiders) make up the phylum Arthropoda, all members of which share the jointed character of the appendages.

MYSIDACEA [OPOSSUM SHRIMPS]. Opossum shrimps are easily recognized by two characters: In the female there is a conspicuous transparent pouch beneath the thorax where the young are carried; and the thorax is completely covered by a cloaklike investment of soft tissue called the carapace. Most specimens are females, but males can be recognized by their evident similarity to females in the same sample. The soft carapace is due to the fact that this organ serves as the respiratory organ, being permeated by blood vessels; it is only loosely attached to the thorax and can be lifted up to disclose the thoracic segments beneath. There are no gills (gills are filamentous structures in most crustaceans, carried at the bases of the thoracic limbs where they pass under the carapace). Mysids are eaten by fishes. Some mysids are benthic, living on the bottom. The food of mysids comprises smaller plankton, such as copepods.

HYPERIIDEA. These animals are members of a crustacean order called Amphipoda, characterized by having no carapace (so the thoracic segmentation is visible externally) and by having the whole body compressed (that is, flattened in the vertical plane). The best-known members of the Amphipoda are the flealike sandhoppers of sandy beaches and the intertidal region. Most amphipods are bottom-dwellers, but one suborder, the Hyperiidea, form an important part of the plankton of the open ocean. These animals have very large, unstalked eyes. They live most of their lives in the faintly illuminated zone below 100 fathoms. Most hyperiids parasitize jellyfishes of the zone they inhabit. Hyperiids are eaten by deep-water fishes such as *Alepisaurus.*

Phronima is a genus of Hyperiidea of interest on account of its peculiar ecology. Spending the greater part of its life in the deeper regions, when sexual maturity is reached *Phronima* swims to the surface of the ocean and there captures a doliolid, or sometimes a salp. The captor kills and eats the captive and then creeps inside the floating transparent barrel-shaped test of the dead doliolid and takes up residence there. In the surface waters the mating and shedding of sexual products occurs. The young hatch out at the surface, later sinking down to the depths normal for the species. Thus *Phronima* has a *vertical migration cycle.*

EUPHAUSIACEA [LANTERN SHRIMPS]. Euphausians are strictly planktonic organisms, unlike mysids (which they somewhat resemble). Their characteristic features are the feathery gills placed at the bases of the thoracic limbs, and *fully exposed to view,* because the carapace in the euphausians does not reach all the way down the sides of the thorax. This character, by the way, distinguishes the euphausians from the true shrimps (*Decapoda natantia*), where the gills are similarly developed but are completely covered by the sides of the carapace. Phosphorescent organs, which give the euphausians their distinctive name of lantern shrimps, are visible in preserved specimens as red spots, two pairs on the thorax and four pairs on the abdomen. In life lantern shrimps emit brilliant flashes of light visible at night when they rise to the surface. Euphausians are represented in the oceans by not many species. Each species prefers particular types of water mass, defined by temperature and salinity. Several species girdle the earth in the southern Pacific Ocean, each band lying at a particular latitudinal range. They constitute a major item of food for the large filter-feeding pelagic vertebrates, such as baleen whales. Euphausians feed on copepods.

Larval Stages

Larval stages of both planktonic and benthonic animals are found in the plankton, usually at or near the surface. This circumstance is probably attributable to the following: Larval stages are small and therefore require a plentiful supply of small particles of food; but small particles are rare in deep water and at the deep bottom, because small particles fall slowly, in accordance with Stoke's law, and therefore are more likely to be eaten up by small organisms. Therefore organisms should increase in size within limits as depth increases, the limits being set by the total availability of food. These deductions lead to the inference that larval stages of deep-water animals must rise to the surface in order to survive the risk of starvation at lower levels. When large enough, they can return to the region where their parents lived. The life history of euphausians includes a *Furcilla* larva, as it is called, somewhat resembling a copepod, but differing in having two long flotation rods, one in front of the thorax and one behind. Similar larvae occur in the life history of lobsters and crabs, where the larva is called a *Zoaea.* The *Furcilla* larva grows into a *Cyrtopida,* characterized by the remarkably long eyestalks. Each stage changes into another by means of shedding the skin. Numerous *ecdyses,* as these sheddings are called, occur in one animal's lifetime. The youngest stages feed on diatoms, gradually changing to copepods as they grow older and larger. The bizarre appearance of larval stages such as these puzzled the early oceanographers, who supposed them to be adult animals and who gave the special names they bear, originally in the sense of generic names.

IANTHINIDAE [VIOLET SNAILS]. Violet snails belong to the molluscan genus *Ianthina,* mainly restricted to tropical and subtropical waters. Their delicate shells are swept by currents on to distant shores, including Cape Cod. They inhabit the sea-air interface, held there by rafts of bubbles

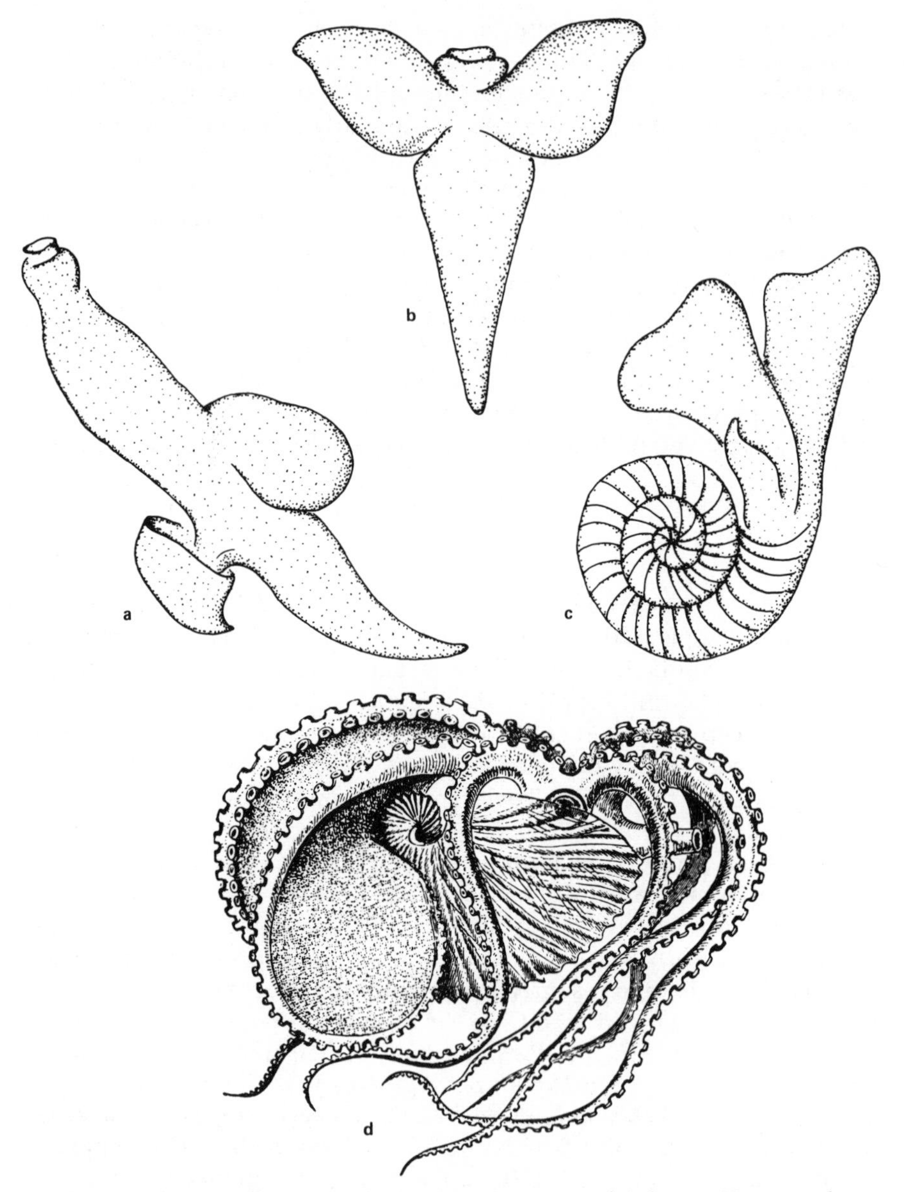

Figure 15. Pelagic mollusks. (a) *Carinaria*, X 0.5, tropical Atlantic, Heteropoda; (b) *Clione* and (c) *Limacina*, X 4, tropical Atlantic, Pteropoda (23); (d) *Argonauta* (female), X 0.5, tropical oceans, Cephalopoda (15).

and slime that they secrete. *Ianthina* lives in swarms, and it is a predator, attacking similar swarms of *Physalia*. Each snail creeps on board the bladder of a *Physalia* colony, abandoning its own raft. It remains on board the captured vessel until it has killed and eaten all the zooids in the colony. The snail then spins a new raft and takes off again in search of another *Physalia*. Siphonophores have existed for the past 500 million years, as the fossil record discloses, and *Ianthina* ranges back for at least 100 million

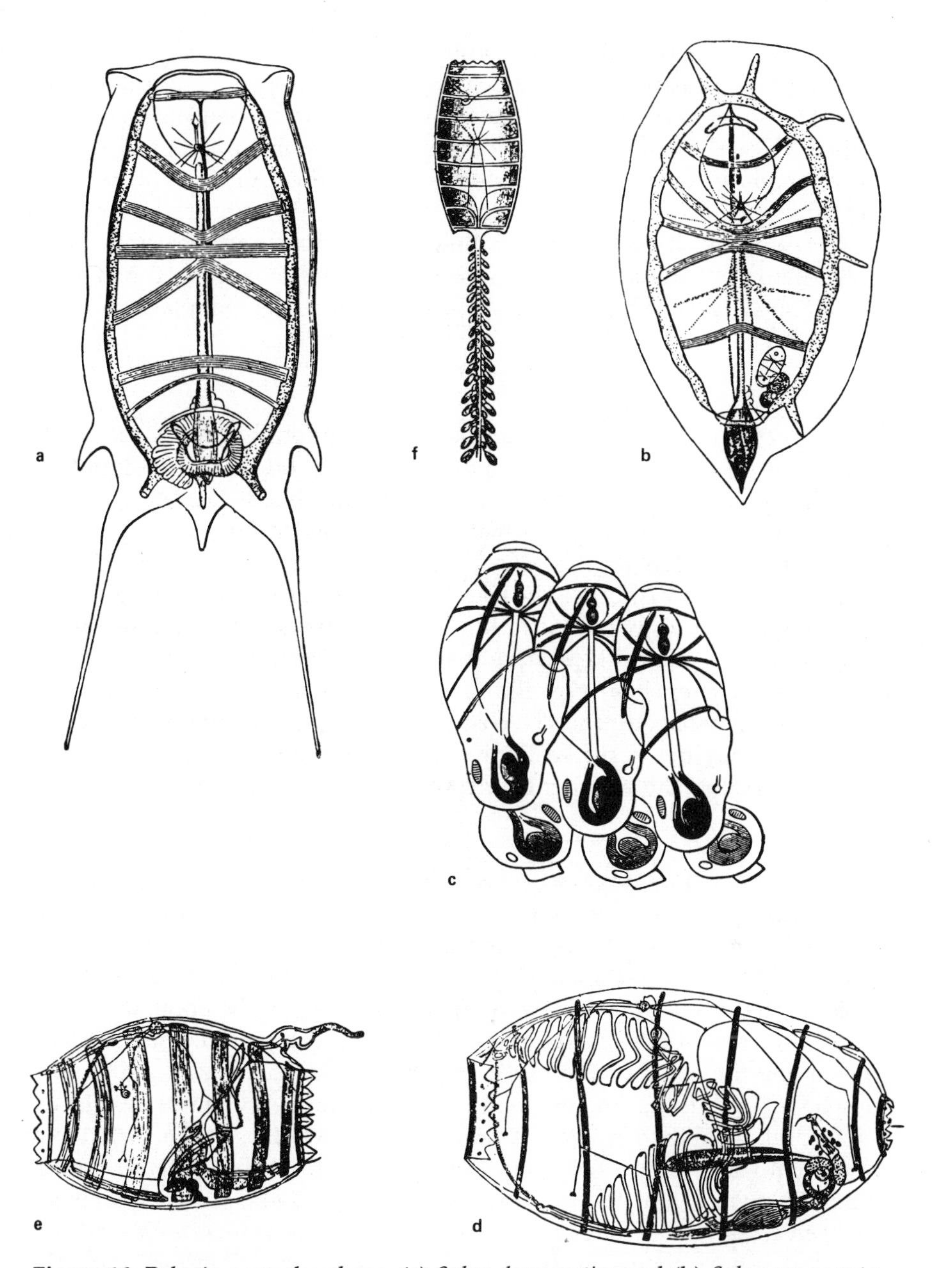

Figure 16. Pelagic protochordates. (a) *Salpa democratica* and (b) *Salpa mucronata*, both X 0.5, world-wide, Salpida; (c) part of young chain of *S. democratica*, X 1; (15); (d) *Doliolum*, X 0.5, world-wide, sexual stage; (e) nurse stage; (f) nurse stage trailing colony (15).

years. Thus peaceful navigation preceded piracy. The violet snails, together with the heteropods and pteropods are classified in the class Gastropoda of the phylum Mollusca.

HETEROPODA. The large sea snails constituting this order have a glassy transparency and a pair of very conspicuously pigmented eyes. The small shell, shaped like a Trojan cap, is carried on a similarly shaped visceral hump. Heteropods are mainly found in tropical waters, but apparently they are carried northward by the Gulf Stream, for they have been collected near the Azores, at 40° north latitude. Heteropods swim rapidly with the aid of muscular contractions and by using finlike lateral flanges of the body. They swim upside-down, and capture and eat small fishes, also jellyfishes. They are seldom taken in plankton nets, presumably because they can swim out of the net.

PTEROPODA [SEA BUTTERFLIES]. Pteropods are really sea snails, and they are classified therefore among the gastropods of the phylum Mollusca. They do not all have shells, but where the shell is lacking, the body still has a general similarity to that of the shelled forms, so they are not difficult to recognize, for the pair of winglike fins is characteristic. Usually the body is brightly colored. The species live in swarms in the ocean, somewhat resembling flights of butterflies as they pass by ships. Some species have the silvery mirrorlike surface that fishes of the lower zones adopt (in conditions of even low-level illumination, a mirror becomes invisible). Shellbearing pteropods characterize warmer regions of the ocean, where their shells may lie on the seafloor in such profusion as to give the name pteropod ooze to the sediment concerned.

CHAETOGNATHA [ARROW WORMS]. The arrow worms have no known close relatives, save only some Cambrian fossils about 600 million years old. They are therefore placed in a phylum by themselves. They are highly predatory animals, ranging the oceans from the polar regions to the equator. They have a pair of jaws at the anterior end of the body, and a pair of tail flukes at the opposite extremity. There are paired horizontal fin folds along the sides of the body, reminiscent of stabilizers, which they may be. Only the tail fins are motile. They swim feebly, but they are voracious and are able to swallow a vertebrate fish whole, even though it may be several times their own bulk. Examples are the deep-water and polar genus *Eukrohnia,* and *Sagitta* in warm, shallow waters. Arrow worms are themselves eaten by vertebrate fishes. Some arrow worms are associated with the sonic echoes produced by schools of organisms at particular depths. In the course of the day, arrow worms rise to the surface at dusk and at dawn and occupy deeper zones at noon and midnight.

SALPIDA. Salps are elongate torpedo-shaped planktonic organisms of a glassy transparency, usually colorless. Some specimens are asexual; these are called nurse stages. The nurse produces babies by the convenient device of growing a cord of undifferentiated tissue out of her body, the cord (called a genital stolon) subsequently dividing up into a chain of babies, all joined side by side like mass-produced siamese centuplets. As her babies grow larger they form a train in which the patient nurse is the locomotive, trailing her progeny behind her as she swims by the usual muscular contractions of planktonic creatures. Eventually the young salps become sexually mature, and they then tend to break up into smaller chains of 10 or 20 individuals, each one of which is sometimes male, at other times female. They shed their ova and sperm into the sea water, and the resultant abandoned progeny grow up to be the sexless nurses. Thus there is an alternation of generations, nurse stages and chain stages appearing in turn. The internal structure of a salp shows that it is related to other animals with archaic features of the ancestors of vertebrates; so salps are classified in the phylum Chordata. *Doliolida* is related to Salpida but has even more extraordinary reproductive aberrations. Here the nurse doliolid grows a tail along which the outer two rows become food gatherers and community stomachs; the members of the middle row do nothing but reproduce sexually, while the nurse degenerates into a tractor for the whole of her varied progeny. The babies that the middle row of zooids (gonozoids) have turn out to be tadpoles, which, after swimming around for a while, redesign themselves into proper doliolids and assume the sexless state of the nurse.

PYROSOMATIDA. Members of this group are related to the salps and doliolids, but differ in having a permanently colonial habit and no alternation of generations; also phosphorescent properties occur. There is no vernacular name in general use, but in New Zealand the fishermen called them mermaids' socks; they look like cones made of gray flannel, more suited to be ice cream cone covers, though occasional giant specimens might fit a very small mermaid. The individuals in the colony are arranged side by side so that their mouths all lie on the outside of the cone. The common species is *Pyrosoma atlanticum atlanticum,* which, despite the insistence of the name, is actually widespread throughout the Pacific. It is considered good eating by a few fish, including *Alepisaurus,* but most fish ignore it.

FISH LARVAE. Plankton hauls and the stomach contents of plankton-eating fishes commonly yield some tadpolelike young stages of various vertebrate fishes. Adult fishes can swim against the current and thus they are classified as nekton, not liable to passive dispersion. Their young stages, however, lack the mobility of the adults and drift with currents. They probably constitute a substantial part of the diet of other, larger plankton, for the infant mortality of fishes must be astronomical. For example, a female mackerel produces 500,000 eggs at one spawning, yet the mackerel population is not known to be increasing; almost all of the young probably die at a very early stage. Since mackerel are plankton eaters, and their progeny are eaten by the planktonic organisms that adult mackerel eat, the recycling of materials could lead to high concentrations of the nonexcretable toxic pollutants in these fishes.

ENERGY TRANSFER IN THE SEA. Since the phytoplankton alone captures radiant energy from sunlight, and can exist only in the illuminated upper waters of the ocean, it follows that the deeper regions, in darkness, acquire energy needed for vital activities of deep-water animals by stealing it from above—that is to say, by feeding on organisms whose food chains lead back to the surface of the sea. But surface organisms do not wander far from the surface, and the animals of deep sea do not visit the surface. The manner in which the energy transfer is effected is discussed in Chapter 6.

4

EXPLORING THE CONTINENTAL SHELF

The *continental shelf* is a fringe of relatively shallow seabed surrounding the major land masses to a depth of about 200 m, or 100 fathoms (600 feet); beyond the edge of the shelf the seafloor plunges more steeply to a very much greater depth. The relatively steeper slope, or *continental slope* to give it its full name, penetrates downward to a depth of about 3 miles, where the seabed evens out into the *abyssal plain,* which underlies most of the world oceans. The average depth of the continental shelf through most of the world is about 132 m, or 70 fathoms (420 feet). The upper part of the continental shelf is accessible to divers using standard types of scuba equipment, such as is familiar to many young men and women to-day; many interesting features of the continental shelf as an abode of life can now be observed directly by students of marine biology. In addition, by observing seabeds and land forms beneath the surface of the sea, one may often form a clearer idea of the processes at work in shaping the coasts and shelf habitats of the world. Thus the modern student may see things that his predecessor of a generation ago might only envisage as a mental picture, laboriously developed from bottom sampling devices, or later by remotely controlled underwater photography.

Types of Coastline

Two main kinds of coast are discernible, known respectively as coasts of *submergence* and coasts of *emergence*. The distinction, of course, seems to imply that coasts have either been sunken beneath the sea or raised above it—and in general it seems that such changes have in fact occurred very widely in recent geological time and probably also throughout most of former periods of earth history. A few coasts seem to be stable, but most show some evidence of the heaving that periodically afflicts parts of the earth's crust.

COASTS OF SUBMERGENCE. These coasts are often recognizable by the extreme degree of indentation of the land-sea margin, with sometimes more or less complexly shaped peninsulas separated by intervening complicated arms of the sea. Shores in this category have evolved from a former land surface on which rivers and streams had carved out a dendritic pattern of confluent valleys, lying between equally complex ranges of hills; then, at some later geological date, the whole area had suffered down-warping, or uniform depression without warping, so that the sea has been able to flood the lower portions of the original valley systems. In some parts of the world submergence of a coastline has occurred within historic times, with the result that buildings or other works of man, even whole ports or cities, have been swallowed up by the sea and now lie wholly or partly beneath the waves. For example, the coast of Lebanon near the ancient Phoenician city of Sidon has undergone a partial submergence during the past 1000 years. One result has been that fortresses built by the marauding crusaders in the Middle Ages now have their lower floors awash by the sea; we may be sure that the architects of these forbidding castles would not place them in situations subject to daily inundation by the tides. Submarine archeologists have found still older remains of the classical port of Tyre, now completely submerged; and the same is true of Sidon and of some other ancient ports. It is a matter of historical record that as long ago as the fourth century B.C. it had become clear to thinking men that the sea and land may change their relative levels. The transient nature of coasts and of river valleys is noted by Aristotle, for example, in his *Meteoritica*, and the poet Ovid of Roman times also comments on such matters, which evidently were thought of as quite reasonable interpretations by the ancients. In medieval times such ideas became heretical. The Persian astronomer Omar Khayyam in the twelfth century observed that the Caspian Sea had changed its level; however, his deductions were found in contempt of the Koran, and he was exiled for the rest of his life to central Asia, where he found time to write the Rubaiyat. In Europe the course of events finally forced on men the realization that

Figure 17. Coast of submergence (Picton, New Zealand), with drowned river-cut valleys continuing out on to the continental shelf.

land and sea interchange. Most startling of the evidence for observers in Britain were such events as the disappearance beneath the sea of the Godwin pasture lands (belonging to the Saxon earl of that name) in the time of Edward the Confessor and of William the Conqueror, in the eleventh century, and recorded in *Domesday Book* (indeed, the very inferences that cost Omar his liberty were simultaneously recorded as observed facts by contemporary thinkers some thousand miles to the west). More alarming still for England was the total loss of her greatest port, Dunwich, reported to be in flourishing condition when *Domesday* tax men spied out William's new territory. Yet, 700 years later, all that remained were a few hundred houses on the edge of the sea, their church towers emergent from the seabed further out; then, one night in the mid-seventeenth century the last 200 houses were swallowed up, and Dunwich became only a memory. Meantime a new port, London, had taken the place of Dunwich and, to judge by the increasingly severe tidal floods now periodically lapping the lower reaches of the Thames, it would appear that the days of London are numbered.

COASTS OF EMERGENCE. In marked contrast to such coasts are shorelines that have suffered upward warping and that have therefore recently emerged from the sea, or shores that have undergone gradual accretion by the action of the tides or current, such that sand is piled up along the beach, and the beaches advance in the seaward direction. Such coastlines generally have relatively simple contours, in the form of gentle curving strandlines or almost straight strandlines. It should be noted, however, that in desert regions that have no rivers, the coasts tend to be of simple contours, irrespective of whether there has been any change in the relative elevation of the sea and the land.

The typical aspect of an emergent coastline includes long sandy beaches, which stretch seemingly without limit into the distant horizon. Hills, if present, do not make direct contact with the sea. Instead, there is an intervening plain or *piedmont* between the strandline and the foot of the hills, with sand dunes defining the seaward margin of the plain. As one travels inland from the coast itself older dunes can be observed, now covered over by a veneer of grass or other low vegetation; and further inland still one encounters scrub and woodland covering the land. If the soil is examined it may often be found to consist largely of marine sands, as can be demonstrated by the presence of subfossil seashells, especially of clams and such organisms as inhabit sandy seafloor. Observations of this kind, and inferences that we now see to have been correct, were made by Fracastori in Italy in 1517, but unfortunately his ideas were considered to be contrary to the biblical revelation of the Deluge, so they were dismissed and forgotten. Leonardo's drawings show that he, too, understood these matters; but since he did not publish his thoughts, they went to the grave with him, only to surface when the English monarchs began to publish the archives of Windsor, with their notebooks and drawings of the Italian polymath.

Sometimes, if actual uplift of the land has occurred through a series of sudden jolts with accompanying earthquake, distinct terraces may be observed, becoming progressively higher as one passes inland. Such *stepped profiles* are common on the restless Pacific margin. Usually the lowermost terraces, or *raised beaches,* still carry wave-cut boulders, dry rock-pool basins, and even the remains of seashells left by the former denizens. Where human settlement has extended over a sufficient space of time historical records often are available from which one can relate a particular raised terrace to a particular record of earthquake uplift. Often the raised terraces are seen to be tilted, showing in such cases that the uplift has been differential. In the 1855 earthquake of Wellington, New Zealand (the first case to be studied by the British geologist Charles Lyell), a differential uplift took place such that tilted wave-cut platforms now stand above the present strandline, raised about 3 m above mean sea level where the uplift was most severe and falling gradually to zero where the fulcrum of the tilting was located.

COMPLEX COASTLINES. Peculiarly complicated situations occur in some parts of the world where earthquake and volcanic activity is common. An example is the central and northern margin of the Mediterranean. In the island of Sicily we find archeological evidence of alternating uplift and submergence. Striking evidence is seen in the columns of Greek and Roman temples, now standing above the sea, yet carrying the unmistakable borings made by marine mollusks into the limestone of the temple columns. The borings for any given temple are found to extend to the same height on the various columns, showing that at some unrecorded epoch *after* the temple was built the land must have sunk beneath the sea, and then, at some later epoch, rose again. The lack of historical records of such events is a reflection probably of the decay and collapse of learning during and after the Dark Ages, when barbarians overran much of Europe. Evidence of retreat of the sea since the eleventh century is seen in the present location of the bronze rings to which the Saracens anchored their ships in Sicily. The rings (cemented into bedrock) are now found at distances up to 3 miles inland from the coast. Evidence of similar changes of this kind can be found from points scattered along hundreds of miles of the Sicilian and Italian coasts.

SEA LEVEL CHANGES. The foregoing examples are from regions of the earth's surface where earthquakes and other seismic disturbances are common, regions where the earth's crust periodically heaves and warps, so that the surface is liable to elevation or depression. Thus consequences of the kind noted are not particularly surprising. But what do we make of cases where we can recognize what appear to be definite wave-cut platforms or former beaches, elevated well above the present level of the sea, and occurring in South Africa and in the southeastern United States, where seismic disturbances are rare and only of minimal character, without measurable effects on the level of the land? In regions of this kind, believed to be extremely stable, it is nonetheless quite common to find well-preserved wave-cut platforms lying at a height of about 30 m, or roughly 100 feet, above the surface of the sea. To follow the unraveling of this interesting problem, as it was resolved by geologists and naturalists during the past century, we may turn aside from the sea briefly, to glance at some instructive features of inland topography and to examine the zonation of plant and animal communities.

THE ALTITUDINAL ZONATION OF BIOTA. Were we to join a party of climbers in, for example, the Great Smokies region of the southern Appalachians, we would note that the lower part of the range is covered by deciduous summergreen forest, but as we ascend the mountains we leave the summergreen trees behind, to pass through a higher zone of evergreen trees, such as spruce and pine. Then, reaching higher elevations, the spruces thin out behind us as we gradually emerge on to an open grass and herbland, the subalpine moor. On these upland meadows are many kinds of flowering herbs that do not grow on the lower levels of the Carolinas; an example is the moss campion, a beautiful carpet plant that bears in season a profusion of red flowers. Yet this same plant, and many others that are found with it on the open windswept tops of the Appalachians, occur also on other, often distant, mountain tops, so that we can recognize the existence of a distinct subalpine flora. More surprisingly, they grow also on the far northern tundra lands of the Arctic region, in Alaska, in northern Canada, in Greenland, and, on the other side of the Atlantic Ocean, in Lapland, in northern Siberia, and in fact all around the North Pole, wherever there is exposed land not covered by permanent ice. Clearly, the distribution of moss campion *(Silene acaulis)* and of the associated subalpine plants is rather remarkable. If we examine other, comparable, data a similar result emerges. The Himalayas are six times higher than the Appalachians, lie much nearer the equator, and are on another continent. Yet here, too, at the appropriate elevations a lower-level deciduous forest gives way to a higher evergreen conifer forest of spruce and pine, replaced still higher by a tundra, then finally a rocky ice desert as the summits are reached. Some of the same mountain plants occur as in North America, and others that are related. Indeed, the phenomenon is general for, if we cross the equator, we meet it again in the Andes, on Kilimanjaro, and in the Southern Alps of New Zealand. However, in the far southern alpine systems, the evergreen mountain forests, though present, do not comprise spruce or related needle-leaf trees—instead, they are the evergreen southern beeches, or *Nothofagus* flora. And just as we found northern mountain herbs in the far northern tundras near the Arctic, so also in the Southern Hemisphere we find mountain herbs in Australia, New Zealand, and South America that are also present in the lowland moors of the subantarctic islets around the margin of Antarctica.

One conclusion we can draw is that as one climbs higher, on land, floras and faunas proper to colder latitudes appear; if you climb, you are experiencing much the same changes in ecology as occur if you travel northward in the Northern Hemisphere without leaving sea level, or if you travel southward in the Southern Hemisphere. So a New Englander can experience Arctic environments simply by ascending the mountains of New Hampshire, for Labrador plants and animals grow on the upper reaches of Mount Washington.

THE BATHYMETRIC ZONATION OF BIOTA. Now, the interesting and relevant aspect of the foregoing rule is that it also applies to the continental shelf, only the sign of the equation is reversed. If you dive deeper, you will encounter progressively colder water, for cold water is denser than warm water and tends to sink. This means that in the deeper parts of the shelf live animals that also inhabit the more northern and colder seas around the North Pole. So the quickest route to the Arctic Sea for a Bostonian or Californian is simply to dive. For the Bostonian, the dive can be well within the capacity of a scuba diver of 60 or 80 feet certification; for at that level live animals that may be found immediately beneath the polar ice. In 1953, as a visitor from the Southern Hemisphere in Lapland, I well remember the interest and excitement aroused when, on a visit to Narvik Fjord, specimens of such Arctic sea snails as *Buccinum undatum* and others of the same northern family first appeared in our collections. They were carried back home so that New Zealand students might see what Arctic animals look like. Years later, on a tropically hot day of the Massachusetts summer, when shallow-water dredging was in progress on the New England shelf, I was astonished to see my old friends of the Norwegian Arctic coming up from below, a vivid lesson on how near at hand are Arctic denizens off our local coasts.

THE ORIGIN OF RAISED BEACHES. With this preliminary diversion behind us, I now return to the core question posed earlier; namely, what brought into existence the elevated wave-cut platforms that we so often find on stable coasts at heights of about 100 feet above sea level?

About 100,000 years ago the climate of the whole earth seems to have been decidedly warmer than now is the case. At that time the swamp cypress of the southeastern coasts ranged as far north as the latitude of Philadelphia, whereas its present northern limit is on the Chiptant River, Dalmaria Peninsula. The coastline lay further inland than its does at present. This was because the higher temperature of the planet had caused the polar ice to retreat, in fact probably the whole Arctic ice cap had melted, and some part of the greater Antarctic ice cap had also melted. So the general level of the sea stood higher than is the case today, for so much ice-melt water had been added to the world's oceans. Thus all over the world where stable crust exists, we may expect to find remnants of the old beaches cut by the waves of that ancient high-level ocean. But in restless regions of the earth's crust the old beaches no longer necessarily remain at a height of 100 feet above the sea, for some of them have since been warped upward, others downward, and others again tilted at various angles, or arched or troughed; so our best estimates of their original heights have to come from the shores of more placid regions, such as have been mentioned.

STILLSTANDS. The sea must have remained at this elevated level for a long period, for there was time enough for cutting quite extensive wave-cut platforms. The term *stillstand* is applied to any such prolonged episode of constant sea level in earth history, when sets of corresponding platforms are cut by the sea at matching levels in different parts of the world. In the case instanced, the sea seems to have remained at about the same level until about 80,000 years ago. Then something happened to the earth's climates, and our planet began to grow colder. We can infer, from other evidence, that ice reformed at the North Pole, and the ice sheet gradually spread outward until it crossed the northern coasts of the northern continents of Europe, Asia, and North America. This spread of the ice 80,000 years ago is known as the Wisconsin advance (because the glacial deposits associated with it are best known from Wisconsin examples; the corresponding term for European deposits is Wuerm). From various data elucidated by geologists we now believe that the rate of southward advance of the ice front averages roughly 50 m (160 feet) a year. As the ice crept southward, it overwhelmed the tundra region, then the taiga, or boreal evergreen belt, overturning the spruce trees, grinding across the tops of the shattered tree trunks, and burying the remains of forests beneath it. Meanwhile a new and more southerly tundra zone became established south of the southern margin of the ice, and polar animals ranged the lands of a desolate belt that stretched westward from the latitude of New York as well as through much of France and Germany and other lands in similar latitudes.

CARBON-DATING. Our ability to date the event and to measure the rates of advance of the ice stem from the circumstance that fragments of the old forest woods were preserved beneath the ice, and these prove to be suited to use in the method of age estimation known as *carbon-dating*. The carbon bound up in the lignin and other organic substances in the wood was obtained (at the time the tree was actually growing) from the carbon dioxide of the earth's atmosphere of that epoch. But the carbon dioxide of the earth's atmosphere is of a dual nature, one part of it constructed from a stable molecule of carbon called C_{12}, with a second portion, of known original percentage in the atmosphere based on an unstable isotope called C_{14}, and derived from atmospheric nitrogen atoms that have been struck by cosmic rays. The C_{14} slowly disintegrates into C_{12} at a known rate, such that half of the C_{14} atoms decay to C_{12} in a little more than 5000 years; if we assume that cosmic ray bombardment of the earth's atmosphere has occurred always at a stable rate, the proportion of C_{14} to C_{12} in the air will always have been the same. In that case, the initial ratios of the two carbons in the wood must have matched that of living wood today. Therefore any difference in the ratios in the fossil wood from that of living wood must be due to the disintegration of the original C_{14}, at the known rate; from this inference derive the equations that permit the determination of the age of the fossil wood.

The Wisconsin advance proved to be the last of a series of ice ages that we know to have occurred at various times over the past 3 million years. This last glaciation episode lasted from about 80,000 years ago until about 12,000 years ago, that is, until about 10,000 B.C. And then something happened again to the earth, and its climates became warmer. The region of New York State had remained all this time under ice or in a general tundra zone, so when the earth began to grow warmer, the initial effects were not very pronounced, for the ice sheet seems never to have transgressed the northern shore of Long Island Sound. But further north, in the Boston area, the changes that now ensued must have been much more dramatic. During the Wisconsin glaciation the coast of New England would probably have resembled that of Antarctica today. The greater glaciers discharging their ice into the Atlantic near Boston would have released icebergs from the ice tongues floating out into the Gulf of Maine, and a polar fauna, presumably dominated by polar bear, would have roamed the New England coasts. Inland would have been herds of musk oxen, with caribou grazing the tundra lands to the south of Long Island Sound.

RETREAT OF THE WISCONSIN ICE SHEET. With the amelioration of climates about 10,000 B.C. the ice sheet, which had reached to the south coast of Connecticut, now began to retreat. Within about 1,000 years the Boston region must have been ice-free, and the tundra had moved in. The newly exposed lands would now have looked something like the present aspect of Prince Edward Island, where tundra still lies at sea level. Man now began to occupy the lands, following the ice front as it retreated north. By 9,000 B.C. roving bands of Indians were crossing coastal Massachusetts and traveling the coastal lands as far north as New Brunswick. One such roving party is known to have encamped on the north shore of Boston Harbor, building cooking fires whose charcoal survives to yield carbon, dating the event. They also left behind stone arrow and spear points whose style recalls that of other contemporary Indian peoples further west, whom we know to have hunted buffalo. Perhaps the Bull Brook visitors to Boston were also following buffalo, or musk oxen. After the tundra time followed the period of development of boreal forests of evergreen conifers, and then later still followed the broadleaf deciduous forests, leading to the state in which the lands are today, or in which they were before the farming and housing development occurred.

MODERN RISE IN SEA LEVEL. Ever since 18,000 years ago the polar ice caps have been melting, sometimes slowly, sometimes faster, with occasional reversals of the warming trend. On the average during the twentieth century the ice melt has caused the sea level to rise at an annual rate of 1.5 mm, and this is probably about the rate for the last few hundred years; if so, then Boston Harbor today has a sea level about 1 m (3 feet) higher than when the pilgrims landed three-and-a-half centuries ago. Over the past 25 years the rate of sea level rise has averaged 2.5 mm per annum, or about an inch in 10 years. So, during the 20 years since the great North

Sea floods, when a spring tide, a barometric low (which causes sea level to rise), and an onshore storm wind all combined to make the sea rise to an unusual height, the general level of the sea has risen by another 2 inches all over the world. But 2 inches was that narrow margin of safety by which London escaped by the skin of its teeth during the last abnormal tides, so the margin of safety of the underground railways of that city is now exceedingly narrow. So, as already noted, it seems that London must ultimately suffer the same fate as Dunwich. The sea has still another 100 feet to rise before it reaches the high levels of the period before the Wisconsin glaciation; at present none can predict how fast this rise will continue, or if the sea will return to its former height. But inevitably, should the polar ice melt more completely, all the city ports of the world will be submerged. Although it does seem that we are presently headed toward the Deluge, it is only fair to say that some geologists have a crystal ball that predicts a return of the ice sheets rather than continued ice melt, and a colleague who read the manuscript of this book and expunged some inaccuracies from it would have me maintain an equivocal posture in this matter.

It follows, of course, from what has been noted about the higher sea levels during warm, or *interglacial,* periods, that there must have been former lowered sea levels during the cold, or *glacial,* periods. So there must exist somewhere beneath the present sea level some invisible wave-cut platforms that were formed during the last ice age, when the sea level was lower. This is a topic on which scuba divers can throw light, by watching the seabed for signs of submarine wave-cut platforms or former storm beaches, marked by horizontal bands of boulders or pebbles, and for evidences of the submergence of coastal strips of land that once lay above the sea when the ocean was at a lower general level. Evidences of such things come in the shape of bones and teeth of ice-age animals, such as mastodons, terrestrial by habit; yet their bones are dredged in fishermen's nets from shelf seafloors that were once dry land. Similarly, from the submerged Dogger Bank of the North Sea, fishermen have for generations fished up the stone axes and other signs of material culture that ancient Teutons left behind, men who once hunted the coastal lands before these were swallowed up by the rising level of the world ocean. Explorer divers working on the sides of coral reefs have discovered features that resemble wave-cut platforms beneath the sea. These are now interpreted as former wave-cut platforms, dating from various standstills during the general progress of the return of milder climates after the end of the Wisconsin glaciation.

Substrates on the Continental Shelf

We may now glance at the various kinds of bottom, or *substrate,* that a scuba diver will encounter as he explores the continental shelf. These different types of bottom provide different kinds of habitat for the animals and plants that inhabit the shelf. But some animals and plants are better adapted to live on one particular kind of substrate, which they therefore prefer or which natural selection ensures that they have better survival chances on.

SAMPLING METHODS. It is not the purpose of this text to describe the mode of operation of oceanographic sampling equipment. Suffice it to note here that standard collecting equipment for bottom animals, or *benthos,* includes various types of *dredges* and bottom trawls, various types of automatic devices for biting pieces of the seabed and returning it to the surface, called *grabs,* various types of *core samplers* for obtaining vertical sections through the seabed, and automatic *cameras* for photographing the seafloor by remote control. Instead of considering the aspects of the substrate disclosed by these devices, it will be more convenient here to summarize the main facts as they might be determined by a scuba diver who is moving across the continental shelf, and observing, collecting, and photographing the habitats as he encounters them. This, in practice, is now becoming a standard procedure among younger marine biologists. For the deeper parts of the continental shelf we must still, of course, rely on the older methods and on the remotely controlled cameras of the oceanographer, still the most important technique in studying the ecology of deep-water faunas beyond the edge of the shelf. However, knowledge of man's physiology under great pressures is increasing rapidly, and the probability now appears great that before very long the entire continental shelf will become the range of the professional scuba diver.

HARD SUBSTRATES. This type of substrate commonly occurs on coasts where there are wave-cut cliffs overlooking wave-cut platforms carrying masses of emergent bedrock and also on more gently profiled coasts where a hard bedrock yields large rounded boulders or cobbles. Examples of the former are very frequent around the margins of the Pacific oceans; examples of the latter are found on most ancient igneous or metamorphic foreshores of the world, or in places where hard ancient sandstones encounter the sea. If a diver-explorer enters the sea from such a coast, he will first encounter an intertidal zone of hard slabs or rock reefs or boulders, covered by such plants as the large brown seaweeds and by such animals as barnacles and mussels, all of which can withstand the heavy surf commonly generated in these localities. Then, as he enters the water, he crosses boulder bottom, passing gradually into masses of smaller rounded stones called cobble, and finally into coarse gravel. Foreshores and inshore bottom of this type are common in cold climates but can also occur in the tropics if for some reason coral is not able to gain a permanent footing. On parts of the Virgin Islands, for example, cobble bottom occurs near coral reefs and is invaded and occupied by the same kinds of hard-bottom denizens as also frequent coral reefs. This circumstance also illustrates the fact that organic structures such as *coral reefs* are themselves a type of hard bottom: and a logical extension is the realization that a mass of shellfish, or the dead shells of shellfish, can also be considered as hard bottom for, even though the shellfish rest on mud or sand, they themselves offer a firm substrate to any organisms that live on them, such as sea urchins or seaweeds. Such *shellbanks* may often constitute *temporary hard bottom,* for a casual disturbance may cause silt to accumulate over a shellbank, or the shellbank may be overwhelmed by a submarine silt flow or turbidity current, converting the region back to soft substrate and also stressing the temporary nature of some hard bottom.

ANGLE OF SLOPE. If you enter the sea from a coast such as that of New Brunswick, or of Nova Scotia, where the intertidal zone is of a kilometer or so in breadth, you are obviously going to encounter similarly gentle-sloped conditions for a considerable way out under the water, gentle slope with a great deal of sand and fine gravel, pebbles, and shells. Sun stars, such as *Crossaster*, favor these substrates, and quite similar fossil starfishes are known from sedimentary rocks of ancient seabeds up to 300 million years old; this is just one of many hints we have that the underwater seascapes seen by divers are probably among the most ancient scenery to be found on earth, so that a diver glancing about himself on the seabed may often, in effect, be peering down the corridors of time and seeing the world as it was and as it has remained during vast intervals of elapsed prehistory. Farther out beyond the pebbly region is a finer bottom of sand. With the decrease in the size of the inorganic constituents of the bottom there is an increase in the organic proportions, such as shells of dead animals, and also the less visible decaying soft parts, or the bacteria that feed on them.

On some coasts the slope of the bottom may be so steep as to be almost vertical, and perhaps the entire distance from the seaward margin of the forest, across the splash zone, the barnacle zone, and the green and brown alga zones down to low-tide level may occupy no more than a vertical distance of 3 or 4 m. A shelf margin of this kind occurs on the seaward edge of mountain regions that were heavily glaciated during the ice ages—for example, the fjords of Norway, or of Alaska, or of the southwestern part of New Zealand. A fjord is a very deep inlet of the sea, formerly occupied by the lower part of a great glacier, capable of eroding below sea level (on account of the weight of the ice), unlike a river. Generally in fjord country the depth of water inshore is of the same order of magnitude as the height of the mountains that rise directly out of the sea; in the case of southern New Zealand this means a water-depth of 1 mile within a few hundred yards of land. In such localities the continental shelf may be said to be nonexistent.

As one proceeds out across the shelf for the most part only soft sediments are encountered, for only the finest particles of silt brought down by rivers are able to float in suspension far out from the coast before sinking to the bottom. However, hard cobble may occur where melting icebergs have dropped moraine rubble or where submarine volcanoes have erupted lava. As we cross the edge of the shelf, and pass down the continental slope and on to the abyssal plain, only very fine soft sediments, or *oozes*, are found, and the sea stars and sea cucumbers that live under such conditions.

SUBDIVISIONS OF THE PHYLA. To permit a more precise designation of the groups of organisms that live on the continental shelf, Table

Table 9. Some conspicuous groups of shelf invertebrates.

PHYLUM	SUBPHYLUM OR CLASS	SUBCLASS OR ORDER	INCLUDED MEMBERS
Coelenterata	Class Hydrozoa		Hydroids
	Class Scyphozoa		Jellyfishes
	Class Anthozoa		Sea anemones and corals
Ctenophora			Comb jellies
Platyhelminthes	Class Turbellaria		Flatworms
Nematoda			Roundworms
Annelida	Class Polychaeta		Bristle worms
	(Other classes)		Sipunculids, and others
Arthropoda	Subphylum Crustacea	Copepoda	Copepods (see Chapter 3)
		Cirripedia	Barnacles
		Isopoda	Gribbles, sea lice
		Amphipoda	Sandhoppers and *Phronima*
		Stomatopoda	Mantis Shrimps
		Euphausiacea	(See Chapter 3)
		Decapoda	Crabs and lobsters
Mollusca	Class Amphineura	Polyplacophora	Chitons or mail shells
	Class Gastropoda		Sea snails and sea slugs
	Class Bivalvia		Clams
	Class Cephalopoda		Octopuses and squid
Echinodermata	Class Holothuroidea		Sea cucumbers
	Class Echinoidea		Sea urchins and sand dollars
	Class Stelleroidea	Asteroidea	Starfishes
	Class Stelleroidea	Ophiuroidea	Brittle stars

9 shows the names of the principal classes, subclasses, and orders of major phyla involved. Only names in frequent use are included, hence the blanks in parts of the table.

Shelf Communities of Rocky Substrates

1. THE AUTOTROPHS. The autotrophs comprise, as already stated, those elements of the ecosystem that carry out the primary production of living tissue from inorganic materials of the environment, utilizing the energy provided by sunlight. All are therefore plants, and on the continental shelf the plants concerned are chiefly algae. The particular algae

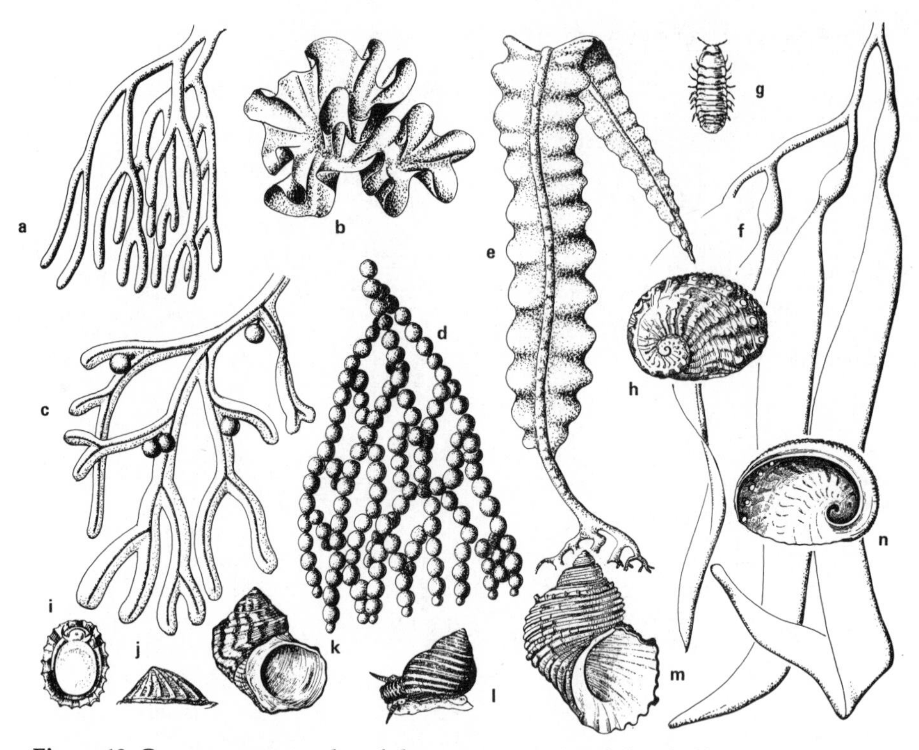

Figure 18. Common autotrophs of the upper margin of the shelf, with associated primary heterotrophs. (a) *Codium*, X 0.5, subtropical and tropical coasts, Chlorophyta; (b) *Ulva* (sea lettuce), X 0.5, all seas, Chlorophyta; (c) *Fucus* (sea wrack), rocky coasts of Northern Hemisphere seas, Phaeophyta; (d) *Hormosira* (mermaid's necklace), rocky wave-beaten coasts of southern Pacific, where it replaces *Fucus*, Phaeophyta, X 0.5; (e) *Laminaria* or kelp, X 0.25, north Atlantic rocky coasts; (f) *Macrocystis* (kelp), rocky coasts of southern hemisphere and northern Pacific, Phaeophyta; (g) *Limnoria* (gribble), X 1, boring into drift wood, Isopoda; (h, n) *Haliotis* (abalone), X 0.1, Indo-Pacific and Mediterranean, Gastropoda; (i, j) *Patella* (limpet), X 0.5, most temperate seas but lacking from America, rocky coasts, Gastropoda; (k, m) *Turbo* (turban shell), X 0.5, and (l) *Littorina* (periwinkle), rocky coasts of most seas, Gastropoda (23).

involved differ according to the depth at which they grow, the stronger more wave-resistant forms predominating at shallow depths, while more delicate and more light-sensitive types occur in the quieter waters at greater depths. Some differences occur also according to geographic location, though the general aspect of the floras does not change as much as might be expected from terrestrial models. Thus the intertidal or subtidal floras of the Southern Hemisphere present many close analogies to the corresponding floras of northern seas, and a naturalist traveling from one hemisphere to the other notices far fewer differences than those he sees in the vegetation of the forests or grasslands.

On north Atlantic shelves the most conspicuous green algae, or Chlorophyta, in the midtidal region are members of the family Ulvaceae, or sea lettuce, forming crinkly thin green expansions of thallus reminiscent of lettuce leaves. Two genera occur, a slightly thicker form with a double layer of cells in the thallus, called *Ulva* (commonest species *Ulva lactuca*) and a thinner, more delicate form with only a single layer of cells, now placed in a separate genus *Monostroma*. Among the genera of brown algae, or Phaeophyta, genera common on New England rocky coasts, are the following intertidal forms: *Punetaria,* with a flat straplike brown thallus, about 5 to 20 cm long and about 1 to 2 cm wide, usually tapering at the tip, sometimes with the thallus bifurcated, growing attached to the shells of mussels in intertidal rock pools; *Scytosiphon lomentaria,* forming very slender grasslike filaments up to 20 cm long and a few mm wide, also growing on mussel shells. Best-known of all the north Atlantic brown seaweeds are the fucoids (family Fucaceae), with two common genera, *Fucus* and *Ascophyllum.* The latter forms long slender cords about 30 or 40 cm long and about 0.5 cm wide, with hollow floats developed as part of the fronds at intervals of 10 cm or so. *Fucus* is commonly represented by two species, *F. edentatus* and *F. vesiculosus.* Both species have branching, bifurcating thalli, like fronds, about 15 cm or more across, the latter with conspicuous bubblelike floats developed in the thallus at intervals on the tips of some branches. *Rhodophyta,* or red seaweed, grows mainly at greater depths, but one mosslike dark purple form, *Polysiphonia lanosa,* is found in the midtidal zone where it grows on *Ascophyllum.* A plant that grows adhering to another plant is termed an epiphyte.

In the lower intertidal zone, below the level of the average low tide, there are some genera of Rhodophyta sometimes conspicuous at this level where the bottom is only occasionally exposed at low tide. These include *Rhodymebia,* with a red fan-shaped expanded flat thallus, subdivided into about half a dozen bifurcated flat lobes, somewhat ragged at the extremity, the whole fan about 20 cm across, but the shape rather variable, sometimes more elongated and straplike. In the lower intertidal zone, and also in rockpools, occurs a much smaller red seaweed, about 6 cm across the frond, bifurcated several times, and of a purple-red color: this is *Chondrus crispus,* the edible sea dulse. Another red seaweed of this zone is the grasslike *Dumontia incrassata,* dark red when alive, but fading to white on preservation, or when cast ashore and dried up. A mosslike feathery red seaweed is *Chondrai* spp., found at a somewhat higher level, usually in rocky pools; also *Porphyra,* closely resembling sea lettuce in shape and size, but differing conspicuously by its red-purple color.

On the rocky parts of the shelf beyond the tidal zone occur larger brown seaweeds known as laminarians. They are conspicuous on the upper shelf region, just below the lowest low-tide zone. Chief of these in the north Atlantic are species of *Laminaria,* such as the large (1 m) brown fans of *L. digitata,* and the elongate frilly undivided thalli of *L. agardhi;* also the curious fronds of *Agarum cribrosum,* growing in deeper water and sometimes cast ashore by storms, with the thallus perforated by hundreds of round holes, as if peppered by grapeshot. The lower extremities of these large algae form tough holdfasts, adhering usually to a boulder. The north Atlantic species lack the numerous floats of the Pacific and Southern Hemisphere laminarians, so they do not form ocean-going rafts when they are broken from the seabed by storms.

The most conspicuous difference of north Pacific shelf floras from those of the north Atlantic rests in the much greater development of the laminarian algae on Pacific coasts, especially on those with rocky reefs that provide a firm attachment for the holdfast. The large bladder kelp *(Macrocystis)* dominates the flora. Its species includes *M. pyrifera,* an alga that has a circumpolar distribution in the southern oceans. There can be little doubt that the genus originated in the southern oceans and was dispersed across the tropics during glacial stages in the Pleistocene, when equatorial waters cooled by some 6°C, to a level tolerated by the alga. Other large laminarians include *Nereocystis,* in which a single algae float carries about 6 or 8 straplike divisions of the thallus and is itself carried at the end of a long flexible stem with a basal holdfast; and *Postelsia,* a palm-shaped kelp with a cluster of straplike segments developed at the tip of a robust trunklike stem. There is also the same development of smaller Phaeophyta, with Clorophyta and Rhodophyta, as elsewhere in northern seas.

In southern oceans a brown kelp called *Hormosira banksii* more or less replaces the northern *Fucus* on Australian coasts. Its appearance is highly distinctive, the thallus comprising strings of small spherical hollow floats, each about 1 cm across and strung in linear series up to 20 cm long or so, bifurcating at frequent intervals. This olive-brown alga occurs at the level of low neap tide, or just below the average midtidal level; here it forms very precise horizontal curtains on rocky shores, around the more sheltered coasts of southern Australia and New Zealand. One vernacular name used for the weed is mermaid's necklace. The veritable giants of the world's brown algae are the species of the great brown bull kelp *(Durvillea),* with *D. antarctica* and others. In these kelps the actual thallus incorporates the floats as honeycomblike cells forming a sort of pith inside the leathery exterior layers of the giant straps of the thallus. The straps may be up to 30 cm wide, about 1 cm thick, and 100 m long. *Macrocystis pyrifera* in the southern oceans is a kelp of rocky exposed coasts, growing in ropelike masses up to 60 m long, the central axis anchored by a rootlike holdfast at the lower end. The free part of the axis slopes at an angle, like a fisherman's line, the angle determined by the direction and strength of the current. At intervals along the axis occur elliptical bladders, one at the base of each straplike branch of the huge thallus. The combined flotation power of all the hundreds of cysts serves to hold the whole plant in its erect posture. If a violent storm tears the plant away from the seabed, then the thallus floats to the surface, to constitute a raft, capable of drifting thousands of miles to sea, and of carrying miniature ecosystems of benthic and epiphytic organisms, plus a cloud of kelp fishes underswimming the whole. Sometimes the alga becomes entangled with a log or logs, and with other species of kelp, in which case great rafts up to 15 m in diameter are produced. Such rafts are believed to be responsible for the widespread dispersion of some shelf organisms in the Southern Hemisphere.

Green algae in the southern oceans include *Ulva* similar to that of northern seas and also the genus *Caulerpa,* somewhat resembling twigs of juniper.

2. PRIMARY HETEROTROPHS. This assemblage, the second tier in the trophic pyramid, comprises the herbivorous animals. Apart from some plant-eating fishes (discussed in Chapter 7), the principal herbivores of the rocky shelf are sea snails and bivalve mollusks, and some sea urchins.

On north Atlantic coasts several species of the periwinkle snail *(Littorina)* are encountered in vast numbers on intertidal rocks, where they feed on encrusting algae. These snails are semiamphibious herbivores of temperate and cold seas, with some species in warm temperate mangrove sloughs, climbing the stilt roots. *Littorina littorea* occurs on both sides of the Atlantic but seems to be a recent immigrant to eastern North America; it extends south to New Jersey and ranges northward through eastern Canada, Greenland, and into northern European waters; it was eaten by Viking explorers, who seem to have carried stocks of live periwinkles in their ships; however, subfossil *Littorina littorea* in eastern Canada implies that the Vikings did not introduce the species there.

Filter-feeders may be included under this head, particularly the common blue mussel *(Mytilus edulis)*, living in great banks, or sheets, attached by the spun byssus threads secreted by the otherwise functionless foot; and the intertidal barnacles, *Balanus* spp. (Crustacea, Cirripedia). The term filter-feeder is applied to any animal that produces currents of water by ciliary action, passing the currents through some sievelike mechanism so as to trap the minute plankton organisms, which are then diverted into the mouth. Sponges also feed in this way, chiefly trapping bacteria.

In the zone below low-tide level occur sea urchins of two genera, namely, *Echinus* and *Strongylocentrotus*, which are the most conspicuous herbivores of the zone together with another sea urchin *(Arbacia)*, mainly a North American genus, though it has species in Europe, too, in the warmer temperate waters. *Arbacia punctulata* ranges the eastern North American coast northward to Cape Cod. In the colder waters *Strongylocentrotus droebachiensis* occurs on either side of the Atlantic, generally distinctive from its greenish color and denser spine cover. *Echinus*, with *E. esculentus*, is conspicuous in Europe; like most species of the genus, this one is red in color. No shallow-water species of *Echinus* occur on the American side of the Atlantic. The regular sea urchins are more or less omnivorous, but most of the food they encounter in the intertidal zones and subtidal seafloor comprises large algae; such smaller invertebrates as happen to fall their way are also taken, and sea urchin stomachs contain varied assortments of worm tubes, small mollusk shell fragments, bits of paper and string and bitten fragments of the substrate, also occasional bus tickets and other man-derived rubbish. Other primary heterotrophs of the zone include limpets *(Acmaea)*, slipper limpets *(Crepidula)*, and barnacles of the genus *Balanus*.

Filter-feeders in this zone commonly include oysters, such as *Crassostrea virginica*, attached to rocks, and some holothurians, such as *Thyone* (with tube feet scattered over the surface of the body) and *Cucumaria* (with the tube feet in five long series). In the latter cases, plankton and organic detrital particles are captured from the water by the sticky frondlike ten-

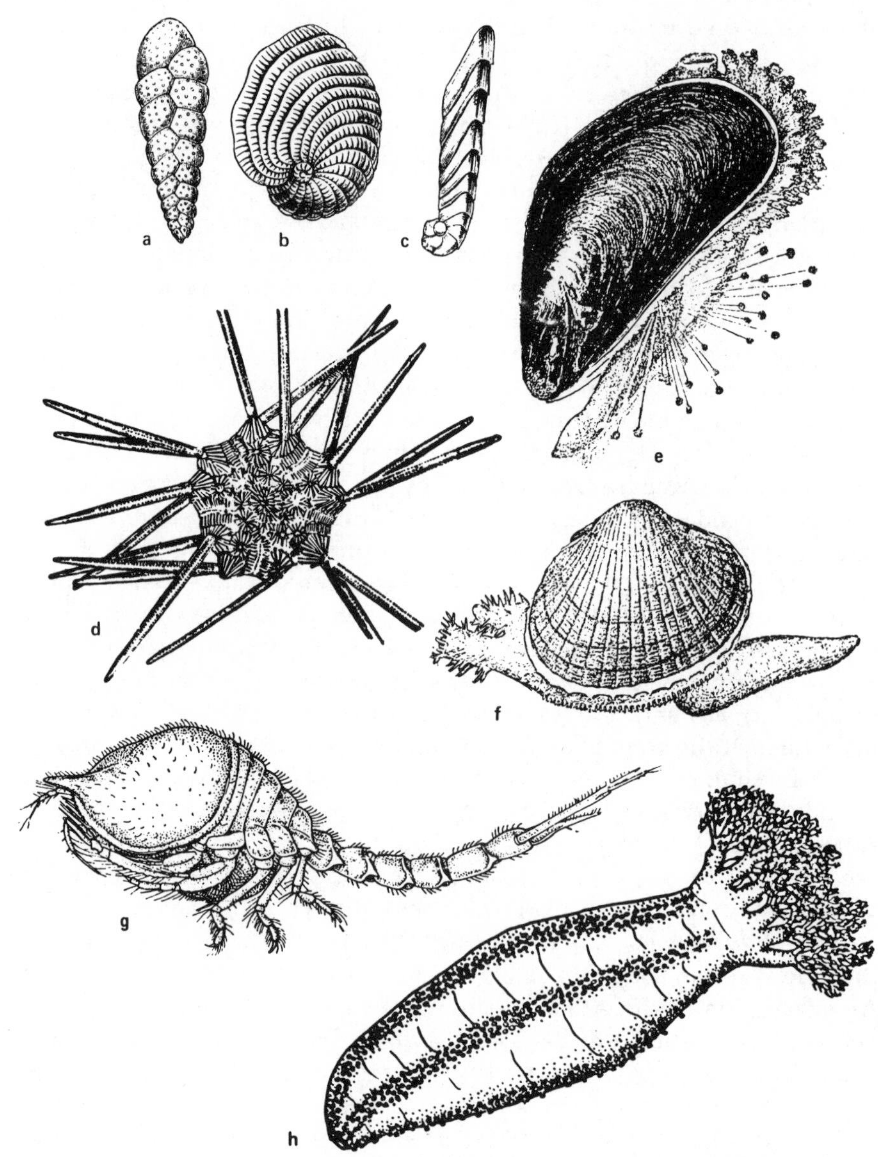

Figure 19. Primary heterotrophs, detrital feeders (a–c, g, h); filter-feeders (e,f), and gnawing herbivore (d). (a) *Textularia*, (b) *Penerophis*, (c) *Cristellaria*, all benthic forams, X 20 (39); (d) *Cidaris*, X 0.5, deep shelf of Atlantic, similar genera on tropical reefs, Echinoidea (23); (e) *Mytilus* (blue mussel), X 0.5. all coasts, shallow water, a common bivalve mollusk attached by byssus threads to substrate (24); (f) *Cardium* (cockle), X 0.5, soft substrate imbedded in sand (infaunal), with siphons (on left) projecting, bivalve mollusk of warmer seas (24); (g) *Diastylis*, X 5, small burrowing crustacean of shelf, soft bottom, of order Cumacea, Crustacea (29); (h) *Cucumaria* (sea cucumber), X 0.5, shallow and deep seafloors all oceans, class Holothuroidea, phylum Echinodermata, drawn by D. L. Pawson.

tacles around the mouth. In the deeper laminarian zone much of the same assemblage of herbivores is found.

On the north Pacific shelf, with the great development of autotrophs there is a corresponding abundance of herbivorous animals. The two most conspicuous groups of these are the alga-eating regular sea urchins and the abalones, which here reach their maximum size.

The abalones are ancient sea snails of a flattened oval form and include such massive species as *Haliotis rufescens,* the red abalone of California, up to 30 cm long and 20 cm broad, and *H. fulgens,* the green abalone, up to 20 cm long—the largest known species. Other north Pacific abalones include *H. kamtschatkana,* of Japan and Siberia as well as western North America. The inner, or nacreous, layer of the shell is well developed, though the colors of the northern species do not equal the opaline and sapphire-like quality of *H. iris* of New Zealand. The Californian abalones are subject to legal conservation, following the drastic depletion of stocks in the last century; but the Mexican beds are fished for the oriental market and headed for rapid extinction.

The sea urchins that make up the other main herbivore assemblage feeding on the kelp beds of the north Pacific nearly all belong to the genus *Strongylocentrotus,* with about 12 north Pacific species, about half a dozen on the Asian side, the rest on the Alaskan, Canadian, and American coasts. The genus has its major development in the north Pacific, and it is evident that the two species of the north Atlantic are relatively late emigrants from the north Pacific, by way of the Arctic Ocean, around whose coasts the emigrant species also occur. The Californian species include some very large and spectacular long-spined sea urchins, notably *S. franciscanus* and *S. purpuratus,* these occasionally yielding specimens some 30 cm across the spine span, among the largest regular urchins in the world. Indeed, this predilection to large size characterizes many of the Californian marine invertebrates and is probably due to the stability of the environmental temperatures throughout the year, no winterkill or dieback occurring, in contrast to the violent climatic extremes of the New England coasts, where the winterkill is so severe that stocks are renewed each spring by migration from deeper water.

An interesting food chain exists, connecting the kelp, the sea urchins, the abalones, and the sea otters. During the recent period of extermination of sea otters, the sea urchins increased in number (having no natural predator any longer), attacked the holdfasts of the kelp, which drifted out to sea, thereby reducing the abalone population. The return of the sea otter to Monterey has meant that the sea urchin population is now being controlled, the kelp beds are regenerating (so far as oil spills permit), and the abalone stocks are increasing.

Other primary heterotrophs of the zone match those of the north Atlantic. The common north Pacific littorinids are *Littorina scutulata* and *L. planaxis;* the filter-feeding mytilids are here *Mytilus californianus* (ranging from Aleutians to Socorro Island, Mexico). In addition species of the turbinid *(Tegula)* are extremely numerous, and chitons are conspicuous on west American coasts. The largest and most spectacular chiton is *Amicula stelleri* (formerly placed in the genus *Cryptochiton*), reaching 30 cm in length and having the valves hidden in a leathery reddish-brown skin; it ranges Japan, Alaska, and south to southern California. Chitons of the genus *Mopalia* reach a length of about 5 cm and usually have bristles or hairlike structures on the girdle that surrounds the central 8 valves of the shell. Another large and striking member of the herbivorous mollusks is the great keyhole limpet (*Megathura crenulata*), up to 10 cm long; and there are other species. Of the true limpets, family Acmaeidae, the most notable west American genus is *Lottia*, with *L. gigantea* (the owl limpet), to 10 cm long, ranging the rocky foreshores of most of California. Several species of *Acmaea* range from Alaska to Baja California. Of the top shells the genus *Calliostoma* is notable. This genus also occurs in Southern Hemisphere communities, Several species of the snail (*Lacuna*) represent the Heritidae on the west American coasts.

The primary heterotrophs of southern oceans present much the same aspect as those of the north Pacific, though the species and genera differ. *Haliotis* and various herbivorous limpets are conspicuous. The sea urchins belong to different families, especially to the family Echinometridae, with representative common genera such as *Heliocidaris* on the kelp beds of Australia, *Evechinus* on New Zealand rocky coasts, and *Loxechinus* in South America.

3. SECONDARY HETEROTROPHS. This division comprises all the predators that capture and feed on other animals, occupying therefore the upper tier of the trophic pyramid.

In north Atlantic seas the dominant predators of the midtidal zone comprise mainly gulls such as herring gulls (*Larus argentatus*), oyster catchers (*Haematopus* spp.), and other coastal waders; also man, who, since Paleolithic times, has systematically hunted the intertidal zone for food, and still does. Associated physically with *Mytilus* is commonly found a commensal crab, *Pinnotheres*. A large five-armed sea star (*Asterias*) is characteristic as the dominant invertebrate predator of this zone in the Northern Hemisphere, with *A. forbesi* and *A. borealis* on American Atlantic coasts, and *A. rubens* on European coasts.

In the subtidal zone *Asterias* is also conspicuous and another sea star, *Henricia*, occurs. The brittle star (*Ophiopholis aculeata*), characterized by having rows of minute platelets around each of the main arm plates, and usually with a variegated brightly colored body, is distinctive in northern faunas; it is, like most ophiuroids, a scavenger. A predatory snail (*Thais lapillus*) occurs here but also spends a good deal of time exposed on upper tide-level rocks, apparently in this way escaping the unwelcome attentions of a snail-hunting crab (*Carcinus maenas*); both the snail and the crab range all northern Atlantic rocky shores, on both sides of the Atlantic. The common edible crabs, of the genus *Cancer*, also range this zone. Two common species of *Cancer* on the American east coast are *C. borealis* and *C. irroratus*, the former with relatively larger chelipeds. Most of the larger invertebrate predators, also the sea urchins, are liable to attack by gulls; the starfish (*Asterias*), however, seems to be immune from attack. Man as predator in this zone takes oysters, sea urchins, and crabs. *Urosalpinx* is a predatory gastropod attacking oysters and other mollusks. Man, in order to conserve oysters from attack by *Urosalpinx*, has in recent years become a government-sponsored predator on this snail.

Out on the shelf proper much the same predators occur as in inshore areas. The same sea stars, and the same crabs, occur here as in the lower rocky tidal zone. In addition the lobster *(Homarus)* is, or was, a conspicuous predator of northern Atlantic rocky seafloors. Lobsters migrate seasonally between shallow and deep-water habitats, so they may be trapped in depths beyond the edge of the shelf.

Notable sea-snail predators of northern seas include the family Buccinidae, a group lacking from the southern oceans. The Buccinidae with large trumpet-shaped shells are carnivores of northern sea floors. Among the conspicuous genera of the North Atlantic are *Buccinum*, with *B. undatum* in northern European and northern American waters, also in the Arctic Ocean; *Neptunea*, with *N. decemcostata*, having the same distribution; and *Colus*, with *C. stimpsoni*, in American waters. These three genera are represented in the offshore communities in the latitude of Boston but become progressively more shallow-water in distribution as one travels northward, until in the region of northern Maine the species of the genera occur as littoral elements. These are evidently stenothermal forms, restricted to cold Arctic water and rising into the littoral zone in far northern waters where the temperatures are low enough.

As might be expected from the exuberance of the primary heterotroph elements, the carnivorous predators of the north Pacific rocky coasts are similarly conspicuous and varied and also very abundant. Of the Buccinidae, *B. undatum* does not occur on the west American coast, but other species of the same genus range from the Arctic to Washington state, for example, *B. plectrum*. Similarly, *Colus spitzbergensis*, *Neptunea lyrata*, and *N. pribiloffensis* are western American species of genera already noted in the Atlantic from other species. The north Pacific also has its restricted buccinids; for example, *Kelletia* ranges California and west Mexico, but not the north Atlantic. *Busycon* and *Melongena*, on the other hand, are totally absent from the west American fauna, though important in the east. Although *Busycon* is lacking, a large left-handed buccinid, *Volutopsis*, occurs in Alaska.

Of other carnivorous sea snails of the north Pacific the genus *Argobuccinum* occurs from the Bering Sea to California; species of the genus range southward to New Zealand. The Naticidae include *Polinices reclusianus* among the west coast carnivores matching *P. heros* of the east coast. Among the sea stars the family Asterinidae is represented by a fine species, *Patiria miniata*, recognizable by its pentagonal form. The Echinasteridae include *Henricia*, and the Asteriidae include the large many-armed sunflower starfish, *Pycnopodia helianthoides* and *Pisaster ochraceous* and *P. gigantea*, the two latter both five-armed and covered by granulelike spines. All these are active predators, with a particular predilection for *Tegula*. Other genera occur in deeper water, particularly *Mediaster;* and further south in the subtropical waters *Linckia* and *Oreaster* appear, matching the species of the same genera that occur in the southeastern United States.

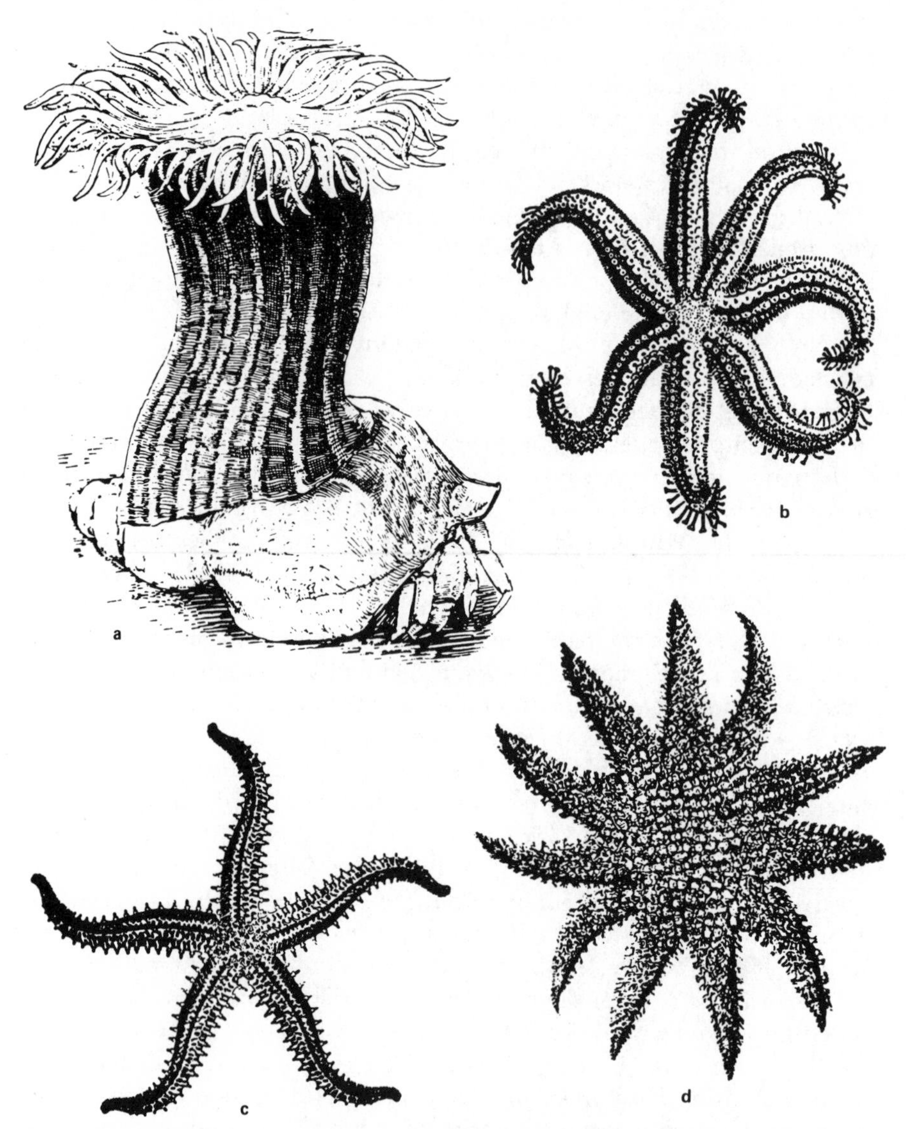

Figure 20. Secondary heterotrophs of hard substrates. (a) *Sagartia* (sea anemone), X 1, growing on discarded shell of whelk (*Buccinum*), now occupied by the predatory hermit crab (*Pagarus*); the crab is protected against octopus attack by the anemone, whose stinging tentacles paralyze octopus, the anemone benefits by obtaining fragments of food from the hermit's victims; the hermit seeks and transfers the anemone to its shell, also transferring it when a new shell is occupied (24); (b) *Astrostole*, confined to south Pacific rocky shores; (c) *Sclerasterias* and (d), *Crossaster*, range widely on soft shelf bottoms; three predatory genera of starfishes, class Asteroidea, phylum Echinodermata, all X 0.25; the genus *Asterias* resembles (c) and ranges shallow rocky coasts of the Northern Hemisphere as a major predator of bivalve mollusks (23).

Shelf Communities of Sand and Silt Substrates

1. THE AUTOTROPHS. In general the communities on substrates of sand or silt tend to be less varied and less abundant than those on hard substrates.

In the midtidal zone a dark bluish-green scum may be found covering tidal mud flats or adhering to occasional boulders or pilings—this is generally due to the presence of blue-green algae (phylum Cyanophyta).

Much-branched green algae called *Codium* may occur, with cylindrical lobes of the thallus, soft and velvety to touch. Filamentous green algae, such as *Bryopsis,* may occur on soft substrates (as also on hard ones) and creeping green algae *(Caulerpa)* with erect strap-shaped thallus lobes may occur.

The other algal groups are generally inconspicuous or absent. Some flowering plants, such as eelgrass *(Zostera marina)*, occur on muddy bottom, including estuaries and brackish-water situations, notably the Baltic Sea. Beds of eelgrass continue into the subtropical and tropical region. They often form a shelter for young or adult stages of fishes. As everywhere, diatoms make up a considerable part of the biomass, especially during the spring. The flora of the lower tidal and subtidal zones is similar to that of the midtidal zone.

Further out, on the floor of the continental shelf, the autotrophs are commonly restricted to the microscopic forms, unless there are scattered buried pebbles or boulders in the soft substrate. In the latter case quite extensive algal growth may develop, even including brown laminarians; but if such scattered hard substrate is missing, little algal colonization can take place. Where substrate includes boulders or cobble partly imbedded in mud, encrusting red algae often cover the exposed portion of the boulders, and sessile invertebrates (sponges, hydroids, anemones) are likely to be associated.

2. PRIMARY HETEROTROPHS. In keeping with the reduction in the biomass of plants, there is a general reduction in the plant-eating species of animals. A few scavenging sea snails may be noted as taking vegetable materials, among other matter—notably the small ceriths such as *Cerithiopsis* and *Seila,* occurring on eelgrass and other soft-bottom vegetation from Massachusetts southward; and also one of the periwinkles, namely, *Littorina irrorata,* which occurs on vegetation between tides, in New England and southern states (this latter species also occurs in hard bottom communities, as on the New Hampshire coast at Hampton Beach). Much more important in the primary heterotroph communities of soft-bottom habitats are the various filter-feeding bivalves, which live mostly buried in mud or sand, save for the protruding siphon when the tide covers the bottom. One family, however, the Mytilidae, have the ability to live on both hard and soft bottom and do not bury in the mud in the latter habitat; instead, the spun byssus threads form a tangle-anchor in the mud, holding

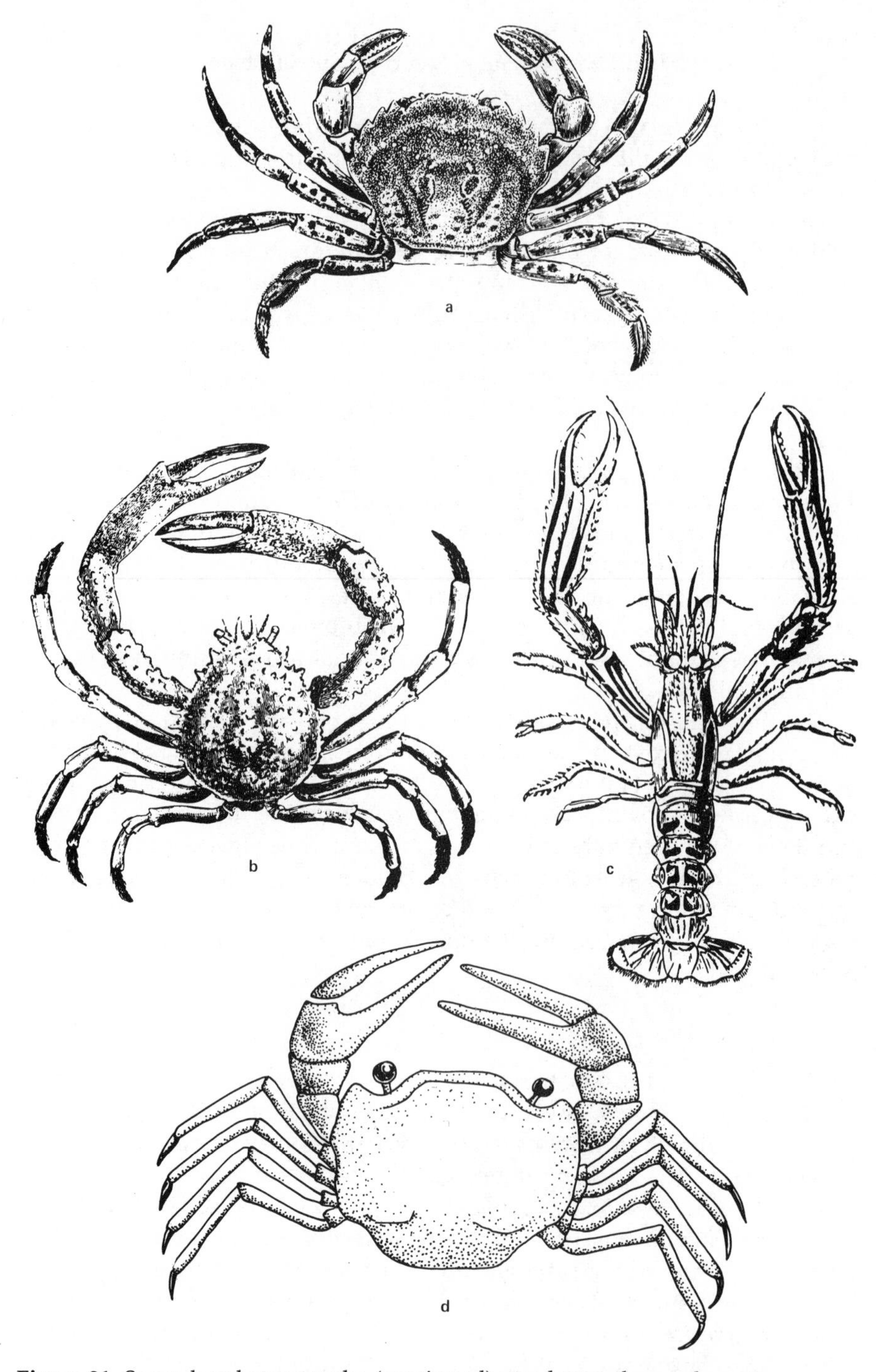

Figure 21. Secondary heterotrophs (continued), predatory decapod crustaceans. (a) *Carcinus* (shore crab), X 0.5, Northern Hemisphere; (b) *Maia* (thornback crab), X 0.5, shallow shelf; (c) *Nephrops* (Norway lobster), X 0.5, deeper shelves of northern seas; a–c (24); (d) *Pachygrapsus,* X 0.5, Mediterranean shores, where it emerges at night on land to forage; this illustration is taken from a Greek coin of 482 B.C. (5).

the shell, with others in great sheets or masses on the surface. The other intertidal families most commonly encountered are infaunal, that is to say, organisms that bury themselves in the soft bottom. Of these filter-feeding infaunal clams the following are examples of commonly encountered genera: (1) the mussels *Mytilus* and *Modiolus*, the former already noted, the latter tending to attach its byssus to pebbles in the sand, and sometimes to bury itself; both genera range these habitats in the temperate oceans, and *Modiolus* also occurs in the Arctic; (2) the colorful sunset shells of the family Tellinidae, with *Macoma* intertidally from the Arctic to Georgia, and *Tellina* more often at or below low tide, same range. The family Petricolidae are infaunal excavators of semicompacted soft materials such as clay. The only conspicuous member of west Atlantic communities is *Petricola pholadiformis*, a dull-colored bivalve, about 5 cm long, found often in peaty substrate, as on the Hampton Beach shoreline at midtide. Note however that another family, the Pholadidae, or piddocks, have similar habits and may be locally conspicuous. Included genera are *Barnea* and *Zirfaea*. These bivalves have as the family character radial ridges making toothlike margins to the anterior margins of the valves, and they gape at either end. Razor clams are recognized by the elongate body. There are several genera, such as *Ensis*, up to 15 cm long and 6 times as long as wide; *Siliqua* and *Solen*, both much smaller and relatively less elongate, occur on sandy bottom, the dead shells often washed ashore. The soft-shell clams (*Mya arenaria* and *M. truncata*) are readily recognized by the spatulalike tooth hinging the shell. The latter, which is the smaller species, occurs in the Arctic (Greenland coasts); both species range the eastern coasts of the United States. As an important seafood for man, these species are often sold under the name of steamers. The surf clams (in these habitats genus *Spisula* and two common species) occur under sand in surf-racked conditions in the lower part of the intertidal zone, also extending on soft bottom offshore, Arctic to southern Atlantic east American states. Often piled ashore in millions after rough inshore storms.

In deeper parts beyond the lowest tidal zone the soft-bottom areas support flattened sea urchins such as the sand dollars (with *Echinarachnius* on west Atlantic shelves, genera such as *Dendraster* on the Pacific coast of America, and *Arachnoides* in the south Pacific). Depending on the time of day, these urchins are sometimes *epifaunal,* that is, found on the surface of the seabed as at night; sometimes *infaunal,* imbedded in the mud, as by day. There are also heart urchins of various families, infaunal elements, imbedded in mud or sand, and extending elongated tube feet from the entrance to the burrow to obtain particulate organic debris and small organisms, diatoms, forams, and so on, as food. The large hard-shell clams of the family Veneridae occur on the upper part of the shelf; these are distinguished by the thick, massive shell, the elaborately interlocking hinge teeth, the powerful hinge and contracting adductor muscles (leaving very strong adductor muscle impressions on the interior of the shell), and the large size and often bright purple pigmentation in the shell interior. Most notable among them are the quahog *(Mercenaria mercenaria)* of west At-

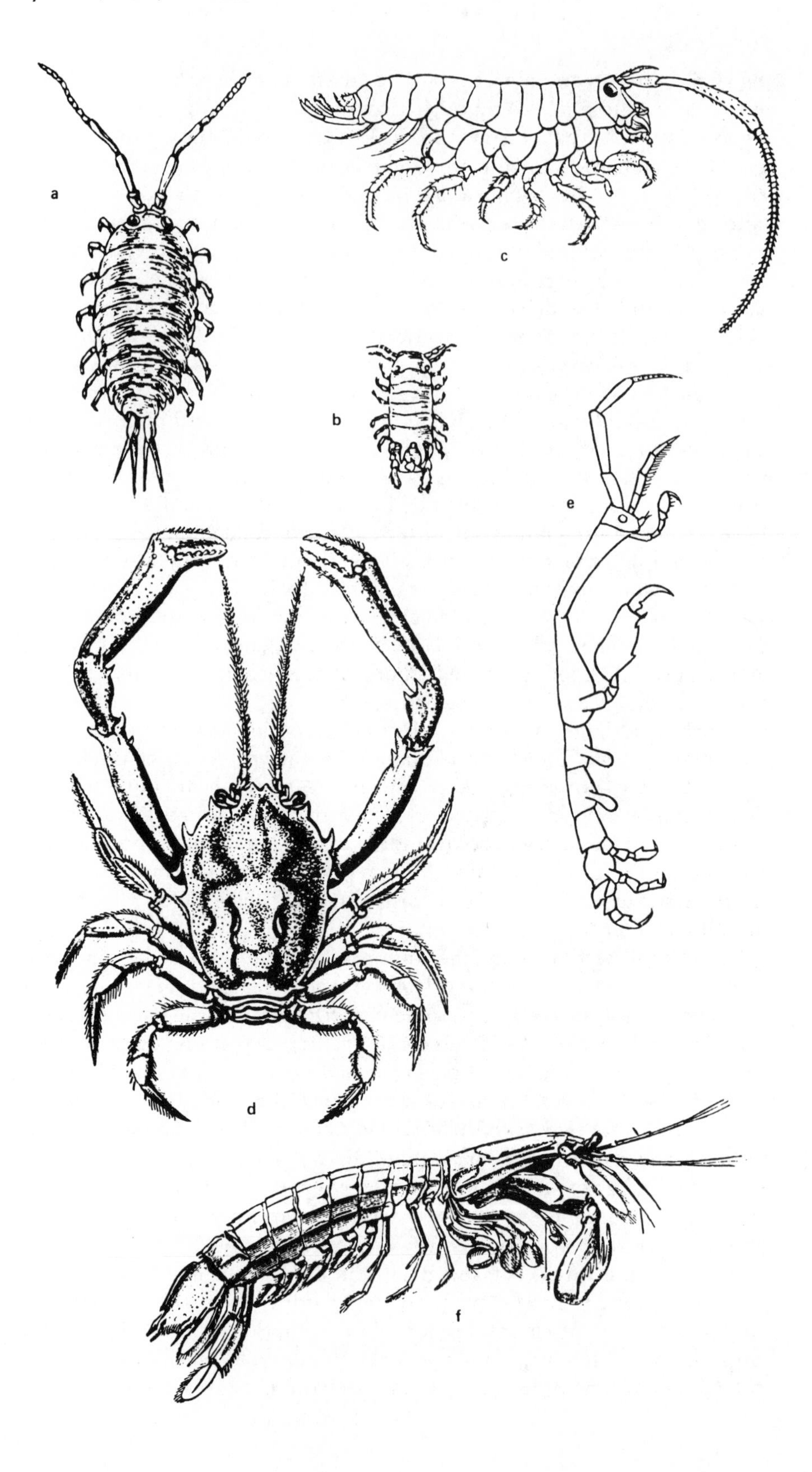
a
b
c
d
e
f

Figure 22. Crustacean secondary heterotrophs of varied habit. (a) *Ligia* and (b) *Nesaea*, X 2, bottom scavengers, Isopoda (24); (c) *Gammarus* (sandhopper), X 4, shallow shelf and intertidal, bottom scavenger, Amphipoda (compare oceanic relative of Figure 14b (24); (d) *Corystes*, X 0.25, shallow shelf, North Sea, bottom predator, Decapoda (38); (e) *Caprella* (ghost shrimp), X 8, parasite on skin of sea stars, etc., Isopoda (38); (f) *Squilla* (mantis shrimp), X 0.5, shallow shelf over soft bottom, swimming form, which also lurks on bottom in a burrow, Stomatopoda (15).

lantic coasts, the object of the hard-shell clam fishery and also used formerly as the raw material from which the American Indians cut the beads used in the manufacture of wampum. Harvard records show that, during early currency shortages in the infant colony of Massachusetts, the professors were for a while paid in wampum (and overpaid too, I bet, adds a disrespectful friend who read this chapter).

3. SECONDARY HETEROTROPHS. Since most predators are roving animals, and adapted therefore to encounter a variety of substrates, there is no great difference between the predators of the various substrates. The dominant carnivorous mollusks on the northern shelves tend to be buccinids, and *Colus*, *Buccinum*, and *Neptunea*, which favor both hard and soft substrates. Asteroidea are more selective, since forms such as the Astropectinidae and Luidiidae, with nonsuctorial tube feet, of course have an advantage on soft bottom; hence these two families tend to supply the genera of main carnivorous starfishes on such soft substrate. Many ophiuroidea favor soft bottom, in which the disk may be buried while the arms alone seek food material; others again creep over the surface of the bottom and rely on their hard skeletal plates to discourage predator attack. Various burrowing anemones occur, and some burrowing worms are also predators. Another scavenger, feeding also on live invertebrates in the substrate, is the horseshoe crab *(Limulus)*. This animal in the breeding season in early spring comes up from deeper water to deposit the sexual products in shallow water. The crabs *Uca* and *Sesarma* are very mobile, and so they commonly enter the intertidal region at times when the beach is deserted, in order to scavenge, especially under cover of darkness. These crabs normally inhabit salt marshes.

The north Pacific lobsters belong to the family Palinuridae, lacking the claw chelipeds of the Atlantic *Homarus* and having conspicuously enlarged antennae. The common west American species is *Panulirus interruptus*, and related to it are several Japanese species of the same genus. The Asian lobsters also include a smooth-carapaced form, *Linuparus trigonus* (note that genera of this family have anagrammatic names formed from *Palinurus*, the langouste of European waters). Other crustaceans are the prawn-killers, represented by the genera *Scyllaridaes*, *Thenus*, and by two other genera. Of the crabs, the genus *Cancer* is represented by species such as *C. antennarius* on the American west coast, the purse crabs by *Randallia*,

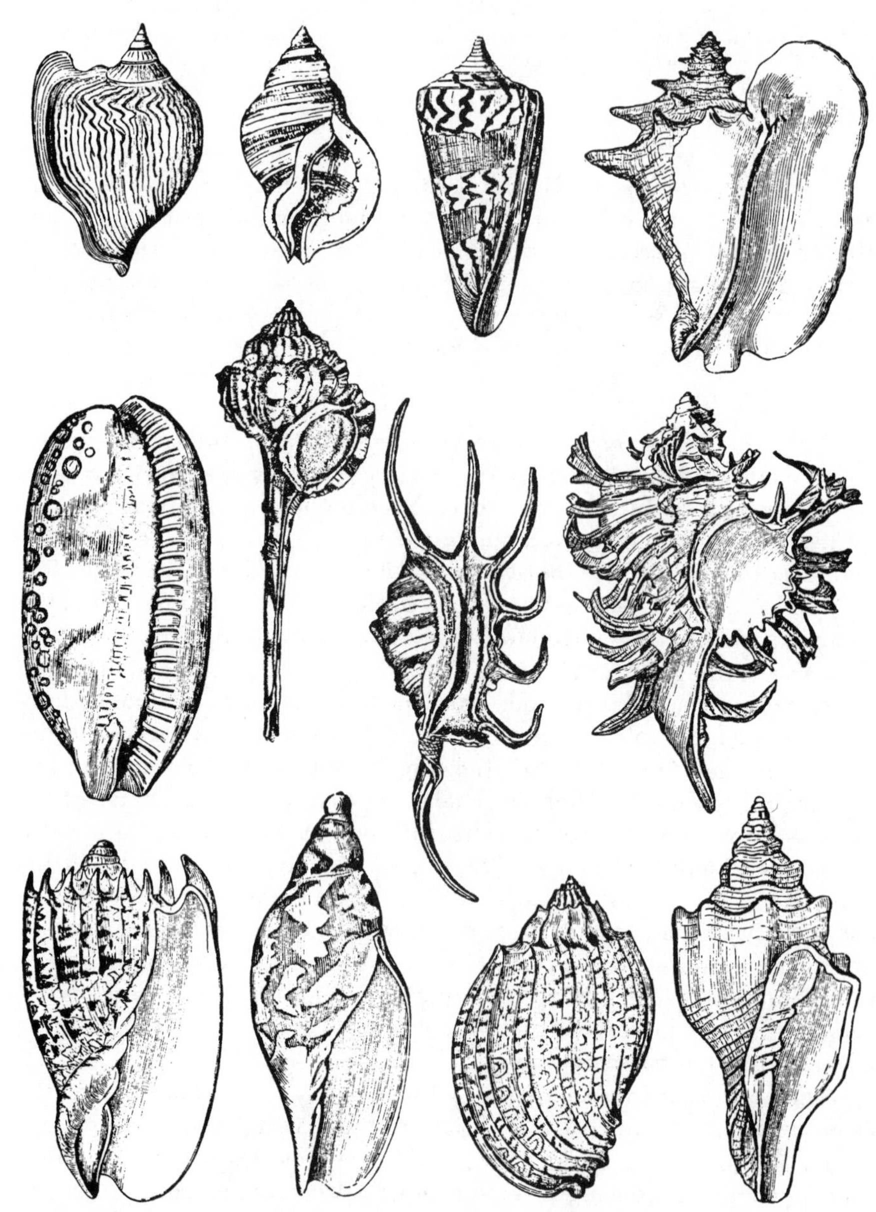

Figure 23. Molluskan secondary heterotrophs; with the exception of *Thais* (top row, second from left, of Atlantic rocky coasts), all the genera illustrated are active predators on tropical reefs. In all cases the animal is a snail with suctorial or biting mouthparts. All are X 0.5. Top row, left to right: *Strombus, Thais, Conus* (venomous, some species with a lethal bite), *Tricornis*. Second row, left to right: *Cypraea, Murex, Lambis, Murex*. Third row, left to right: *Voluta, Voluta, Harpa, Xancus*. From Woodward, *Mollusca*, in Duncan (21).

and the spider crab by *Libinia*. Giant spider crabs, or king crabs, characterize the north Pacific and include the huge *Paralithodes camtschatica* of Alaska, presently the object of a destructive fishery. The crab fauna of the north Pacific is very rich and cannot adequately be characterized in the limited space available here.

Other carnivorous invertebrates that are conspicuous on west American coasts are the shell-less sea snails, or nudibranchs, which are upper trophic level forms, feeding mainly on sea anemones (themselves also secondary heterotrophs); some nudibranchs are brilliantly colored, and some arm themselves with the nematocysts that they acquire from their anemone victims. The sea anemones are also highly differentiated predators on west American coasts. In general they resemble ecologically the southern Pacific anemones.

Chief predators of the intertidal region are the vertebrates, including man especially, and the shore birds. Below the shoreline fishes are the major predators. These are reviewed under nekton in Chapter 7.

Coral Reefs and Bioastronomy

Corals are organisms related to sea anemones but differ in that they secrete a lime skeleton. In the tropics, where corals are abundant in shallow waters on the continental shelf, a special kind of hard substrate develops from the dead coral skeletons. These, accumulating over millions of years, build up massive coral reefs on which a rich benthos lives. At the present time reefs extend to about 30° north and south of the equator, where the surface water of the sea is never colder than about 18°C. So the reefs form a girdle around the tropics. Fossil reefs are found on some lands on whose coasts corals do not now live, and this is taken as evidence of changing climates, a matter to which reference was made in Chapter 1. Some further consideration of this question may now be given.

FLUCTUATING WIDTH OF THE REEF ZONE. During the glaciations of the past 2 million years (Chapter 16) the overall temperature of the oceans fell, with a resultant contraction in the width of the coral reef belt. This means that the more northern and more southern reefs must have died at that time. That they are now living again is evidently due to subsequent recolonization of the dead reefs by living larvae from more equatorially placed reefs; the larval stages float and can therefore be carried far and wide by ocean currents. There are still some dead submerged reefs along the northern and southern margins of the existing belts, so apparently these are old reefs not yet reactivated, presumably because the oceans are not yet as warm as they have been in the past. During the Miocene period, about 20 to 25 million years ago, the whole earth was warmer than it is today, and living reefs extended as far south as New Zealand, at 42° south latitide. These data, and other similar distribution data on terrestrial plants and animals, tell us of periodic warming and cooling of the earth, such that tropical animals sometimes spread far beyond the limits of the present tropics.

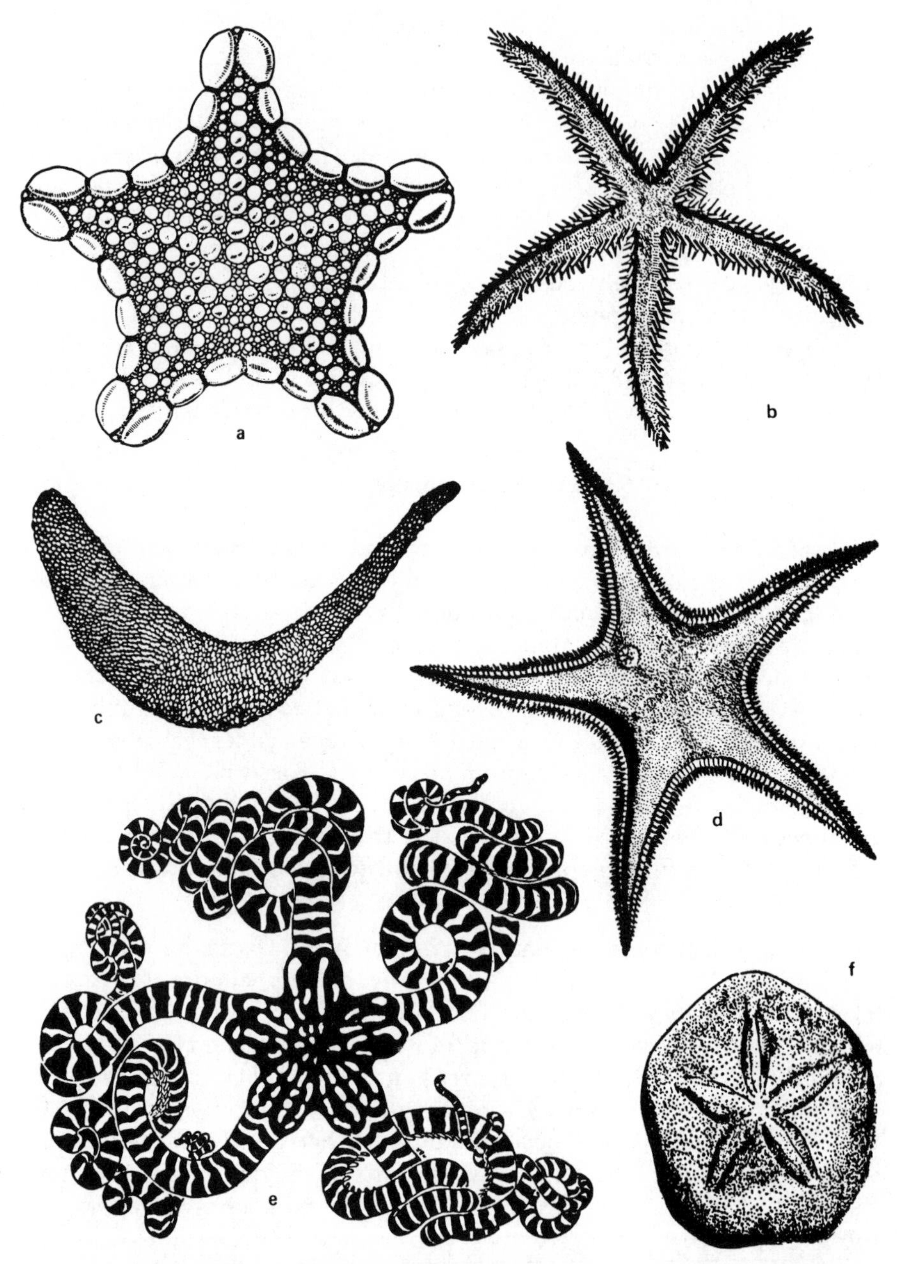

Figure 24. Echinoderms of soft shelf substrates. (a) *Pentagonaster*, X 0.5, New Zealand, detrital feeder; (b) *Luidia*, X 0.2, tropical and subtropical shelves, voracious predator; (c) *Placothuria* (infaunal sea cucumber), X 0.5, New Zealand (drawn by David L. Pawson); (d) *Plutonaster*, X 0.25, world-wide, deep shelf and beyond; (e) *Astroceras* (brittle star), X 0.5, preying on corals of deep shelf, Pacific; (f) *Laganum* (flattened sea urchin, or sand dollar), X 0.5, warm Pacific shallow shelf, infaunal detrital feeder (23).

POLAR WANDERING THEORY. If we carry the investigation further back in geological time a different kind of variation becomes discernible. Instead of simple northward and southward fluctuations in distribution, we find that the east-west parallelism varies too. In the Cretaceous period, for example, some 100 million years ago, the coral reef belt of North America and of Europe extended northward to about 50° north latitude, and warmth-demanding animals such as crocodiles also ranged much further north than they do today. Yet in the southern part of the Western Hemisphere the coral reef belt seems to have extended only to about 20° present south latitude. So the belt of coral reefs seems to have been about as wide as it is now, only displaced toward the north. On the opposite side of the world the reverse seems to have been the case, the displacement being toward the south. Many investigators believe that facts such as these imply that the earth's axis of rotation then lay in a different position from its present one, the North Pole lying on the north coast of Siberia, and the South Pole near the Palmer Peninsula, south of Magellan Straits. If facts like these are interpreted in a similar manner for still earlier geological periods we find, for example, that in Ordovician times, some 500 million years ago, the corals seem to form a belt tilted so steeply with respect to the present belt as to girdle the earth by swinging along the east coast of North America, passing through Greenland and northwestern Europe, then south through India to reach Australia. This suggests a North Pole near Hawaii and a South Pole off west Africa! Geologists such as Strakhov in Russia find that desert sandstones and other climate-related rocks also seem to girdle the earth in positions such as to suggest poles in the positions indicated by the corals. Unfortunately we lack data from much of the Pacific (where no large land surfaces occur to yield fossil reefs of such great age). Interestingly, the idea that the earth's poles may have changed their positions in the course of time is by no means new, and a truly scientific exposition of the matter was given by Robert Hooke nearly three centuries ago.

HOOKE'S CROCODILES. After the Great Fire of London in 1666, Sir Christopher Wren and Robert Hooke were commissioned by Charles II to design and build the major architectural monuments of London. One of Hooke's responsibilities was to inspect the stones sent from the Portland quarries for the construction of Saint Paul's Cathedral. He records that one day his attention was drawn to the impression on one of the blocks of stone of what looked like a giant sea snail. He traveled to Portland and when the quarrymen showed him their exposed rock face Hooke soon realized that the rock was indeed the hardened remains of a former seabed, with the remains of sea animals imbedded in it as fossils dating from a time when the rock was soft sediment. Further inquiries among the quarrymen brought to light bones that Hooke recognized as similar to those of turtles and crocodiles in the collection of tropical specimens of the Royal Society of London. In a paper that he read to the Royal Society Hooke theorized that England had once ". . . at a certain time for Ages

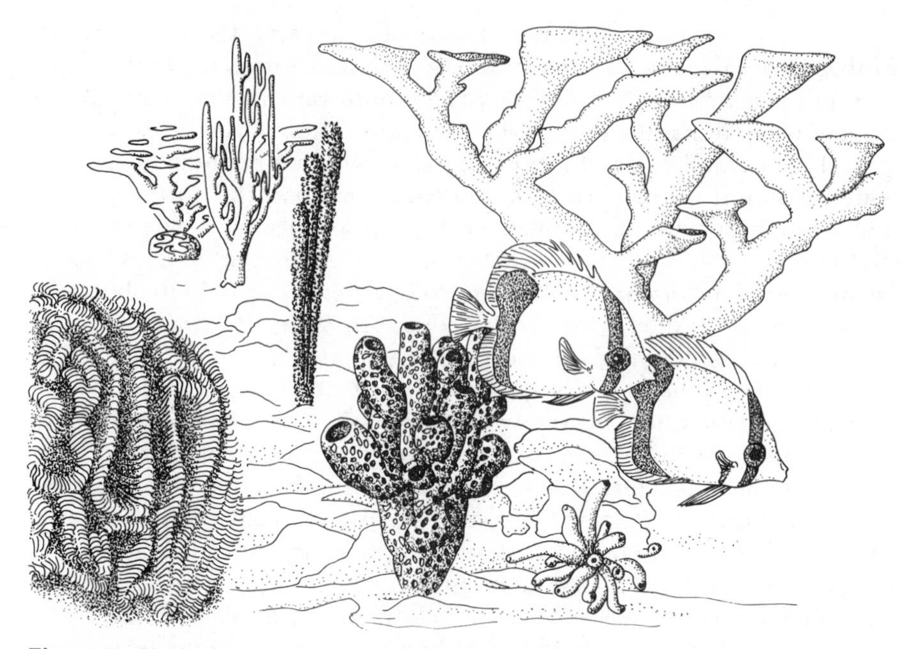

Figure 25. Underwater view of coral reef in Caribbean. Lower right, colonial coral *Meandrina*; center foreground, a tube sponge; and behind this to left, a young gorgonian coral (*Muricea*) with sea snail (*Cyphoma*) feeding on coral polyps; right, two butterfly fishes (*Chaetodon ocellatus*); and in front of them a sea anemone. The staghorn coral behind is *Acropora*. Drawn by David Moynahan.

past lain within the Torrid Zone . . . I think it not improbable that there may be some such motion of the Earth's Axis as may alter the Latitude of Places. . . ." At a later meeting of the Society in 1696 Hooke set up a heavy sphere of lignum vitae suspended by a cord. Spinning it, all present could see the vertical axis of rotation. Hooke then attached an iron weight to one side of the sphere and set it spinning again. He was able to demonstrate that the change in the center of gravity of the globe caused its axis of rotation to change its position in the direction of the weight; the rotating ball wobbled around its new axis instead of spinning smoothly as before.

Now, Leonardo da Vinci (who did not publish his speculations) had entered in his notebooks a century before Hooke's experiment an observation to the effect that rivers constantly sweep sediment from land into the sea and that therefore the position of the center of gravity of the earth must change with time. A similar idea must have occurred to Hooke, but unfortunately for his ideas, Sir Isaac Newton resolutely declined to believe that the axis of the earth might move. So Hooke's ideas were forgotten until 1830, when the British geologist Charles Lyell drew attention to them.

THE CHANDLER WOBBLE. In 1885 Seth Chandler of Harvard Observatory noticed that some measurements of the altitudes of stars were discrepant with other measurements he had made some months earlier. On examining the matter he eventually discovered that all stars vary slightly in their apparent positions, the variation being such as to imply that the earth slides about or wobbles on its axis of rotation. In effect, the latitude of Cambridge was found to be varying, and hence the apparent altitude of stars varies. Observations repeated in other parts of the world eventually showed that when the apparent altitude of a star is at a maximum on one side of the world, it is at a minimum on the other side, so Chandler's interpretation was proved. Chandler went on to discover that the variation is complex and is made up of the summation of two variations; one variation is annual, and apparently reflects changes in the distribution of air and water masses as the Northern and Southern hemispheres exhange water and air as a consequence of ice melt in their respective summers; the other variation had a period of 14 months, and Kelvin showed that this is the natural vibration period of a body the size of the earth if it has the elasticity of steel (most of the earth is composed of a steel alloy of iron and nickel). The lowest common multiple of 12 and 14 is 84, which means that every 84 months, or 7 years, the two variations reinforce each other, with intervening periods when they counteract each other. This effect, known as the Chandler wobble, causes the earth's poles to move irregularly over an area about the size of a football field.

However, taking into account Leonardo's observation, and also recent measurements of the effect of major earthquakes on the earth's rotation, it becomes apparent that after a long lapse of time the slowly moving center of gravity of the earth is bound to cause the poles to wander outside the area they presently traverse; after a very long span of time such wandering can become highly significant.

What is really happening is that the earth's axis itself remains fixed at an angle of 23½° to the plane of the earth's revolution about the sun, but the *whole body of the earth* is slowly sliding with respect to that fixed axis. For an observer on the earth this must mean that the poles appear to move across the surface of the planet over a long span of time. If the inferences as to the position of the poles are correct for the Ordovician period, apparently during the 500 million years since the Ordovician the North Pole has moved from Hawaii northwest (by modern reckoning) across the Pacific to pass through Kamchatka about 300 million years ago, and thence on to the north coast of Siberia by the Cretaceous period, and after that across the Arctic Ocean to assume its present position. If the inferences are correct, the future motion of the North Pole will take it into Greenland and southward toward the northeastern part of the United States, though that would not occur till some hundreds of millions of years after now.

BIOLOGICAL ASTRONOMY. It may be thought strange to include material such as the foregoing paragraphs under the heading of coral reefs. Nonetheless it is an example of how these organisms can throw light on astronomical problems, a topic that will be further discussed in Chapter 5. There seems to be a likelihood that the fossil coral reef belts of our planet may ultimately prove to be a reliable set of datum points, or rather datum zones, for other geophysical and astronomical measurements. The subject, however, is still in initial stages of study.

5

PLANETARY RHYTHMS AND THE MARINE ENVIRONMENT

The biota of the continental shelf, the benthos especially, inhabit a relatively stable and nearly uniform environment in comparison with the conditions imposed on organisms that inhabit land surfaces. There are several reasons for this.

1. Random or seasonal variations in rainfall have virtually no effect on bottom-dwelling organisms protected by an overlying water mass.
2. Exposure and desiccation rarely occur, for mean sea level is determined by the mass of water in the entire ocean, and local fluctuations or irreversible changes only happen as a result of rare events such as earthquakes or great hurricanes.
3. The aqueous medium protects its denizens against the effects of violent short-term fluctuations in the temperature of the overlying air masses, such as commonly occur during a single day over land.
4. Wind variations affect substantial bodies of water only if they are prolonged, for there is a considerable delay in the transfer of air motions to water.

As a result of these immunities to random short-term environmental fluctuations, the marine environment tends to be dominated by planetary rather than local factors, and planetary factors are dependent on the long-term properties of the part of the solar system occupied by the planet earth.

THE NATURE OF PLANETARY FACTORS. Two main categories of influences may be distinguished: (1) *scalar,* that is, effects that can be represented by quantitative measurements, such as temperature, light intensity, duration of light or darkness, and so on; and (2) *vector,* that is, effects that involve directional components and require a special type of notation such as rotating arrows—for example, wind flow and the course of ocean currents. The scalar factors will be discussed here. The vector field will be discussed in Chapter 8.

Three categories of scalar rhythms may be distinguished, each associated with a celestial body:

1. Diurnal (or terrestrial) cycle: the repetitive 24-hr cycle of feeding, growth, activity, and rest, imposed by the daily rotation of the planet earth
2. Annual (or solar) cycle: the repetitive yearly cycle of growth, activity, migration, reproduction, and hibernation, imposed by the annual orbital revolution of the earth about the sun and accentuated by the fact that the earth's axis is inclined to its orbital plane
3. Monthly (or lunar) cycle: the 30-day cycle of sexual activities related to the phase of the moon and hence to the orbital motion of the moon in revolution about the earth

THE DIURNAL CYCLE. This cycle, which has a present frequency of 365 per annum and a wavelength therefore of 24.0 hrs, is so well known as to require little comment. Organisms tend to conform to the geophysical frequency quite precisely, performing corresponding activities on consecutive days at the same time of day. The rhythms are obviously superimposed by the environment and, in the same species, will differ therefore according to the latitude. In subpolar regions, where the winter nights are long and the winter days are short, the biological rhythms are correspondingly adjusted; while in the same latitudes in summer the biological rhythms conform to the long day and short night. Nearer the equator, where seasonal variations in the length of day and night are very slight, the biological rhythms show a corresponding degree of near uniformity throughout the year.

But there are significant differences between terrestrial biota and marine biota. On land, a week of bad weather or a prolonged major storm may cause the biota to take shelter, to refrain from hunting or feeding activities, perhaps to sleep—growth, of course, ceasing during such periods of enforced inactivity. On the other hand, the relative immunity of the marine environment from atmospheric variations ensures that the biota carry on with their normal diurnal cyclic activities, whether the weather above the air-sea interface is fair of foul. This means that growth continues much more regularly, with a daily increment almost without fail. Some marine organisms, such as mollusks, corals, and similar shell-secreting forms, add a marginal zone of calcium carbonate around the edge of the growing part of the shell every day.

Successive *growth lines* may remain visible months or years after they were originally secreted, thereby providing a permanent record of the successive sizes and stages of growth of an individual. About 2 percent of the time a marine invertebrate may fail to perform its normal growth increment; perhaps this may be due to some internal factor such as bacterial infection or other disease, to some external factor such as a temporary food shortage, to an accidental injury from predator attack or other natural cause, or to an unusually severe storm, disturbing the seafloor. Whatever the case, there is a slight nonconformity between diurnal activities and the diurnal planetary cycle, but the discrepancy is only a matter of a few omitted days' growth in the course of a year. Panella (1971) has recently demonstrated that daily growth lines are readily observable in the bone of the inner ear (otolith) of certain fishes, thus bringing chordates into line with nonchordates in respect to diurnal growth cycles.

Growth lines of essentially the same kind as in existing organisms of the sea can be recognized in fossil skeletons of extinct mollusks, corals, and other biota, as far back as the early Paleozoic, thus proving that diurnal cycles of growth and behavior are very ancient features of living organisms.

Some organisms, such as the shore crab *(Carcinus)*, pattern their activities on the semidiurnal cycle of the tides. Successive high tides occur at intervals of 12.4 hrs, so that about 700 cycles occur in the space of a year. Such activities, however, are moderated by light and darkness, and it appears to be a general rule that a single growth line forms each day.

THE ANNUAL CYCLE. With the exception of the central belt of the tropics, where the elevation of the sun, and the consequent mean annual temperature, vary little in the course of a year, seasonal variations in the number of hours of daylight and darkness, and a seasonal rise and fall of environmental temperature, are quite general. It is a general rule that metabolism (biological activity of tissues and cells) occurs faster at higher than at lower temperatures (within the viable range). It happens therefore that corals and other marine organisms that live on or near the equator grow at a nearly uniform rate throughout the year. On the other hand, corals that inhabit the northernmost or southernmost limits of the coral-reef zone, in latitudes of about 30° north and 30° south, or thereabouts, experience a pronounced seasonal variation in water temperature. In such cases the rate of growth in winter is much less than that in summer, when the water temperatures are higher. For example, species of the genus *Fungia* have been found to grow in a sequence of about 7 annual installments, the radius of the corallum increasing by about 10 or 11 mm each year; the periods of rapid growth (corresponding to summer) are marked by thicker growth lines, and the winter lines are discernibly thinner. Similar observations have been made on corals in the Caribbean and Floridian reefs. On account of the differences in the thickness of the summer and winter growth lines, it becomes possible to make counts of the growth lines and to classify them into successive sets of about 360 rings, representing successive annual increments. This is not as easy as it sounds, for the rings are produced in a thin layer of the coral skeleton called the *epitheca,* and unfortunately the epitheca is attacked by boring worms and other organisms, as well as by the abrasive influences of the inorganic constituents of the environment; so that it is quite difficult to make accurate counts of growth rings. Genera where detailed counts have been carried out with some success include *Manicina* and *Lophelia.*

The observations noted in the preceding paragraph seem relatively straightforward and predictable. However, in 1963 John W. Wells, an authority on fossil and living corals, drew attention to a hitherto unnoticed fact, namely, that annual sets of growth lines can also be recognized on some fossil corals and that the number of growth lines in an annual sequence *increases with antiquity* of the coral fossils. He found that whereas living corals such as *Manicina* produce about 360 growth lines per annum, a coral of the Pennsylvanian period (about 300 million years ago) will have between 385 and 390 rings in an annual sequence, and a coral from the middle Devonian period (about 380 million years ago) will show between 385 and 410 diurnal rings. In January of the following year the National Aeronautics and Space Administration was host to an international conference on the history of the earth-moon system, at which Wells was invited to present his data. Like Hooke in London in 1696, Wells presented paleontological data to a meeting of astronomers and geophysicists; the latter, in marked contrast to Hooke's contemporaries, immediately perceived the possible implications of the discovery. Wells's findings have been under study ever since, by biologists, astronomers, and physicists. Although some uncertainties remain (as Wells himself was careful to point out), particularly because of the difficulty in making precise counts of growth lines, most of all on worn fossils, nonetheless a general body of literature now exists on the subject, and the following is a summary of the main points.

THE STABILITY OF THE ANNUAL CYCLE. Accepting the probable fundamental truth of the observations made by Wells, we are left with several initial possibilities:

1. The year may once have lasted longer and therefore comprised a greater number of days than now is the case.
2. The day may have been shorter, so that more days elapsed in a year.
3. Both the length of the day and the length of the year may vary.

Fortunately the first and the third of these possibilities may be dis-

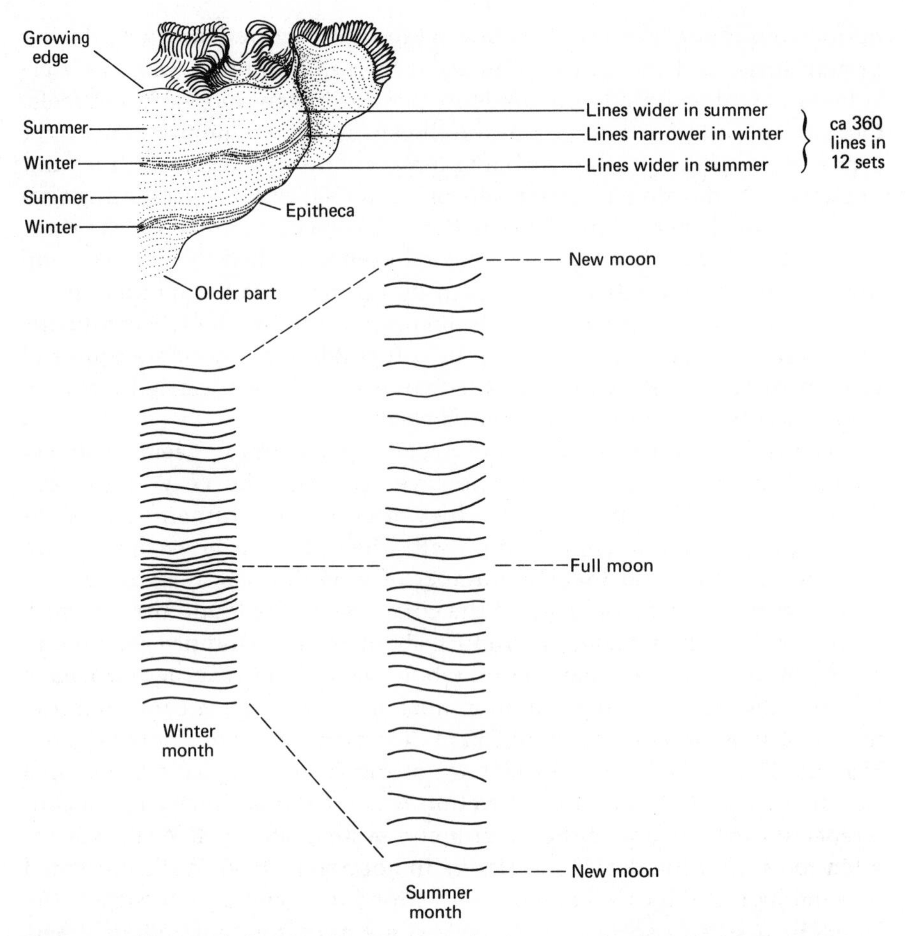

Figure 26. The coral clock exemplified by the living genus *Manicina*. In the upper figure the outer layer of the skeleton, or *epitheca*, is laid down in daily increments, the annual bands being produced through the varying thickness of each band. Under higher magnification (lower diagram) the daily bands are seen to vary also on a 30-day cycle, both in summer and in winter, the summer lines being wider (23).

missed. A long series of classical investigations—beginning with those of Laplace on the properties of the earth's orbit; continuing with those of Leverrier (1843) on the mutual perturbations of the planets, leading in 1846 to the discovery of Neptune; and concluding with the further development of Leverrier's equations by later investigators—have established that the parameters of the orbit of the earth are such that the eccentricity varies slowly between a maximum and minimum value, extremes separated by hundreds of thousands of years, but the period of revolution about the sun is virtually stable. This means that the absolute length of the year is a fixed quantity. If therefore the number of days constituting an annual cycle in the Carboniferous or the Devonian periods was greater than is the case today, then the absolute length of the day must have been increasing through geological time. In other words, the earth is gradually rotating more and more slowly.

When Wells plotted his data on the number of days in a year against the elapsed time since the observed values for Devonian and Pennsylvanian periods based on radiometric determinations, he found that the points for the Recent, the Pennsylvanian, and the Devonian all lay on the same straight line. This suggests a steady change; and astronomers further noted that the slope of the line was such as might have been expected from observed variation in the earth's rate of rotation within historic times. Assuming the correctness of these inferences, we could interpolate and extrapolate from the points given by Wells and so obtain a set of expected values for other geological periods. Such interpolation and extrapolation must still be treated with reserve, but the present state of the art is such as to encourage us to think that these are reasonable deductions. (See Table 10).

Table 10. Number of days in the year at various epochs.

PERIOD	ELAPSED TIME IN MILLIONS OF YEARS	NUMBER OF DAYS	LENGTH OF DAY	
			HOURS	MINUTES
Recent	0	365.24	24	00
Pleistocene	1	365.4	23	59.5
Pliocene	11	366.3	23	56
Miocene	25	368	23	50
Oligocene	30	369	23	46
Eocene	50	371	23	38
Cretaceous	100	377	23	16
Jurassic	150	383	22	53
Triassic	200	389	22	32
Permian	250	394	22	16
Carboniferous	300	400	21	55
Devonian	380	410	21	23
Silurian	420	415	21	08
Ordovician	470	420	20	53
Cambrian	550	430	20	23

NOTE: These estimates are tentative and derive solely from biometric data, as explained in text.

THE CAUSES OF THE LENGTHENING OF THE DAY. In 1897 George H. Darwin pointed out that the earth's rate of rotation is being retarded by tidal friction, by the delay in the time at which high tide occurs. The mechanism operates in this way: tidal friction (see Chapter 7) causes delays in the transmission of the tidal crest across an ocean; so the high-tide crest arrives *after* the moon has already crossed the meridian. This means that the tidal bulge lies always to *the east* of the sublunar point. The tidal bulge may be thought of as a protuberance of the earth. As such, it suffers gravitational attraction by the moon, which therefore tends to pull it toward the west. But the earth is rotating from west to east, so the earth's motion of rotation is carrying the tidal bulge in the opposite direction from that in which the moon is drawing the tidal bulge. In other words, the moon is exerting a torque on the earth itself, in such a direction as to counteract the earth's rotation, and therefore a brake is being applied. Hence the day is gradually lengthening.

THE MONTHLY, OR LUNAR, SEX CYCLE. The sexual activities of many marine animals are mediated by the nervous system, especially by the photosensitive light-detector organs. At particular thresholds of illumination, or duration of illumination, hormones are released that cause the male and female sex cells to mature, followed by their liberation some hours or days later. Peak illumination at night by the full moon serves as an effective maturation stimulant in the cases of many marine organisms that are normally active at night. In such forms the sex cells are likely to be liberated several days after the night of the full moon. Among the possible advantages of this arrangement is the fact that most or all members of a given species of a community will become sexually active on the same night; thus wastage of sperm or eggs is avoided, because large numbers of gametes of both types are simultaneously liberated. For any given species the time that elapses between the full moon and the mating night is fixed within rather precise limits. The limits may vary according to the season.

The nuptial dances of the polychaete, *Platynereis dumerilii,* for example, are held on the French coasts at the first and last quarters of the lunar month. In *Ceratocephale,* the Japanese palolo worm, the July and August meetings are held 6 to 9 days after new and full moon, whereas the September gatherings are delayed until the fifteenth night after the new and full moon (doubtless on account of slowing maturation as the water temperatures fall). In reef-building corals such as *Stylophora* the reproductive stages (planulae) are released with a lunar periodicity throughout the year. In the Palao islands, the planulae of *Fungia* are released "in the days of the new moon" in every month from September to April. On the other hand, *Rhizopsammia,* a small Japanese littoral coral, liberates its planulae only between August 10 and September 10, the period of maximum water temperatures. Thus corals differ, with lunar periodicity most marked in the cases of species living in the tropics, where water temperatures are high at all times. It is perhaps true to say that in corals the more conspicuous the annual cycle is, the less conspicuous is the lunar cycle, and water temperature factors would explain this relationship.

THE EFFECTS OF LUNAR SEXUALITY ON GROWTH. When corals are actively producing and liberating reproductive stages, there is a reduction in the fraction of metabolic activities directed to normal growth, including growth of the skeleton. This means that the growth lines laid down during the sexually active period are narrower than those produced during the nonsexual part of the month. Consequently, when a long series of growth lines is studied, it is usually easy to distinguish the sets of consecutive lines laid down during the days or week of reproductive activity and the alternating sets of broader growth lines that are laid down on the intervening days. Numerous data on these have now been obtained and published. A variety of other lunar-related factors probably also influence the rate of growth of marine organisms; whatever the factors may be, the net result is a lunar-related cyclic fluctuation in the growth rates. When a large number of such lunar growth lines is measured and analyzed statistically it becomes apparent that the existing marine organisms follow a lunar periodicity that matches the *synodic month*—that is, the time that elapses between two successive matching phases of the moon, from full moon to full moon, new moon to new moon, and so on.

FOSSIL MOONLIGHT. Soon after Wells detected annual periodicity in Devonian corals, Scrutton in England observed lunar periodicity in a number of genera of corals of the Devonian period. In 1964 he published his conclusion that the middle Devonian year contained 13 lunar months each of 30½ days. Subsequent studies by other workers have confirmed this conclusion and have further implied a gradual change in the length of the synodic lunar month from about 31½ days in the middle Cambrian (about 550 million years ago) to 29.53 days, its Recent value. It must, of course, be remembered that the days vary in absolute length according to the geological period to which they relate. Therefore, in order to obtain a true picture of the change in the length of the lunar cycle, we have to convert the units of measurement into uniform standard absolute values.

THE ABSOLUTE MEASUREMENT OF THE LUNAR CYCLE. This measurement is achieved very simply by converting the units of time into fractions of the year, for we know that the absolute length of the year has not varied significantly. Since there are 29.53 days in the present (Recent) synodic month, we can can say that the length—or *period*—of the synodic month at the present time is

$$1/\frac{365.24 \text{ days of 24 hrs}}{29.53 \text{ days of 24 hrs}} = 1/12.4 \text{ year}$$

This value can now be compared directly with Scrutton's value (1964) for the middle Devonian synodic month of 1/13 year. From this comparison it is now clear that the synodic month has increased in length (13 − 12.4)/13, or 4.6 percent during the past 380 million years.

KEPLER'S LAWS OF PLANETARY MOTION. The foregoing paragraph implies an inferred change in the period of revolution of the moon about the earth. The statement acquires a special interest because Kepler's third law of planetary motion tells us that "The square of the period of a planet (or satellite) is proportional to the cube of its distance from the primary body.'' In other words, for the moon to have orbited the earth in Devonian times in the observed period of 1/13 year, its distance from the earth cannot have been the same as is now the case; and the difference can be computed by inserting the known values into Kepler's equation. Before this calculation can be performed, one adjustment is needed, for the synodic month is not the true period of revolution of the moon about the earth.

SIDEREAL AND SYNODIC MONTHS. The *synodic* month is the time that elapses between one full moon (or new moon, or any other recognizable phase) and the next corresponding phase. As already noted, at the present epoch 12.4 such synodic intervals occur in a year. However, it has to be remembered that while the moon is completing one revolution about the earth, the earth itself is traversing its own orbit about the sun. In the space of one month the earth moves through about 30° of its own orbit, with a consequent alteration of the direction in space in which the sun now lies. The change in direction is such that the moon is required to traverse an extra 30° of its own orbit before it returns to a position that appears from the earth to match its former position relative to the sun. In the course of one year, therefore, the moon actually completes *one extra orbit* around the earth. Thus, for the present epoch, 12.4 synodic months correspond to $12.4 + 1 = 13.4$ revolutions of the moon about the earth. The period 1/13.4 year, or 27.32 days, is called the *sidereal month*, and this is the true period of revolution of the moon. Similarly for the middle Devonian the sidereal month would be 1/13 + 1, or 1/14 year.

THE DISTANCE OF THE MOON IN DEVONIAN TIMES. This value can now be calculated from the growth-line data of corals, given that the present distance of the moon is 384,400 k, or 238,857 miles. By substitution in Kepler's equation:

Devonian distance of the moon

$$= \text{present distance } [\,(12.4 + 1)^2 \,/\, (13 + 1)^2]^{1/3}$$

$$= 384{,}400\ (13.4^2/14^2)^{1/3}\ \text{km}$$

$$= 372{,}870\ \text{km}$$

from which it appears that the moon has retreated over 11,000 km during the past 380 million years. If certain modified values for the numbers of days in the year are used, as recommended by Newton (1969), the distance the moon has retreated since Devonian times becomes about 15,000 km.

THE CAUSE OF THE RETREAT OF THE MOON FROM THE EARTH. Although the growth-line data from marine fossils has yielded the first measures of the rate of retreat of the moon from the earth, the inference that this retreat is occurring was already drawn by George Darwin in 1897, in the same paper as that in which he inferred the earth's rate of rotation to be slowing down. He pointed out that as the earth's rate of rotation is reduced, there is a loss of angular momentum. But angular momentum is indestructible and must therefore be transferred elsewhere. The transfer is effected by way of the tidal torque applied to the earth by the moon. Just as the moon is exerting a backward tug on the tidal protuberance of the earth, so also the rotating tidal protuberance exerts a forward tug on the moon. The moon responds like a slingshot, flying ever faster in its orbit and therefore, in accordance with Newton's and Kepler's laws, moving outward into successively larger orbits appropriate for its greater velocity.

Darwin predicted that this process will continue until the day and the month both equal 47 of our present days. The effects of the solar torque will then become predominant. Later still, if the solar system still exists, the transfers of angular momentum will operate in such a manner as to return the moon to the near vicinity of the earth. For a planet of any given mass there is a specific radial distance, known as Roche's limit, within which it is not possible for a satellite to occur without suffering physical disruption under the gravitational attraction of the planet. Jeffreys believes that the final fate of the moon will be to suffer disruption within Roche's limit, to become a ring of fragments around the earth, like the rings of Saturn.

6

FIVE MILES DOWN

Man has ventured on the sea for 10,000 years, yet for nearly all that span of time he had little conception of what lay beneath the waves, or how deep the waters might be. From archeological finds, as also from some surviving documents, we learn that the sounding line was known to the peoples of the Mediterranean in classical times. For example, we have part of the diary of a Roman citizen who made a voyage from Sidon to Rome in the fall of the year A.D. 62. He writes: "One night, when we were two weeks out from Crete, and still being driven up and down the Ionian Sea, the ship's officers began to suspect we were approaching land. They took a sounding, and found twenty fathoms. A little farther on they sounded again, this time finding only fifteen fathoms. . . . " The method, which was still in use throughout the Middle Ages, and on into the eighteenth century for that matter, was to lower a rope weighted with a lead sound, in the base of which was a recess containing beeswax. If the sound touched bottom, it came up with pieces of shell or sand or pebbles adhering to the wax. Curiously enough, there are still today certain parts of the seabed that can be sounded effectively only by this ancient method.

HOW DEEP IS THE OCEAN? The first man to attempt to discover the depth of the Atlantic Ocean was, appropriately, Christopher Columbus. In a letter to Ferdinand and Isabella, which is still extant, Columbus cites from his log: "September 20, Thursday (1492)—A flat calm came, so the Admiral, believing in an early landfall, ordered the lead hove. They found no bottom at 200 fathoms. . . . " Columbus, who always refers to himself in the third person as Admiral of the Ocean Sea, was actually sailing over 2300 fathoms on that Thursday. About 40 years later Magellan, in mid-Pacific, over similar depths, made a similar trial, and he too found no bottom at 200 fathoms.

The mystery of the ocean's depth remained unsolved for another 250 years. Then, in 1775, on the eve of the American Revolution, a French aristocrat announced from his study chair in Paris that the Atlantic Ocean has an average depth of about 13,000 feet, or 2½ miles. Another century elapsed before experimental proof was forthcoming that this statement is in fact correct. How was it obtained? By solving an equation! The man who plumbed the Atlantic was Pierre Simon, Marquis de Laplace, a brilliant mathematician who taught at the Ecole Militaire in Paris.

Accepting Newton's explanation of the tides as being due to the attraction of the moon and the sun, he engaged in a detailed study of the theory of propagation of a tidal wave across an ocean. Ideally, if Newton's explanation is correct, one would expect the high tide crest to lie beneath the moon and sun, when these bodies are in the same direction in space, as is the case at the time of the new moon. On the equator, where the surface of the earth rotates at a surface speed of 1000 miles per hour, the tidal wave should sweep across the ocean at the same speed, so as always to lie beneath the heavenly bodies whose attraction raises it.

But a study of ships' logs disclosed to Laplace that these circumstances by no means hold. He found that the ships' officers reported that on the west African equatorial coast, the high tide occurs *2 hrs after* the moon passes the meridian, whereas on the eastern coast of Brazil, on the other side of the Atlantic, the equatorial high tide occurs *6 hrs after* the moon passes the meridian. Evidently, therefore, something delays the tide as it sweeps across the Atlantic from east to west, following the apparent motion of the moon (that is, the real west-to-east motion of the rotating earth). Since the earth's rotation causes all points on the equator to move at 1000 miles an hour, if there were no delays, the tidal wave would cross the equatorial Atlantic in 3 hrs, for the Atlantic here has a width of 3000 miles. As there was an observed delay of 4 hrs, the total time taken by the tidal wave to cross 3000 miles must be $3 + 4 = 7$ hrs.

Laplace now sought an explanation for the observed delay by investigating the way in which a long-period wave is transmitted through water. He discovered that there is a limitation on the speed with which a long-period wave can travel, the parameters being set by the value of the constant of gravitation, g, and the depth of the fluid transmitting the wave, such that

$$v = (gd)^{1/2}$$

where v is the velocity of propagation, g is the constant 32 feet per sec^2, the acceleration produced by gravity at the earth's surface, and d is the depth of the fluid, in this case the sea. Thus, for a tidal wave to travel at its expected maximum speed, so as always to crest beneath the moon, moving therefore at 1000 miles per hour (1467 feet per second), we have, by rearranging the terms of the equation above,

$$d = (1467)^2/32 = 67{,}000 \text{ feet, or 13 miles}$$

Evidently, therefore, the Atlantic Ocean must be shallower than 13 miles. To find its average depth, Laplace inserted into the equation the observed speed of the wave that, as noted above, crosses 3000 miles in 7 hrs. This yields a speed of 637 feet per second, whence we have

$$d = (637)^2/32 = \text{approximately } 13{,}000 \text{ feet, or 2.5 miles}$$

OTHER MEANS OF FATHOMING THE SEA. During the nineteenth century experimental oceanography developed as a separate discipline, following the need for seafloor surveys before the laying of the first transoceanic cables. Long tapered cables and wires were produced in lengths great enough to reach the deepest parts of the ocean. The invention of a meter for measuring the tension applied to a suspended cable permitted direct fathoming, for the touchdown on the seafloor could now be detected by the fact that the tension on the cable became stabilized when once the free end began to pile up on the bottom. A clinometer measured the angle of the cable, permitting solution of the vertical side of the triangle whose three angles were known, and whose hypotenuse was the cable of known length. These methods disclosed depths of 5 miles in certain parts of the sea.

Modern soundings are made with the aid of echo-sounders, which give a continuous recording along any transect followed by a ship. Thus very detailed topographic maps of the ocean bed are now becoming available at relatively little cost.

A DARK, LIFELESS ABYSS? Before the improvements mentioned in the last paragraph had been devised, some thinking men had begun to ponder the implications of Laplace's astonishing discovery. It was now realized that if the average depth of the ocean is 2½ miles, then by far the greater part of the seabed must lie in utter darkness, under pressures that had hitherto been considered appropriate only to the interior of a planet. Joseph Priestley and his fellow chemists in England, France, and Germany demonstrated that sunlight was essential for the production of sugars and other organic carbohydrates in plants and that animals lack this power to make organic substances from inorganic ingredients. It seemed, then, that the floor of the ocean must be a lifeless desert. Calculations disclosed estimated pressures (see Table 11), and in face of such estimates, few naturalists or other scientists would dare to predicate the existence of life.

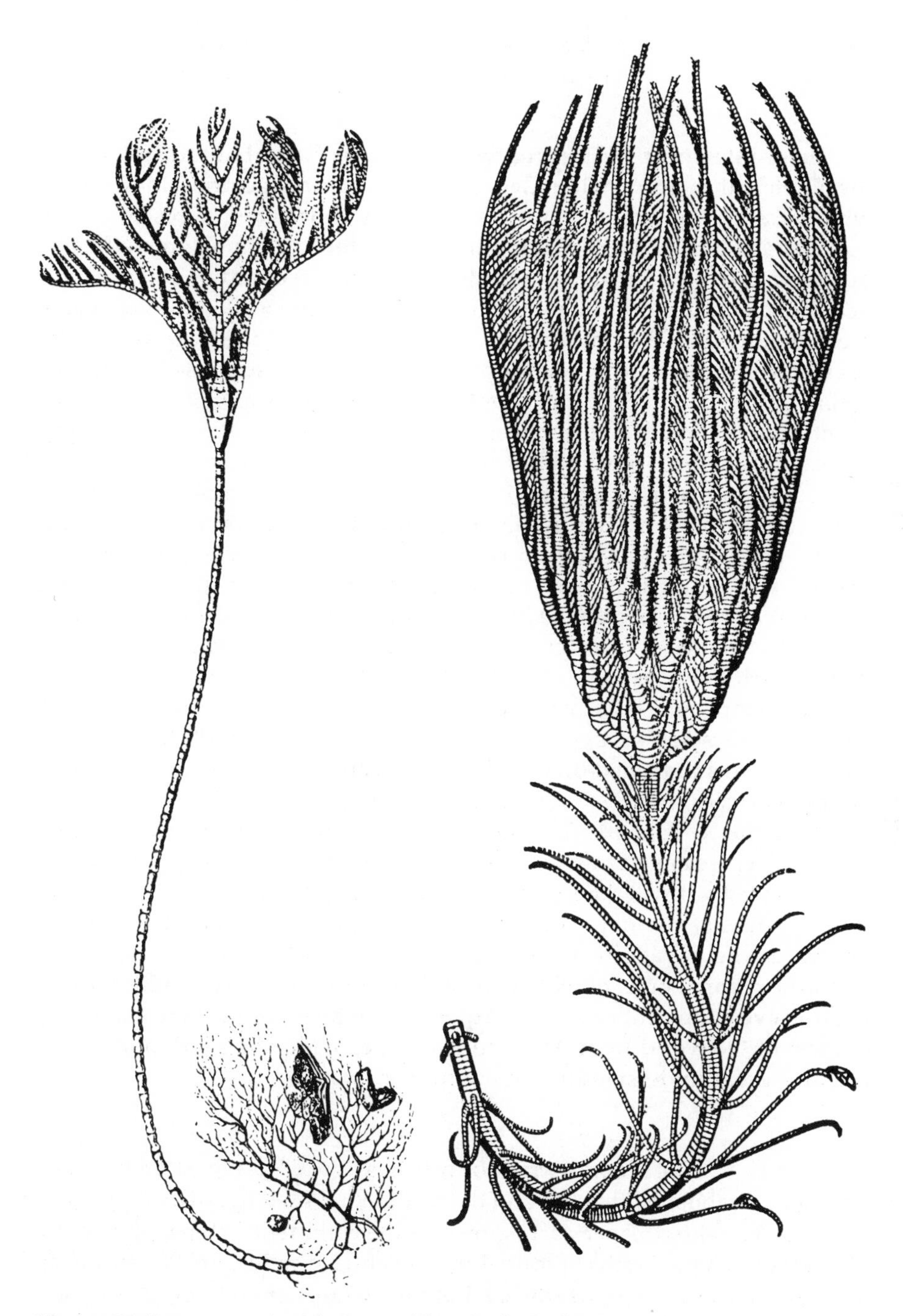

Figure 27. Deep-sea crinoids, or sea lilies. Left, *Bathycrinus*, from Sars; right, *Metacrinus*, taken by the *Challenger*, after Carpenter (38).

Table 11. Ocean depths and estimated pressures.

DEPTH	PRESSURE
1000 m (3300 feet)	3/4 ton per square inch (110 atmospheres)
2000 m	1 1/2 tons per square inch (220 atmospheres)
3000 m	2 1/4 tons per square inch (330 atmospheres)
4000 m	3 tons per square inch (440 atmospheres)

These inferences were abruptly shown to be false about the middle of the nineteenth century, when Georg Sars, son of a village priest in Norway, who had seen the local fishermen's catches and become interested in the sea, devised a means of dredging in deep water and almost immediately discovered living animals at great depths. Soon afterward he further startled the scientific world by producing from the deep seafloor living specimens of sea lilies, thereby triggering speculation that the deep sea might prove to be a place of refuge for organisms that had supposedly become extinct millions of years ago. The latter hypothesis was not justified by events, but the important discovery had been made that the deep ocean is indeed the abode of life, and not an empty desert. Other curious and mistaken ideas were dispelled one by one. Some scientists spoke of a mysterious zone in deep midwater where all sunken ships, bodies of drowned sailors, and so on, were supposed to remain suspended in water whose density was so great under pressure as to equal that of the items mentioned. Water is in fact almost incompressible, and objects sink to the floor of the deepest seas; research ships not uncommonly dredge from great depths such items as beer cans thrown overboard from their own refuse the day before, and less commonly objects thrown or fallen from passing ships 200 years ago or more—all lie on the seabed, and with them lie the teeth of sharks that shed them millions of years ago and the hard petrosal bones of whales that lived and died 10,000 years past.

After the work of the Sars, father and son, became known in Britain, a major deep-sea exploring expedition was planned in considerable detail and provided with all the investigatory equipment and scientific staff that the resources of the day could encompass. This expedition of the 1870s, known as the voyage of H.M.S. *Challenger,* resulted in a great series of scientific reports, most of which are still in constant use, new editions appearing even in recent years. The expedition, which sampled the deep seas of most of the world from Arctic to Antarctic, marks the solid foundation of our present understanding of life in deep waters.

THE STRATIFICATION OF BENTHOS. The *Challenger* and later expeditions established that beyond the continental shelf the seabed plunges more steeply; this steeper part, known as the continental slope, plunges to about the average depth computed by Laplace, 2½ miles, or 2000 fathoms, and then becomes more nearly level. Actually there are numerous submarine ridges and depressions. In some restricted regions, usually near land, the earth's crust seems to be thrown into great vertical folds, with deep trenches extending to depths of 5 and 6 miles below sea level. The benthos lying on, or imbedded in, the floor of the ocean is classified by depth in Table 12.

Table 12. Vertical stratification of benthos.

DESCRIPTIVE TERM	DEPTH AT WHICH BENTHOS OCCURS (IN METERS)	EXAMPLES
Littoral	0–5	Biota of beaches, reefs, mangrove sloughs
Sublittoral	5–20	Region commonly entered by scuba divers
Neritic	0–200	Benthos of the continental shelf
Archibenthal	200–1,000	Benthos of upper continental slope
Bathyal	1,000–2,000	Benthos of lower continental slope
Abyssal	2,000–4,000	Fauna of the general oceanic floor
Hadal	4,000–10,000	Fauna of the deep trenches

NOTE: For *hadal*, the term *ultrabyssal* (Russian, *ultraabyssal*) is also used.

SAMPLING PROCEDURES. Samples of deep-water benthos are obtained by what are essentially the same methods as those for the continental shelf. The collecting instruments are dredges, beam trawls, and grabs, as well as sledges and corers. The principles are the same, the manner of operation demanding more sophisticated controls and more experienced or more ingenious operators. Underwater photography has become an indispensable technique in both qualitative and quantitative investigations, either by means of remotely controlled cameras lowered from ships or hand-operated cameras viewing through the port of a deep submersible. Besides showing the life in its natural state, photography enables more random sampling of the epifauna than do methods such as dredge- or trawl-sampling. To obtain maximum information photographic and more traditional collecting methods should be used together, each supplementing the other. In this way much can be learned about the ecology, including feeding habits, modes of locomotion on abyssal substrates, reaction to light and temperature change, community formation, and population densities of underwater life.

To be most effective, seafloor photographs should be taken at frequent intervals along the track of any ship that is trawling or dredging for biological specimens. If possible, photographs should not be taken without also securing actual biological samples, for identifications of images in photos are always difficult, often impossible, without comparative material in the form of specimens of biota from the same or nearby stations. Neither should biological samples be taken without seafloor photographs, unless it is unavoidable; for photographs nearly always throw much extra light on details of the environment and the ecology of the animals collected. Photographs may reveal unexpected circumstances, such as animals in an environment previously thought inappropriate for them, an example being Hurley's discovery (1959) of *Pyrosoma* lying on the seafloor.

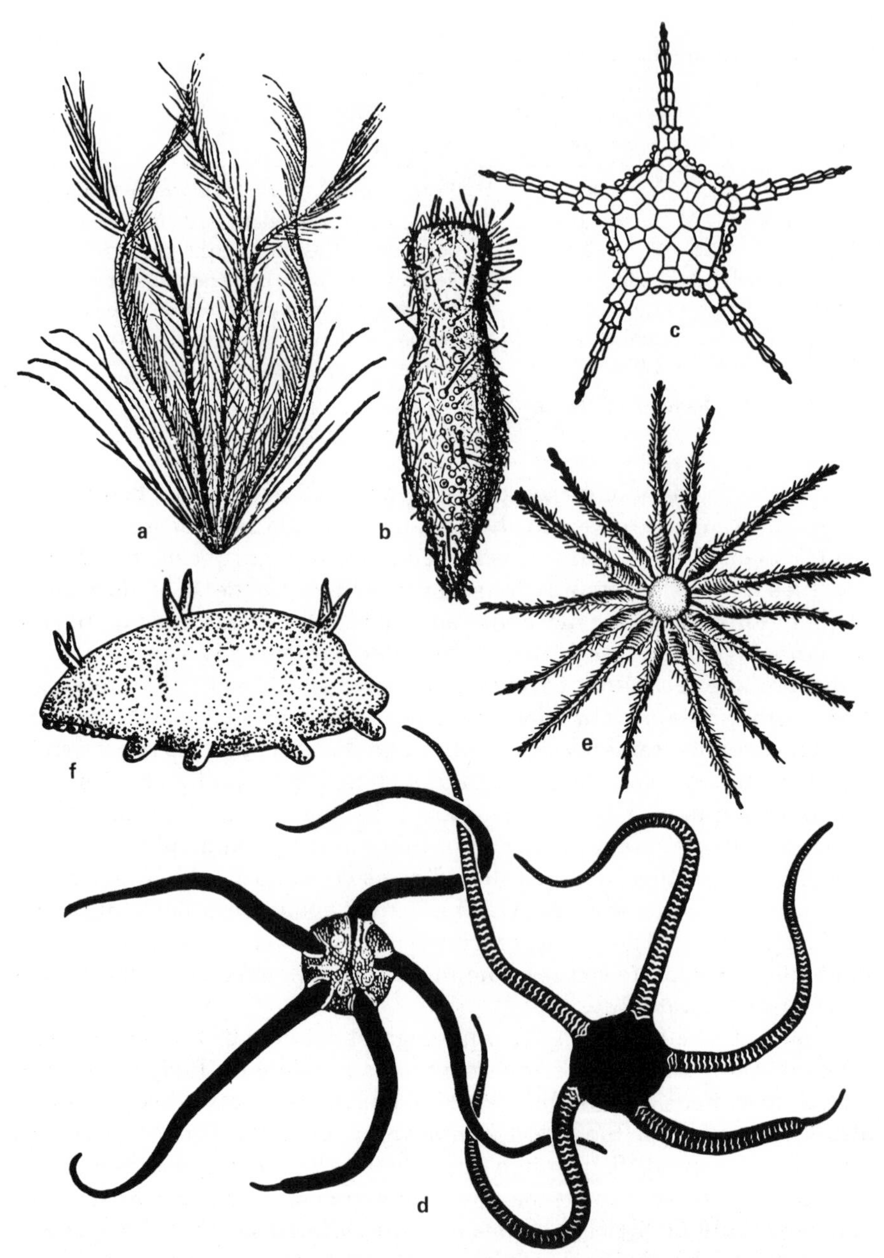

Figure 28. Echinoderms from beyond the shelf. (a) *Pentametrocrinus* (feather star), X 1, Crinoidea; (b) *Echinosigra* (sea urchin), X 2, Echinoidea; (c) *Ophiomisidium*, X 3, and (d) long-armed brittle star (various genera occur), X 0.5, Ophiuroidea; (e) *Brisinga* (long-armed sea star), X 0.2, Asteroidea; (f) *Scotoplanes* (sea cucumber), X 0.5, Holothuroidea (23).

THE RELATIONSHIP OF ORGANISM TO SUBSTRATE. While some benthic organisms tolerate a wide range of seafloor conditions, many are restricted to particular types of bottom. This is not always evident from the collections taken from trawls and dredges, because these instruments may traverse a variety of bottoms during a single transect. Here a useful adjunct to the collecting procedure is provided by taking periodic bottom photographs during the progress of the transect. Bottom may vary from naked, or thinly silted, rock to coarse or fine gravel, or soft silt or ooze. On the harder substrates many of the organisms will be anchored to the hard objects permanently, as in the cases of sponges or corals and similar sessile forms, or will be facultatively mobile, as in the cases of crinoids with clasping organs such as cirri. The epifauna on a soft substrate may itself provide a hard substrate for such mobile elements as crinoids or ophiuroids.

Soft substrate may form flat submarine plains, or the floor of a submarine canyon, carrying a characteristic population of organisms so constructed that they either do not sink into the soft material or else have means of extricating themselves from it. A typical population of this type in deep water will comprise in large part echinoderms, especially ophiuroids and holothurians. The ophiuroids do not sink into the soft substrate because their weight is distributed by their long, horizontally disposed arms. The holothurians are usually very delicately constructed, consisting mainly of watery fluids in fine membranes. Photographs taken from the U.S. Navy bathyscaphe *Trieste* show that they can actually walk across the sediment on the tips of the lateral rows of tube feet. Some benthic animals such as free crinoids and echinoids develop very elongated cirri or spines. The purpose of these is evidently to provide a stable support by sinking just so far into the sediment as is needed to reach a state of equilibrium, without submerging vital organs in the substrate. Sea urchins of the family Aspidodiadematidae possess exceptionally long, hollow, curved spines that are very delicately constructed and so do not add appreciably to the weight, yet are very effective in supporting it in its favored soft environment. These echinoids, almost invariably shattered in the process of bringing them to the surface, are excellent subjects for photography, which alone provides an undistorted record of their ecology.

DISTURBANCES OF THE SUBSTRATE. A soft substrate is not always a flat substrate. Photographs in submarine canyons sometimes disclose relatively steep profiles, resembling hillsides. If the substrate is soft, then crevasselike openings are sometimes visible, suggesting that periodic slumping must occur. Doubtless many fossil assemblages, in which many animals of the same species and age group are found preserved in the same bedding plane, may have been entombed through sudden slumping of soft substrate of this type.

Feeding habits are related to substrate, and one consequence of the feeding habits of holothurians on soft substrates is relevant to the study of many bottom photographs; this is the production of copious castings

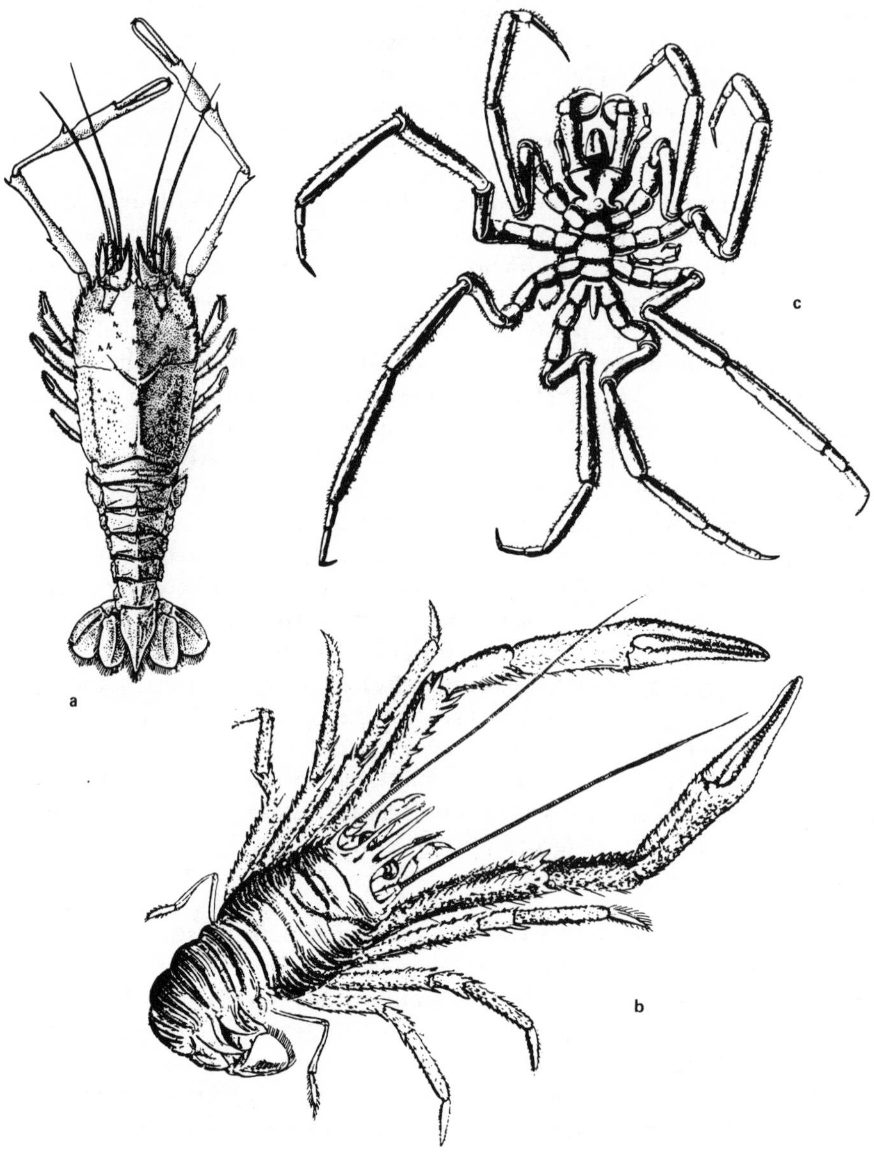

Figure 29. Deep-water Crustacea (a, b), and a pycnogonid (c). (a) *Polycheles*, 800 fathoms, Panama, a deep-sea decapod remarkably similar to a Jurassic genus *Eryon*, X 0.5; Alcock, from (29); (b) *Munida*, continental slope, Decapoda, X 0.5 (38); (c) *Nymphon* (sea spider), Arctic seafloors, Pycnogonida, X 0.25 (38).

of egested materials. Probably the main source of nutriment in the substrate itself in the deep-sea environment is the bacterial content, together with its by-products. To secure enough nutritives from such a source holothurians swallow relatively immense quantities of mud, thus continuously disturbing the upper layers and preventing any very precise stratification of matter falling from above. Hence considerable care is needed in distinguishing the evidence of living bottom communities *(biocoenoses)*, from that of secondarily associated evidence of materials post mortem *(thanatocoenoses)*. This warning particularly applies to grab or core samples, where quantitative treatment is often accorded the samples. Seafloor photography warns us that it is highly improbable that annual increments of sediment and microfossils can survive for long in undisturbed horizontal layers on the bottom of the sea, except under unusual abiotic conditions. Indeed, the castings of holothurians are among the commonest objects on the seabed disclosed by photography, and these imply the constant dislocation of sediments. Castings fall to pieces at a touch, so they are never taken in a recognizable state in sampling devices operated by remote control; here then is an area of study where seafloor photography has proved critical.

THE DETECTION OF BOTTOM CURRENTS. Bottom photographs often disclose a nonrandom orientation pattern in bottom communities, more particularly the orientation of sessile members of the community. Fixed animals, and also those errant forms that adopt a sedentary habit when environmental conditions permit, may sometimes be seen directing the mouth or feeding organs upward. Presumably this means that they are collecting fragments of organic materials, or small organisms, which are falling from above. This would imply relatively little movement in the water mass. Under such conditions normally mobile wandering holothurians may adopt the posture of a sea anemone, standing upright with the body anchored by the posterior tube feet and the oral tentacles at the upper, free end, held out in a ring. A random orientation of such erect organisms would suggest the presence of micronekton or swimming plankton, capable of being captured from various directions, the water mass remaining still.

Parts of the Antarctic seafloor have proved to carry communities in which the dominant member is a tubicolous polychaete. Photographs of the community frequently show a definite orientation of the polychaetes such that all will have the free tentacle-bearing end of the body directed the same way. It is inferred in these cases that a gentle bottom current is flowing toward the animals from the direction to which they are pointing. Sponges, which can create their own feeding currents by the vigorous activity of the flagellated cells lining their gastral chambers, generally show quite random orientations; or, in cases of clusters of sponge persons, the latter are mutually divergent, as if to maintain each feeding system independently of the neighboring ones. Polyzoans and sea anemones show current orientation; on the other hand, the mobile benthos such as ophiuroids, asteroids, and mollusks appear in photographs to be creeping randomly. In general, mobile animals disregard current.

POPULATION DENSITIES. Transects utilizing photographic techniques can provide good data on epifaunal elements. But in most areas there are also likely to be infaunal elements, such as burrowing sea urchins and burrowing asteroids. These latter are likely to be missed by any technique wholly dependent on photography. Consequently the photographic method cannot replace traditional dredging procedures. Clearly, the biological samples are needed, not only to provide accurate determinations of the fauna visible in photographs, but also to yield data on the infauna that is not visible in photographs. Conversely, without photography, small but numerous members of the fauna may be lost through inappropriate mesh size, or at least reduced numerically in the samples, thus confounding qualitative analysis. Highly mobile species escape from the mechanical sampler and can only be taken by lucky accident, whereas the camera registers their existence without difficulty. Again, the camera will record peculiar attitudes—for example, fishing postures, plankton-trapping postures, and so on—that throw light on the bottom ecology. So neither camera nor dredge alone is sufficient, for neither one can replace the other. Population densities cannot accurately be studied without cameras. The high incidence of feather stars in Antarctic seafloor trawl and dredge samples has been interpreted in the past as indicating high population densities of crinoids in the Antarctic. Photographs do not confirm this. It is more probable that feather stars are disturbed by the passage of the trawl, swim slowly above the seafloor, and become immediately entangled in the net, while other smaller or more mobile elements escape the net (Fell, 1967). Thus, populations should be studied from photographs, or better, from photographs and dredge or trawl samples.

THE MAXIMUM DEPTHS OF THE OCEANS. The greatest depths of the oceans occur in relatively narrow elongate *trenches*, usually not wider than about 100 km, and up to about 2000 km long. Table 13 shows the deepest trenches of the Pacific, Atlantic, and Indian oceans.

DEEP-SEA FOOD WEBS. Since there is no sunlight reaching the deep-sea floor, there can be no autotrophic organisms there. The postures of microphagous benthos show that food particles arrive either as plankton-shower from above, as inflowing suspended materials, or as organisms in bottom current. Thus there is a materials input and also energy input from elsewhere. Therefore the deep-sea communities must ultimately be dependent on other communities for their continued existence. Carnivorous members of the deep-sea nekton doubtless feed on the overlying nekton of the midwater masses. The latter, through the diurnal mechanism of the rise and fall of the plankton, and the accompanying vertical migrations of

Table 13. Deepest trenches of the oceans.

			MAXIMUM DEPTH		
OCEAN	REGION	TRENCH	KILOMETERS	MILES	FATHOMS
Northwest Pacific	Marianas	Mariana	11.02	6.85	6028
Southwest Pacific	Tonga	Tonga	10.88	6.76	5950
Northwest Pacific	Philippines	Philippines	10.50	6.52	5740
Southwest Pacific	New Zealand	Kermadec	10.05	6.24	5494
Northwest Pacific	Japan	Idzu-Bonin	9.99	6.21	5460
West Atlantic	Caribbean	Puerto Rico	8.39	5.21	4585
East Indian	Indonesia	Java	7.45	4.63	4074

the associated nekton at successive levels, probably set in motion a continuous vertical transfer of energy and biomass, from above downward.

Thus, the deep-sea biota must be viewed as an extension of the biota of the shallower parts of the oceans, and not as an independent ecosystem. Further strength is given to this view by the observed fact that the youngest stages of deep-water organisms, the larval forms, are obliged to rise into the shallow parts of the sea to obtain food particles of correct size, a circumstance suggesting that the deep sea was colonized by stocks that originated in shallower regions.

It should not be overlooked that the deep-sea environment plays a critical part in the rejuvenation of the planktonic ecosystem by way of the ascending currents that carry dissolved nutrients to the surface from the bottom where the bacterial degraders are active. One is tempted to speculate on the possibility that the rather sudden rise to eminence of diatoms in the epiplankton in the Cretaceous period, 100 million years ago, may perhaps have been related to a possible invasion of the deepest parts of the ocean by significant new biota at that time. For example, most of the groups of deep-sea fishes belong to orders whose earliest known members are of Cretaceous or Eocene age.

THE SYSTEMATIC CONTENT OF ABYSSAL AND HADAL BIOTA. As would be expected from the foregoing discussion, the phyla represented in the benthos and nekton of the deepest parts of the oceans prove to be the same as those encountered on the continental shelf and in the overlying waters. There is therefore no need to review again the characters of the phyla concerned, since cross reference may be made by the reader to the relevant pages of foregoing parts of this guide. Representative genera from the various phyla conspicuous in deep-sea collections are listed in

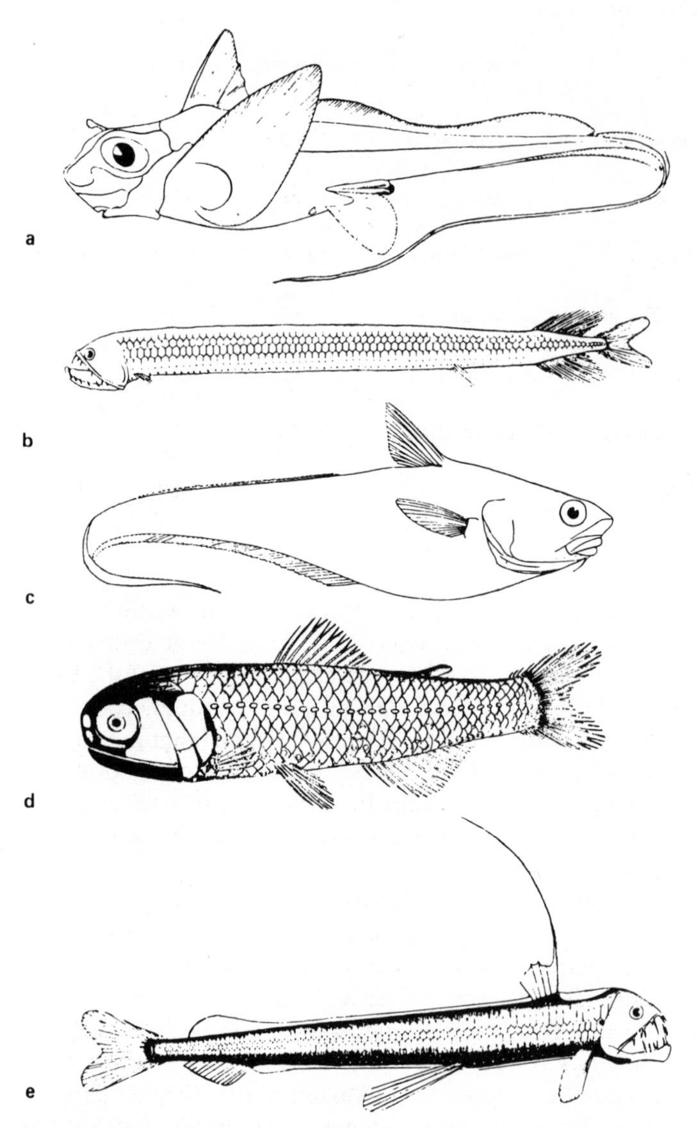

Figure 30. Pelagic and bathyal fishes. (a) *Chimaera* (ghost shark), X 0.1, benthic form, see Chapter 7 for classification, Holocephali; (b) *Stomias* (viper fish), slightly reduced; (c) *Macrurus* (rat-tail cod), X 0.1, benthic, bathyal, see Chapter 7 for classification, Gadiformes; (d) *Myctophum* (lantern fish), X 1, pelagic, Osteichthyes; (e) *Chauliodus* (viper fish), X 1, pelagic, Osteichthyes (38).

Table 14. Figures 27–30 illustrate some of the deep-water animals encountered in waters and on bottoms beyond the shelf. Most belong to phyla or classes already mentioned from shallower water; of groups not hitherto noted, the crinoids, or sea lilies, are members of the phylum Echinodermata and have somewhat the structure of a starfish, inverted and usually

Table 14. Representative genera of animals conspicuous in deep-water environments.

GENERIC NAME	PHYLUM OR OTHER GROUP NAME	DEPTH AT WHICH TAKEN (IN KM)	TRENCH WHERE TAKEN
Hyalonema	Sponge (Porifera)	6.86	Kurile
Nereis	Bristle worm (Annelida)	7.29	Banda
Scalpellum	Barnacle (Crustacea)	7.00	Kermadec
Serolis	Isopod (Crustacea)	3.70	(abyssal)
Nucula	Clam (Mollusca)	6.77	Kermadec
Xylophaga	Log-boring mollusk	6.29	Banda
Elpidia	Sea cucumber (Holothuroidea)	9.74	Bonin
Urechinus	Sea urchin (Echinoidea)	4.48	cosmopolitan
Bathycrinus	Sea lily (Crinoidea)	9.05	Kurile
Amphiophiura	Brittle star (Ophiuroidea)	6.04	Atlantic
Porcellanaster	Sea star (Asteroidea)	7.58	Mariana
Eremicaster	Sea star (Asteroidea)	7.24	Aleutian
Freyella	Sea star (Asteroidea)	6.18	Kermadec
Macrourus	Deep-sea cod (Chordata)	1.00	bathypelagic

attached to the bottom by a stalk (the feather stars have no stalk). The Pycnogonida are spiderlike marine animals, classified in the Arthropoda, and occurring also in shallow polar seas. Of fishes, the classification of which is the subject of the next chapter, note may be taken here of some of the commoner deep-water forms; these usually have a whiplike tail if they frequent the bottom, and some have chemical luminescent organs—for example, the lantern fishes (family Myctophidae). Some of these rise to the surface at night, when they may even be caught and eaten by oceanic birds such as petrels.

7

NEKTON: FISHES AND OTHER MARINE VERTEBRATES

Nekton may be defined as the set of organisms able to disperse in any direction in water. For marine nekton this means dispersion independently of ocean currents and independently of the substrate. Thus swimming organisms such as mysids or scyphozoans are not part of the nekton, for their swimming powers are insufficient to counteract ocean currents, and so they drift with the water masses and are counted with the plankton. Fishes, on the other hand, may drift with ocean currents but, if carried into unfavorable environments, they can facultatively swim against the currents. Similarly, fishes can perform annual migrations northward and southward, though the ocean currents they traverse continue to flow in the same direction. Of the other marine vertebrates, the marine mammals behave similarly to the fishes and obviously fall in the category of nekton. Marine birds such as penguins, lacking wings, also fall in the category. Other marine birds can both swim and fly; as a matter of convenience they are here considered to be part of the marine nekton. A few marine invertebrates, such as squid and some pelagic octopuses, have such highly de-

Table 15. Vertical stratification of the nekton.

DESCRIPTIVE TERM	DEPTH AT WHICH NEKTON OCCURS (IN METERS)	DEFINITION
Littoral	0–5	Frequenting inshore water
Neritic	0–200	Frequenting waters overlying the continental shelf
Epipelagic	0–200	Frequenting the upper waters of the open ocean
Demersal	Any depth	Nektonic, but habitually resting on the seafloor, as, for example, flounder
Mesopelagic	200–1000	Frequenting pelagic waters at depths corresponding to the archibenthos
Bathypelagic	1000–2000	Frequenting pelagic waters at depths corresponding to the bathyal benthos
Abyssopelagic	2000–4000	Frequenting pelagic waters at depths corresponding to the abyssal benthos
Hadopelagic	4000 and deeper	Inhabiting the waters of the deepest trenches

veloped powers of swimming as to place them in the category of nekton. No plants can be classified as nekton (except in science fiction). Table 15 gives the commonly employed categories of nektonic distribution.

STRUCTURE. The structure of the fish body is such as to promote rapid movement, and this is especially true of sharks. However, despite the robust tail, which provides the power for swimming, sharks suffer a constant tendency to sink, because there is no means by which the specific gravity of the body can be matched to the surrounding medium. To overcome the sinking, the anterior parts of the body, especially the snout and the head, are flattened horizontally; also the pectoral fins are held in a horizontal position, slightly inclined downward on the trailing edge. These features, combined with the motion imparted by the tail, maintain the fish's spatial relationship in much the same manner as for an airplane. If the fish stops swimming it will sink gently to rest on the seabed. Rays have probably evolved from time to time from shark stocks as a response to the constant natural tendency of sharks to sink to the floor of the sea. Any fish that rests on the bottom (or imbedded in the sediments) is termed *demersal*. Demersal fishes such as sand sharks and eagle rays and sting rays feed on bottom-living organisms, especially on mollusks, whose shells are crushed by the powerful jaws armed with flattened platelike teeth at the back of the mouth, or all around the jaw.

In contrast to the sharks (which have a cartilaginous skeleton), the fishes called teleosts (that is, fishes with a skeleton of bones) have evolved an organ called the swim bladder. This is an air-filled structure provided with blood vessels capable of secreting gas or absorbing it, and so adding to or subtracting from the amount of air in the bladder.

The acquisition of the swim bladder by teleosts led to change in locomotor activity. The specific gravity of the fish body could now be adjusted to equal that of the surrounding medium, with a resultant disappearance of the tendency to sink to the seafloor whenever forward progression ceased. Thus the head region no longer had to be flattened in the horizontal plane, and the pectoral fins were liberated from the task of serving as stabilizers in the horizontal plane. Stabilization in the vertical plane could be achieved by the body's assumption of a deep laterally compressed form, with appropriate further extension in the vertical plane by means of the unpaired fins. Energy, as before, was provided for forward motion by the tail. The pectoral fins were now available to serve a new purpose, that of braking, and accordingly the attachment of the anterior edge of the pectorals rotates through 90° into the vertical plane and comes to lie just behind the operculum. The rapid braking now possible enables a teleost to reverse its direction in about its own length, a great increase in agility thereby resulting.

Certain teleosts have secondarily lost the swim bladder. These fall in two categories: (1) fishes that have adopted the demersal habit, such as flatfishes and angler fishes and their relatives; and (2) fishes that through body armor (for example, coffer fishes, porcupine fishes, cowfishes) or through venomous properties (for example, puffer fishes) are not susceptible to predator attack and therefore do not demand great agility.

Other variations in locomotor habit are seen in mud skippers *(Periophthalmus)*, where the pectoral fins are converted into levers adapted to walking on damp or dry land and to ascending mangrove roots and branches, in search of insects and crustaceans, on which they feed out of the water. Other modifications that may briefly be cited include pectoral fins adapted for gliding through the air from one wave crest to another, as in the pelagic flying fishes (Exocoetidae); also the conversion of the dorsal fin into an attachment sucker in the remoras, which adhere to sharks and other fishes and are thus transported passively. Morays among the tropical eels have lost both sets of paired fins, adopting instead a snakelike manner of progression on the seafloor, where much of the time is spent in concealment with only the head emergent, on watch for passing prey.

DIURNAL PERIODICITY. Most sharks are nocturnally active, as is the case with most reef organisms. Pelagic species, however, which are occasional visitors to reefs, are active by day. Occasional visitors may include dominant carnivores of the open ocean, of which particular individuals ("rogue sharks") develop a taste for human prey. Nearly all attacks seem attributable to a few individual sharks, which once detected and caught, leave no impress on the behavior of other sharks. There is a diurnal periodicity in shark attacks—most occurring between 2:00 P.M. and 6:00 P.M.; these, however, are the hours when most human beings swim, so the cycle is probably an imposed one, dictated by the availability of prey.

Squirrel fishes (family Holocentridae), morays (f. Muraenidae), and grunts (f. Pomadasyidae) are nocturnally active, hiding near or on the bottom by day; however, a moray can be wakened and encouraged to feed if suitable prey is offered. The butterfly fishes and angelfishes (f. Chaetodontidae) are diurnal and become torpid, with color changes, at night. Wrasses (f. Labridae) are also diurnal fishes, and at night young individuals sink to the soft bottom and conceal themselves as infauna. Squirrel fishes are related to mainly deep-water families and share with them the large eyes of animals that live in near-darkness; hence the nocturnal habit is to be expected. Parrot fishes (f. Scaridae) are diurnal and sleep on the bottom by night, sometimes secreting a nest of mucus about the body.

PROTECTIVE STRUCTURES. Sharks, by their powerful locomotor organs and teeth, require no additional protective structures; but the demersal forms, with the head, trunk, and pectoral fins converted into the disk, are only feeble swimmers, and their teeth are suited more to crushing than to severing; these have become for the most part unaggressive animals, but many of them have developed a protective organ in the shape of a more or less erect and barbed dorsal spine at the base of the tail; an intruder, such as a human being accidentally standing on such a ray, may be severely injured by the animal as it makes its escape. Most wounds are in the ankle region; it is advisable, therefore, to shuffle if walking on the seabed in areas where rays abound, so that the animal may escape before being pinned down by the weight of the person standing on the disk. The electric rays, or torpedos, can administer a sharp shock if trodden on; the electric organs are in this case two lateral bands of modified pectoral muscle, reconstituted in the form of a charge accumulator of parallel plates separated by semipermeable membranes, and energized by part of the hind brain; positive and negative ions, by separating on the upper and lower surfaces of the plates, may collectively build up a charge of several hundred volts, capable of sudden discharge through the body of the intruder.

Paleozoic fishes depended largely on a strong external armor plating of bony scales for protection against predators, chief of which seem then to have been giant invertebrates such as eurypterids and cephalopods. With the acquisition of the swim bladder and the agility imparted by the consequential elaboration of locomotor organs, the need for heavy external armor diminished. At this epoch, agility must be rated as a major protective characteristic of teleost fishes. However, other relevant features occur under this heading; among those that are significant in the coral reef environment and in neighboring areas of the tropical shelf may be listed sharp spines and venom glands often associated with them, sharp anterior teeth, and protective coloration.

The surgeon fishes (family Acanthuridae) are essentially herbivorous and quite aggressive reef fishes that have evolved an effective organ of defense and also offense in the shape of one or more erectile spines, housed in a slot on either side of the base of the tail. According to Randall (1968) mere threat of using the spine is enough to impart dominance to a surgeon fish forming part of a community in an aquarium tank. If provoked, the spine is erected and severe wounds inflicted on other fishes, by violent lashing of the armed tail. Among species listed by Halstead (1959) as capable of inflicting deep and painful wounds on man are *Acanthurus xanthopterus,* of the Indo-Pacific reefs, *Acanthurus bleekeri,* of the Indo-Pacific, exclusive of Hawaii; and *Naso lituratus,* of the Indo-West-Pacific, from Polynesia to east Africa and the Red Sea. The genus first named has a single erectile spine, hinged at its posterior end, and rising like the blade of a pocketknife, sharp edge facing anterior. *Naso* has on either side of the caudal peduncle two pairs of permanently erect spines, attached to plates imbedded in the skin. The Atlantic surgeon fishes all belong to the genus *Acanthurus.* It appears that no venom glands are associated with the apparatus.

The scorpion fishes (family Scorpaenidae) are mainly Indo-West-Pacific reef inhabitants, though there are some representatives in the Caribbean and some also in temperate and Arctic seas. Their outstanding characteristic is the presence of erect spines derived from the unpaired and paired fins, each spine associated with an investing venomous dermal gland. Halstead (1959) distinguishes three main groups according to the nature of the venom organs. These are:

1. *Pterois* type, exemplified by *Pterois volitans,* variously known as the lion fish, dragon fish, or zebra fish (on account of the contrasting bands of black and orange semantic coloration), or turkey fish (from the habit of spreading the feathery fin spines in the manner of a turkey's tail), restricted to the tropical reefs of the Indo-West-Pacific. Here the venom spines are very long and slender, each invested by a thin venomous integumentary gland, with no venom duct. Several species range from the Red Sea through the intervening tropical coasts eastward to the reefs of Polynesia and northern Australia. They occupy shallow parts of the reef, or crevices, or swim openly in unprotected shallows. They occur often in pairs. Their apparent fearlessness leads the incautious to be stung by the needle-sharp spines, and stings are also received when such a fish is hooked and is being taken from the line. The venom organs comprise 13 dorsal spines, 3 anal spines, and 2 pelvic spines. The biologist can readily recognize the nature of fishes of this genus and is only likely to be stung if incautiously groping in crevices with unprotected hands. The effects of a sting include intense pain, nausea, and shock.
2. *Scorpaena* type, in which the venom apparatus is similar to that of *Pterois*, but the spines are shorter. Among fishes of this group are waspfish *(Centropogon australis),* of northern and eastern Australia; the bullrout, *(Notesthes robusta)*, with the same distribution; scorpion fish *(Scorpaena plumieri)*, Brazil to Massachusetts; *S. guttata,* of California; *S. porcus*, of Africa and the Mediterranean; *Scorpaenopsis diabolus* of Australia, Polynesia, and Indonesia; and the lup *(Inimicus japonicus),* of Japan. Effects of stings are like those of *Pterois.*
3. Stonefish type, in which the stinging spines are short and robust,

each invested with a conspicuous dermal venom gland from which a duct runs along the spine to enter the puncture wound. These fishes are the most dangerous species, and include deadly stonefish *(Synanceja horrida),* of Indonesia, India, China, and Australia; the hime-okoze (*Minous monodactylus*), of Polynesia, China, and Japan; and *Chloridactylus multibarbis,* with the same distribution.

Adaptive Coloration

Cott (1940), in the course of an exhaustive and scholarly study of animal coloration, isolated some 144 categories of what appear to be adaptive coloration of evolutionary value or significance in relationship to the mode of life or the environment. These he arranged in three broad groupings, namely, concealment, advertisement, and disguise. Reef fishes were among the numerous assemblages of animals considered; following is a résumé of Cott's ideas, moderated by some more recent studies made possible in part by the development of free diving as an investigatory technique available to biologists.

1. CRYPTIC COLORATION. Obliterative patterns include *countershadings,* as in offshore species such as Spanish mackerel *(Scomberomorus),* feeding at the surface and having the illuminated upper surface tinted bluish green, and the undersurface a silvery color, to match the background according to the viewing angle; *physiological color change,* as in the demersal flounders, with their remarkable power to open or close chromatophore cells in the skin, to match the background; *disruptive coloration,* as in most members of the family Chaetodontidae, in which contrasting bars of light and dark break up the outline of the fish; *coincident disruptive coloration,* in which vital organs such as the eye are concealed by disruptive means, again as in the Chaetodontidae; *general resemblance to medium,* for example, the blue *Chromis,* which is common in blue water above the outer reefs, though highly conspicuous in another environment.

Demersal bottom-dwelling fishes may adopt a mottled color pattern, disrupting the outline of the animal and tending to match it to the variegated bottom. Examples are the torpedos, often with ring-shaped markings, and the mottled pattern of the upper surface of the spotted eagle ray. These would appear to be protective rather than aggressive adaptations.

2. SEMATIC, OR WARNING, COLORATION. This coloration is adopted by many animals that are either poisonous to eat or can inflict a venomous wound. Examples of the former are spotted and chain morays (family Muraenidae), whose flesh is often toxic; and of the latter, the lion fish (*Pterois,* f. Scorpaenidae). The various soft-skinned puffer fishes (f.

Tetraodontidae) frequently have patterns of conspicuous bars or stripes, possibly serving as sematic warnings, an example being the (somewhat tautologically named!) deadly death puffer, or makimaki *(Arothron hispidus)*, which ranges the whole tropical Indo-Pacific reef region from the Red Sea to Panama.

3. DISGUISE COLORATION. Deflective characters include *false eye marks* to deflect predator attack from vulnerable to less vulnerable parts of the body, as, for example, in the four-eyed butterfly fish *(Chaetodon capistratus)*, where the posterior end of the animal is likely to be mistaken for the head; *deflective colors* diverting a predator to less vulnerable members of a species, as, for example, the dull colors of female parrot fishes and younger males, in contrast to the brilliant colors of terminal males; and the sergeant major *(Abudefduf saxatilis)*, a pantropical pomacentrid, in which the male is light yellowish to bluish, with several dark bars, but when guarding eggs becomes a deep blue, and thus much less conspicuous.

TERRITORIAL BEHAVIOR. For some of the foregoing materials alternative (though not necessarily mutually exclusive) explanations have been offered in recent years. Among the chaetodonts, for example, where vertical contrasted stripes are of common occurrence, it has been observed that these are apt not to be schooling species, but rather solitary or living in restricted reef situations as members of bonded pairs. It is probable that the distinctive species colors enable members of the same species to recognize one another, and when an intruder fortuitously enters the territory of another specimen of the same species, he either quits it on recognizing the occupier, or is driven off when the occupier recognizes the intruder. Various porgies, as, for example, *Calamus penna*, may adopt color patterns of dark vertical bars when near the bottom of the sea; this is suggestive of territorial advertisement, too.

COMMENSAL RELATIONSHIPS. The advent of undersea observation by scuba-diving naturalists has brought to light many examples of commensalism or other interspecific relationships. Probably the most interesting of these have been the numerous cases of small species (or commonly the young stages of larger species) serving as *cleaners* (parasite removers) for other fishes, including very large species. A series of 7 color photographs by Faulkner (1970) shows gobies and wrasses performing this function, and the young stages of some chaetodonts are also known to be cleaners. In some cases service stations are set up for visiting clientele, as, for example, the Caribbean neon gobi *(Elacatinus oceanops)*, a pair of 2-inch-long gobies setting up a jointly operated clinic for fishes as large as 60-lb sea bass *(Epinephelus itajara)*. In Queensland waters off Australia Faulkner found small wrasses (*Labroides dimidiatus*) similarly cleaning bass *(Plectropomus maculatus)*, "fearlessly entering its mouth to pick parasites," later to "exit through the gill openings . . . like many fishes, the bass extends its gill rakers to facilitate cleaning." Other Pacific wrasse Faulkner found to operate a cooperative clinic on the Scandinavian style, with more than one species of doctor attending to visiting patients; others again, such as *Labroides bicolor*, operate a fixed clinic in the young stages, but as they grow older and larger, gradually extend the limits of the station until, at adulthood, they are on continuous house call over a territory of up to 10,000 square feet. Schools of oceanic fishes, such as jacks (family Carangidae) apparently visit the reefs in order to be cleaned.

Fishes such as *Fierasfer* and *Carapus* shelter in the cloaca of large holothurians, backing in in much the same manner as a jawfish backs into its burrow. Some fishes such as the gobiesocid *(Diademichthys)* in Indonesia take refuge among the long spines of *Diadema*, an urchin that presumably affords them protection.

PREDATOR-PREY RELATIONSHIPS AND SPECIES DIVERSITY. It follows that since fishes are mostly predators there is a great diversity of predators (and their patterns of predation) on coral reefs and in the associated shallow lagoons. As already noted, the invertebrates of coral reefs are the most diverse and richest of any marine ecosystem, a point stressed by Thorson (1971). The two sets of organisms are intimately related in a complex food web. It might perhaps be supposed that the invertebrate biota would be even more diversified if they were not subject to the severe predation exerted by the rich predator community; but the case is the reverse, and the diversity of both predator and prey is mutually supported. A few years ago Paine (1966) artificially eradicated the dominant predators (mainly starfishes of the genus *Pisaster*) from a delimited stretch of coast in Washington. He then observed the consequent changes in the rather simple food web of the habitat and community. At first, as could be expected, there was a rapid increase in the prey organisms, especially a species of barnacle on which *Pisaster* feeds. But later certain other prey organisms, notably the bivalve *Mytilus*, became more common, eventually displacing much of the barnacle population.

The end result was a less diversified community than was initially the case. Evidently the removal of the predator permitted all the prey species to increase to the maximum carrying capacity of the habitat, followed by intense competition for food and space, and eventual elimination by natural selection of the less vigorous or less efficient members. Thus it appears that the effect of predators is to apply selection to a prey community in such a way as to override the effects of competition based on individual efficiency or vigor, or in effect to randomize the predation pressure. So it seems likely that the more diversified the predator population, the more diversified the prey population. Thus probably the diversity of both predators and prey on coral reefs is an ancient feature, and one in which the two elements are mutually interacting so that either party reinforces the diversity of the other. In the present rather puzzling population explosion of the starfish *(Acanthaster)*, certain reef communities have become saddled with a dominant uniform predator population that consistently destroys a fixed range of coral species. The observed concomitant depauperization of reef populations would seem to be a predictable outcome if Paine's observations are of general application. For example, since many of the butterfly fishes and angelfishes feed largely on the tentacles of tube worms and the tube feet of echinoderms (sea urchins) no activity of *Acanthaster* in destroying living coral would affect them; but coral-eating ophiuroids would be adversely affected and tend to be displaced from the community.

Table 16. The classes of marine vertebrates forming the nekton.

DIAGNOSTIC CHARACTERS	CLASS	EXAMPLES
Vertebrates breathing by means of gills in water, fishlike animals		
Body eel-shaped, with no paired limbs, and a circular suctorial mouth, lacking jaws	Cyclostomata	Lampreys and hagfishes
Body typically fishlike, with paired pectoral and pelvic fins, and jaws; skeleton of cartilage only	Chondrichthyes	Sharks and rays
As above, but skeleton includes bone elements	Osteichthyes	Bone fishes
Vertebrates breathing by means of lungs in air, four-footed animals		
Marine reptiles	Reptilia	Turtles, crocodiles, and sea snakes
Marine birds	Aves	Shore and oceanic birds
Marine mammals	Mammalia	Furred animals, whales, dolphins, and so on

SYSTEMATIC REVIEW OF MARINE VERTEBRATES

Lampreys and Hagfishes (Class Cyclostomata)

These are archaic eel-like animals with relatives in the lower Paleozoic. The living representatives are few and comprise the lampreys (Petromyzontia) and hagfishes (Myxinoidea). The hagfishes are exclusively marine, but lampreys usually have at least the early stages of life in fresh water. All cyclostomes are predators, attacking larger vertebrates and using their rasping, plated tongue to bore through the body wall. Injured and dead whales are attacked by hagfishes in large numbers.

Sharks and Rays (Class Chondrichthyes)

Although more ancient types of shark are known, the first sharks of modern aspect appeared at the end of the Triassic period, about 200 million years ago. The gills open to the exterior by a number of separate slitlike openings, a character described as *elasmobranch* and used to form the names of the suborder to which true sharks and rays belong, as set out in Table 17.

By the late Jurassic the reef shark faunas had come to resemble those of modern seas, and in the following period, the Cretaceous, nearly all sharks prove to belong to the genera that still exist today. Thus shark genera has as great an antiquity as do most of the extant reef invertebrate genera; and the pantropical distribution of such genera in modern seas can evidently be explained by the same mechanism as has been invoked to explain the present dispersion of echinoderm genera, namely, the circumglobal continuity of tropical seas through most of the Tertiary.

The first modern types of rays (order Batoidea) differentiated from the general elasmobranch stock during the latter part of the Jurassic period. They were at first rare, but by the late Cretaceous (about 80 million years ago) rays had assumed as conspicuous a position in reef faunas as is the case today, all major existing families being present, with the exception of the eagle rays and the electric rays, these being Tertiary differentiates.

In present-day seas about 500 species of elasmobranchs are known, most of them confined to shoal waters, though some of the best-known sharks and the largest rays are large pelagic species of the open ocean. A few inhabit the deepest parts of the ocean. Apart from the diagnostic features already mentioned in the key, other characters of the elasmobranchs are these: there is no swim bladder; the skin is not covered by scales, but instead small spinous dermal *denticles* are imbedded in it and impart a rough texture to most species; copulation occurs in all species, the male having clasping organs developed on the pelvic fin for attaching to and holding open the female genital aperture; the development of the embryo may either occur in the uterus, in which case live birth follows, or else fertilized encapsulated eggs may be laid and subsequently hatch independently of the parent.

Table 17. Key characters of the orders of cartilaginous fishes (class Chondrichthyes).

Class characters: vertebrates breathing by means of gills in water; the body of fishlike shape, with paired pectoral and pelvic fins, upper and lower jaws; the skeleton of cartilage, without bone elements. Three extant orders, as follows:

Character	Order	Common name
Several (5 to 7) separate slitlike gill openings on each side of the neck region (subclass Elasmobranchia)		
Body elongate, tapering evenly into tail	Selachii	Sharks
Body flattened, abruptly narrowed behind trunk	Batoidei	Rays
Single gill opening on either side of neck region	Holocephali	Ghost sharks

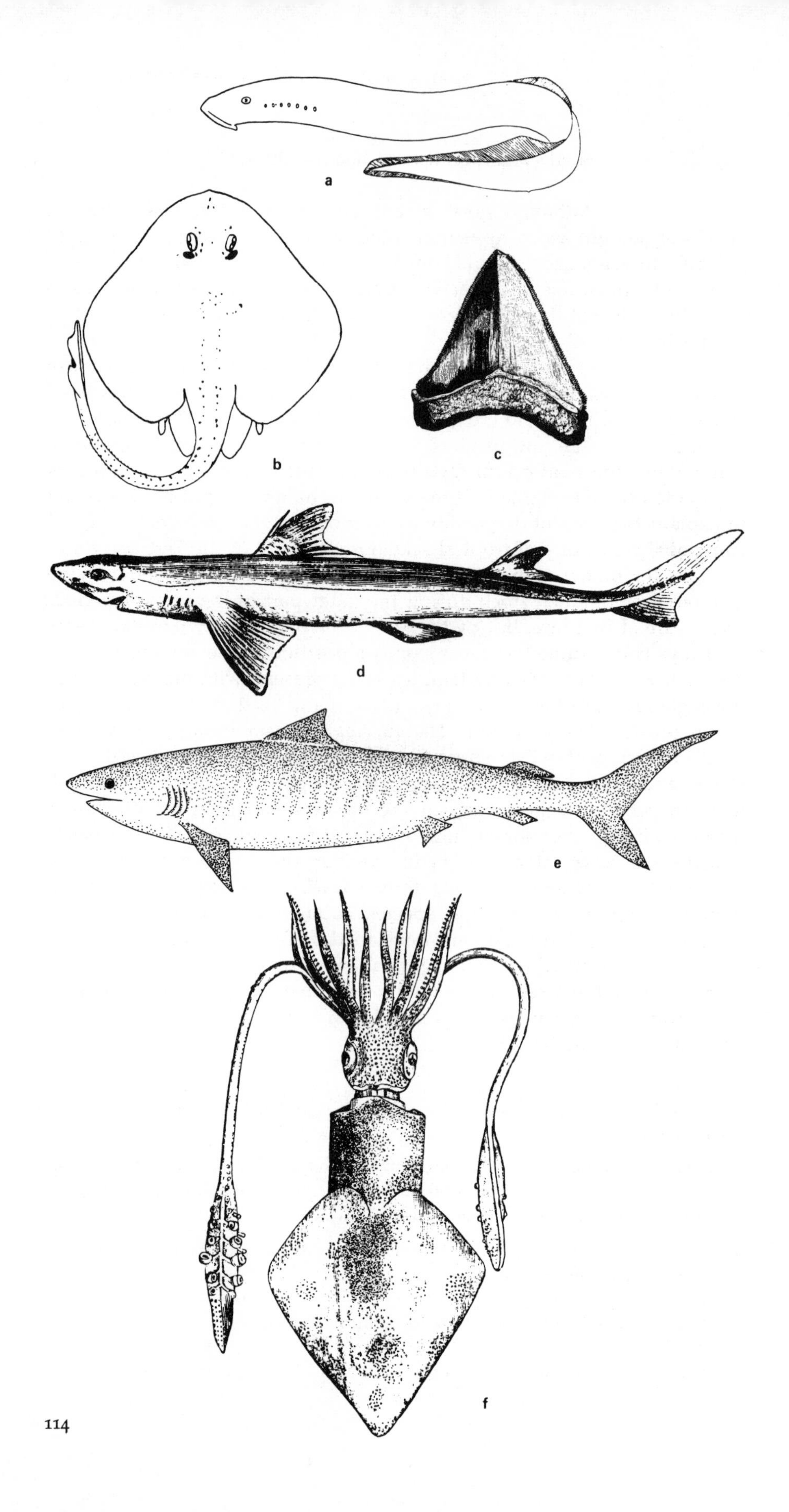
a
b
c
d
e
f

Figure 31. (a) *Petromyzon* (marine lamprey), X 0.2, Cyclostomata (38); (b) *Raja* (ray), X 0.1, demersal neritic and bathyal, Chondrichthyes (38); (c) *Carcharodon megalodon*, X 0.25, tooth of gigantic Miocene shark, Malta, Chondrichthyes (43); (d) *Galeocerdo* (tiger shark), length 18 feet, warm seas, Chondichthyes (23); (e) *Acanthias* (spiny dogfish), 4 feet, north Atlantic, Chondrichthyes (15); (f) *Loligo* (squid), X 0.1, Cephalopoda. Members of this class of mollusks are active swimmers and form part of the nekton (15).

MAN AS PREY TO SHARKS. Of the various large carnivores that are capable of preying on man, sharks alone remain uncontrolled. Although rigorous programs of selective trapping have been instituted on the Australian coasts (where most shark attacks occur) and have brought the problem under control, many fatalities still occur in exceptional circumstances, particularly after shipwrecks and plane crashes. Commercial shipwrecks and passenger-plane crashes are usually reported discreetly by the news media for the sake of good relations with the corporations concerned. Not so, however, is the case in time of war, when most losses involve noncommercial vessels and aircraft. During World War II it became apparent that severe losses of life following wrecks are due to attacks by swarms of sharks attracted to the scene by the presence of men in the water. Injured people are always attacked first, presumably because blood attracts a shark. People floating in life jackets are usually attacked below water first; sharks of the order of size of the tiger shark *(Galeocerdo)*, 3 to 5 m long, attack by dismembering the limbs one by one. The white pointer (*Carcharodon*), more than 6 m in length, can swallow a man whole, but the younger specimens usually bite a man in half first. About 20 species are believed to be potential man-eaters. Other smaller species take an occasional bite if given the opportunity. During World War II tens of thousands of sailors are believed to have been killed by sharks, most while clinging to ropes on the sides of lifeboats, or to flotsam, or while swimming or supported by life jackets. Of the few civilian cases on record in detail an example is an airliner that crashlanded apparently intact on the surface of the sea on November 16, 1959, 120 miles southwest of New Orleans; the 42 passengers and the crew all died, either from sharks or by drowning, 10 half-eaten bodies being recovered. The extensive research on so-called shark repellents has not thus far produced a remedy. The supposedly effective "repellent" issued to crewmen by naval authorities had, in fact, no more than a psychological value. Active research is in progress, with periodic international conferences by shark experts to review results, thus far mostly producing a confusing mass of seemingly contradictory data.

Sharks are widely distributed, mainly in warm waters, though not exclusively so. Many sharks are very large fishes, none are very small. The largest species now surviving are the basking shark *(Halsydrus maximus)* and the whale shark *(Rhineodon typicus)*; both are known to reach lengths of 40 feet (12 m), but records of up to twice that size exist, unconfirmed. However, extinct species reached 24 m. The largest sharks are unaggressive if unmolested, feeding on plankton and on invertebrates on kelp. The white pointer *(Carcharodon carcharias)* occurs in warm seas throughout the oceans. A large specimen measures 11 m (36 feet), and one of that size at Sydney, Australia, had teeth 2 inches long. Fossil teeth of the same genus are known about 6 inches across, suggesting a shark of 80 to 100 feet in length. A 20-foot white pointer swallows a man whole, whereas a 12-foot specimen first bites him in half. One Californian specimen contained a 100-lb sea lion. The tiger shark *(Galeocerdo cuvier)* is another man-eater and ranges as far north as Iceland. A specimen taken at Durban was found to contain the anterior half of a crocodile, the hind leg of a sheep, three sea gulls, and two 2-lb cans of peas and a can of cigarettes; a New Zealand specimen contained a large collie dog, a blue penguin, and two large lobsters. The hammerhead shark (*Sphyrna lewini*) ranges all warm seas and is often seen in midocean. The young stages feed on mackerel. A specimen from Riverhead, New York, contained many portions of a man together with his clothes. Sharks enter rivers and swim far inland, some of the most serious Australian attacks having occurred under these circumstances. Nonetheless, the chances of death by shark bite are much less than the chances of being run over on a street. Sharks are valuable scavengers, and there can be no doubt that many of the land animals found in their stomachs were ingested after their cadavers were washed to sea by rivers.

The thorny skate *(Raja radiata)* ranges the continental shelf around the northern margins of the North Atlantic, on the American side as far south as Cape Cod. It feeds on shrimps, crabs, anelids, and small fishes. South of Cape Cod the species ranges on the continental slope at least as far south as North Carolina, where it descends to a depth of 300 fathoms. Skates or rays occur all over the world. All have a demersal habit. The largest rays are the devil fishes, or mantas, mostly found swimming in open oceanic waters near the surface. A large devil ray *(Manta birostris)* reaches 6 m (20 feet) across, weighing 3500 lbs (1.6 metric tons).

Bone Fishes (Class Osteichthyes)

STURGEONS [ORDER ACIPENSERIFORMES]. Sturgeons are another ancient group of fishes. *Acipenser oxyrhynchus* ranges neritic waters from Hudson Bay to the Gulf of Mexico. It reaches a length of 5.5 m (18 feet). Sturgeons are bottom feeders. The barbels are probably the sensory organs by which they detect worms of the infauna; the snout is used to disturb the bottom in the search for food. When mature sturgeons ascend rivers to spawn in fresh water, the young may remain in rivers and lakes for 1 to 3 years, leaving them for the sea when 1 to 3 feet long. A sturgeon 8 feet long is estimated to be about 12 years old and weighs about 85 kg (190 lbs). Sturgeons were common in American waters when the first colonial settlements were established. They are now rare, and the sturgeon fishery has disappeared. They are captured only occasionally as incidental species associated with other fisheries.

HERRINGS [ORDER CLUPEIFORMES]. The herringlike fishes are also ancient, the oldest fishes with clupeoid structure occurring in Triassic rocks, and more herringlike members occurring in the lower Jurassic, about 150 million years ago. The characters are indicated in Table 18. The most distinctive feature, and one by which a clupeoid marine fish may be

Table 18. Key characters of the more conspicuous orders of bone fishes (class Osteichthyes).

Class characters: vertebrates breathing by means of gills in water; the body of typical fishlike shape, with paired pectoral and pelvic fins, upper and lower jaws; the skeleton principally of bone, cartilage also occurring. Numerous orders, of which the following are important in marine environments:

Key character	Order	Common name
Tail fin with long upper lobe containing tip of vertebral column (heterocercal); Mouth below the snout; bone plates form rows along flanks	Acipenseriformes	Sturgeons
Tail fin not heterocercal, the two lobes of fin usually similar (homocercal)		
Both eyes on one side of head; body flattened, demersal	Pleuronectiformes	Flatfishes
One eye on each side of head		
Dorsal and tail fins continuous; no pelvic fins	Anguilliformes	Eels
Dorsal fin separate from tail fin		
Snout tubular; body enclosed in bone plates	Syngnathiformes	Sea horses
Snout not tubular; body not enclosed in bone plates		
Fin rays of separate segments, not fused into stiff spines		
Chin barbel present; 2 or 3 dorsal fins	Gadiformes	Codfishes
No barbels; pelvic fins set far back	Clupeiformes	Herrings
Similar, but pectoral fins set high on body	Beloniformes	Flying fishes and halfbeaks
Fin rays not segmented, fused to form stiff or flexible spines		
Anterior dorsal fin placed on head	Lophiiformes	Angler fishes
Not so	Perciformes (and so on)	Perchlike fishes

recognized at sight, is the low insertion and wide separation of the pectoral and pelvic fins, the latter lying well back on the body as is the case in sharks. About 30 families are recognized, the best-known being the Clupeidae, or herrings proper, and sardines. Other families include the salmon and trout *(Salmonidae)*, a few of which are marine, most however being freshwater fishes. Several other clupeiform families are encountered in freshwater environments (Chapter 9). One deep-sea family, the stomiatoid viper fishes, and a few other deep-sea families, complete the roster of clupeiform fishes.

The herrings and the salmonids are both essentially northern groups.

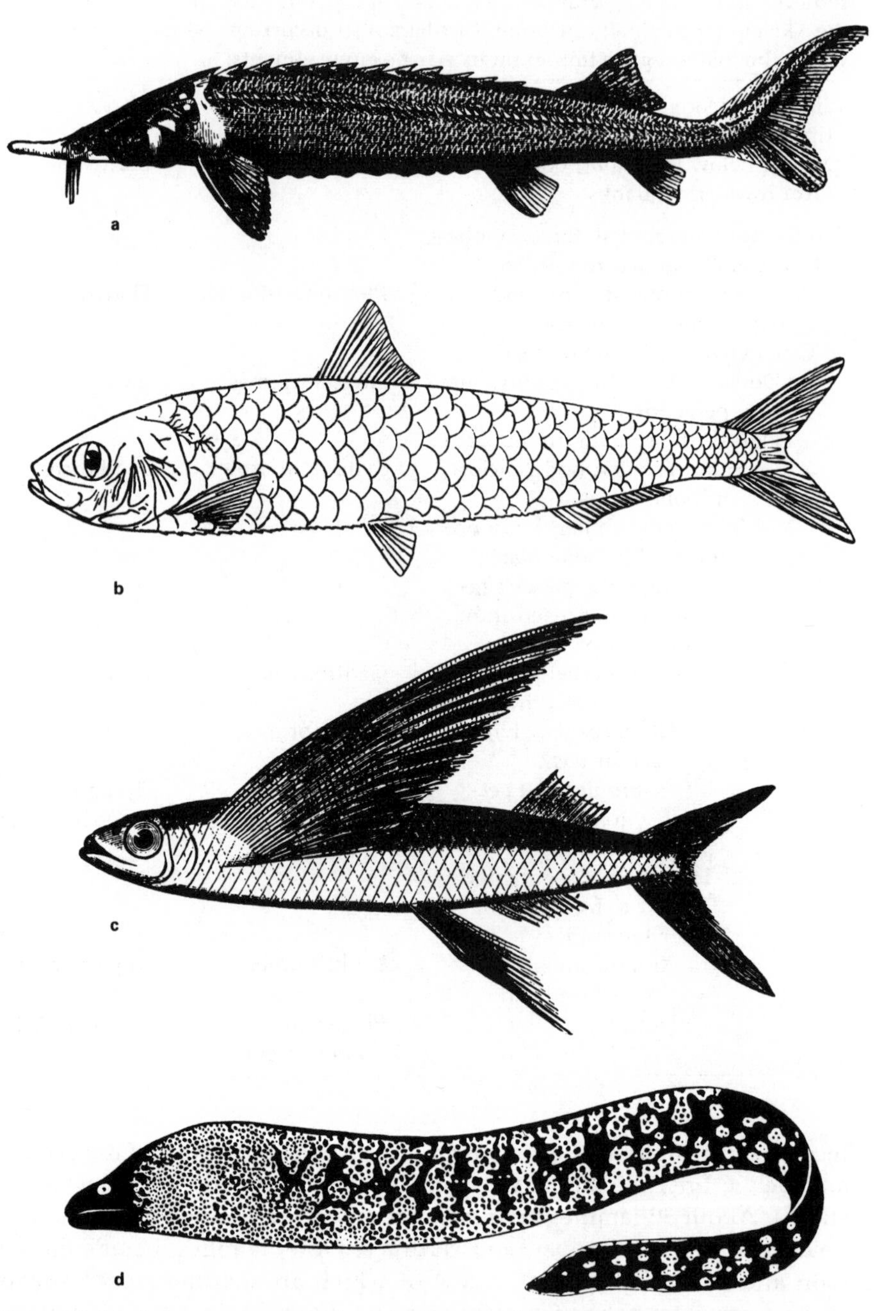

Figure 32. (a) *Acipenser* (sturgeon), to 18 feet, north Atlantic, Acipenseriformes (15); (b) *Clupea* (herring), X 0.2, pelagic, Clupeiformes (38); (c) *Exocoetus* (flying fish), X 0.2, warm seas, pelagic, Beloniformes (15); (d) *Muraena* (moray eel), X 0.1, demersal in warm seas, Anguilliformes (38).

They are valuable food fishes in the Northern Hemisphere, where extensive fisheries are located. The sea herring *(Clupea harengus)* occurs in large schools, especially in open coastal waters. It feeds on plankton and has a life span of about 10 years, reaching a maximum length of about 37 cm (15 inches). It is eaten by larger fishes and by man.

FLYING FISHES AND HALFBEAKS [ORDER BELONIFORMES]. A small group of herringlike fishes, they resemble the clupeoids in the wide separation of the pelvic and pectoral fins, but differ in having the pectoral fins placed high on the body, the pectoral insertion lying at or above the horizontal axis. There are no fin spines; the swim bladder, which usually has a duct in clupeiform fishes, has no duct in the beloniform species. Best-known members of this small order are the flying fishes. Other members include the long-snouted billfish and halfbeaks, the lizard fishes with elongate body, and various other fishes, mostly inhabitants of warm surface waters. Flying fishes have enlarged pectoral fins, by which they plane from wave crest to wave crest, gliding rather than flying; they break the surface when pursued by larger pelagic fishes and are equally disturbed by the passage of a ship's hull through the water, so flying fishes are most often seen taking to the air in shoals from the bow waves of ships. With a strong wind the fish may be carried 5 or 6 m high and fall on the deck. The oldest known beloniform fishes date from the Eocene period, about 70 million years ago.

EELS [ORDER ANGUILLIFORMES]. The most ancient eels seem to date from the Cretaceous, about 100 million years ago. Best-known are the species that enter fresh water, though these are marine at the breeding stage. Among the strictly marine members of the order are the moray eels of shallow tropical seas, particularly common on coral reefs, and the conger eels. The American Atlantic conger *(Conger oceanica)* ranges from Massachusetts Bay southward, perhaps as far south as Brazil, ranging the whole shelf region from inshore shallows to beyond the edge of the shelf (to 150 fathoms). The adults feed on smaller fishes, crustaceans, and mollusks. They do not enter freshwater. In the breeding season the conger migrate into deep oceanic waters, where the young stages (called leptocephali) are found.

SEA HORSES AND PIPEFISHES [ORDER SYNGNATHIFORMES]. These tubular-snouted fishes are mostly found in warm shallow waters, but the American Atlantic species, or northern sea horse, ranges from the Carolinas northward to Nova Scotia. *Hippocampus hudsonicus* conceals itself by clinging to vegetation of any kind in shallow water. It feeds on planktonic crustaceans, as do also the pipefishes. The male carries the developing eggs in a brood pouch from which 100 or more may be born simultaneously. Pipefishes (*Syngnathus* spp.) have elongate tubular bodies, and they lack the prehensile tail, whereas the sea horses have a more inflated trunk and a slender prehensile tail.

CODFISHES [ORDER GADIFORMES]. These are structurally intermediate between soft-rayed fishes and the spiny-rayed (or perchlike) fishes. One character they share with the latter is an internal swim bladder that lacks any connection with the gut; this means that deep-sea members of the order if trapped and hauled too rapidly to the surface suffer evisceration through the mouth on account of the sudden expansion of gases in the swim bladder under the reduced pressure near the surface. The oldest known members of the order date from the Eocene period, about 70 million years ago. There are five families of Gadiformes, the best-known being the Gadidae, or codfishes. Cod and their relatives (pollock, haddock, hake) all tend to be rather shallow-water fishes, distributed in northern seas. *Gadus callarias* occurs on both sides of the North Atlantic, ranging from western Greenland southward to Virginia; but it is commonest from Labrador to Cape Cod, where the main fishery occurs. Cod reach a length of about 2 m and a weight of nearly 100 kg; however, most specimens are smaller. At an age of 7 or 8 years, a cod measures about 1 m. Cod feed on bottom organisms, particularly crabs, lobsters, and mollusks; some squid and smaller fishes are also taken. They swallow bright or unusual objects lying on the seabed, such as stones, rope, and bottle caps; jewelry has been found in their stomachs. The cod are fished throughout the year in the more northern waters, mainly in winter months in the more southern part of their range. The other families of Gadiformes are mainly deep-water fishes.

FLATFISHES [ORDER PLEURONECTIFORMES]. This order of fishes is one of the most easily recognizable, for the included members have partially lost the bilateral symmetry that characterizes vertebrates. Young specimens resemble other fishes, but during the course of growth the eye of one side of the head migrates across the top of the skull and comes to lie beside the eye of the other side of the head. According to which eye migrates, the fishes of this order may be separated into right-eyed and left-eyed groups. Any one species is normally always a member of one group, but occasional reversals occur. The purpose of the eye migration is to permit the fish to lie on the bottom of the sea, on the blind side, without any loss of vision. Fishes that habitually rest on the bottom are termed demersal. Demersal fishes thus become temporary members of the benthos, though they retain the ability to swim and may still be considered as nekton. There are about 500 species of flatfishes, mostly inhabitants of the continental shelves of the world, and mostly in temperate or warm waters, though a few occur in the polar regions. Of the six families, among the best-known are the Pleuronectidae, or right-eyed flatfishes; these forms, which rest on the left side, include the north sea plaice *(Pleuronectes)*, Pacific flounders such as *Pleuronichthys*, and west Atlantic flounders such

Figure 33. (a) *Gadus* (codfish), X 0.1, north Atlantic, epibenthic, Gadiformes (38); (b) *Serranus* (grouper), X 0.1, north Atlantic, Perciformes (38); (c) *Selene* (lookdown), X 0.2, neritic, north Atlantic, Perciformes (23); (d) *Scomber* (Atlantic mackerel), X 0.1, pelagic, Perciformes (23).

as the blackback *(Pseudopleuronectes americanus)* and the yellowtail *(Limanda ferruginea),* both species caught in considerable numbers on the New England coast. Flounder occupy quiet waters, especially in sheltered inlets, at depths from a few fathoms down to about 50 fathoms. Some species enter rivers. The oldest known flatfishes date from the middle Eocene, about 50 million years ago.

ANGLER FISHES [ORDER LOPHIIFORMES]. Members of the order are recognizable by the characteristic translocation of the anterior dorsal fin rays to a position on the head, where they are converted to form a fishing lure in many of the species. Most anglers are deep-sea fish, often of small size, but some species are larger demersal forms of the continental shelf. The best-known species are the east Atlantic angler fish *(Lophius piscatorius)* and the west Atlantic congener, known in America as the goosefish *(Lophius americanus)*. The latter species ranges from Newfoundland south to the Carolinas, reaching a length of over 1 m and a weight of over 20 kg. There is no air bladder in anglers. The goosefish is extremely tolerant of wide variation in depth of habitat, probably because of its immunity to changes of pressure; the species is known to occur in water as shallow as the low-tide zone and as deep as about 350 fathoms on the continental slope. It lies quietly on the bottom but when a potential food animal approaches, it darts forward very rapidly to engulf the prey in the oversized mouth and stomach. In shallow water it is reputed to seize sea birds.

THE PERCHLIKE FISHES. As Table 18 shows, these are spiny-rayed fishes. There are several orders, differing only in respect to certain internal characters (especially in respect to bones of the head); so it is not feasible to make a key to these orders using only external features. However, the great majority of the perchlike fishes belong to the order Perciformes, the largest known order of vertebrates. The Perciformes are the dominant fishes of existing seas, therefore, and they have been so since Eocene times, about 70 million years ago. Although the precise ordinal characters of perchlike fishes are difficult to state without undue technicalities, the included families are often easy to recognize on account of their external familial characters. There are about 1200 genera arranged in about 140 families. A few of these are noted here.

EXAMPLES OF THE PERCIFORMES

1. Jacks (family Carangidae): mainly tropical, but drifting into temperate regions, mainly fishes of the open sea, but some enter reef environments to hunt other fishes. The lookdown *(Selene vomer)* is a strangely flattened North Atlantic example. The permit and pompano are both species of *Trachinotus*, important Caribbean food fishes.
2. Tunas and mackerels (family Scombridae): swift-swimming pelagic fishes in schools, mostly only in offshore blue water. Mackerel *(Scomber scombrus)* feed on plankton, especially copepods. Tuna may reach a length of over 4 m and a weight of 800 kg (1800 lbs). Large tuna do not school; they feed on herring and mackerel.
3. Snappers (family Lutianidae): mostly benthic nocturnal carnivores, frequenting reefs or mud bottom, or in deeper water beyond the edge of the shelf. Yellowtail snapper *(Ocyurus chrysurus)* occurs on the Caribbean reefs. The fishes of this family feed mostly on benthic crustaceans, the adults also taking smaller fishes.

4. Angelfishes and butterfly fishes (family Chaetodontidae): frequenting tropical coral reefs. Usually brightly colored in life, bold active fishes, diurnal, feeding on sponges and algae chiefly. Example: French angel *(Pomacanthus parus)*, Caribbean.
5. Parrot fishes (family Scaridae): colorful herbivorous reef fishes, their teeth modified into strong plates for scraping algae from hard substrate, also for grazing on eelgrass. Example: blue parrot fish *(Scarus coeruleus)*, Caribbean.
6. Scorpion fishes (family Scorpaenidae): inhabiting tropics, over hard bottom, and having venomous spines. Example: the deadly lion fish of the Indo-West-Pacific *(Pterois volitans)*, ranging from the Red Sea to Polynesia. Wounds from this fish may be lethal; no species occur in the Atlantic. The genus *Scorpaena* ranges all warm seas.
7. Sculpins (family Cottidae): north temperate representatives of the scorpion fishes, but lacking the venomous character; have conspicuous fin spines. Example: *Myxocephalus octodecimspinosus*, found off New England coasts.
8. Lumpfishes (family Cyclopteridae): having a ventral suctorial disk by which they attach to the seabed. Example: *Cyclopterus lumpus*, off the Massachusetts coast.

Other perchlike fishes are the squirrel fishes (order Beryciformes, family Holocentridae), which frequent tropical reefs and also deeper water. Squirrel fishes are nocturnal, probably because they are related to deep-sea forms and share their large eyes; by day they hide in crevices. They feed mainly on crustaceans. An example is the red squirrel fish *(Holocentrus rufus)*, found in the Caribbean. Trigger fishes (order Tetraodontiformes, family Balistidae) have the first fin spine of the dorsal fin enlarged. The body is compressed, the eye is high on the head, and the scales are very strong. They are found in outer tropical reefs and tropical open ocean, grazing in shallow water on benthic plants and rising to the surface to feed on plankton. An example is the black durgon *(Melichthys niger)*, which is pantropical.

Marine Reptiles (Class Reptilia)

Reptiles are air-breathing animals primarily adapted for life on land (see Chapter 10). However, a few of the surviving groups have become secondarily adapted to living in the sea. These are the marine turtles, one species of crocodile, and the sea snakes, as set out in Table 16.

MARINE TURTLES [ORDER CHELONIA]. There are five living genera, all of which occur in all warm seas of the world. The oldest known fossil marine turtles date from the Cretaceous, about 100 million years ago, and they seem to have arisen from marsh-inhabiting stocks. Marine turtles come ashore to lay their eggs; the young make their way back to the sea after hatching, and at this stage in their life they are extremely vulnerable to predator attacks. Turtles are a threatened species and should be protected, for they are very ancient animals of great scientific interest. The leatherback *(Dermochelys coriacea)* is the only known representative of its family. On the other hand, the family Cheloniidae is represented by four genera, the green turtles *(Chelonia mydas)*, the hawksbill turtles *(Eretmochelys imbricata)*, the loggerheads *(Caretta caretta* and *C. gigas)*, and the ridleys *(Lepidochelys olivacea* and *L. kempii)*. The green turtles are mainly herbivorous; the hawksbills are omnivorous; the loggerheads are mainly carnivorous (on fish, crabs, conchs); the ridleys eat crabs, sea urchins, and mollusks; and the leatherbacks are omnivorous.

MARINE CROCODILES [ORDER CROCODYLIA]. The only marine crocodile is *Crocodylus porosus* (note the spelling!), of the Indo-West Pacific. It ranges from India to northern Australia and Fiji. It inhabits estuaries but also is found at sea, far from land, and it has reached many oceanic islands. The 1-m-long skull at Harvard was taken from a 9-m-long specimen killed in the Philippines by a Boston merchant nearly 150 years ago; this animal was reputed to have eaten many people. Specimens today seldom exceed 6 m, but the species is dangerous when 3 m long. Adults feed mainly on mammals approaching water to drink; the newly hatched young crocodiles feed on water beetles, changing to fish as they grow older. They live to about 20 years of age. The males fight during the breeding period. The females come about 60 m inland to lay the eggs, 25 to 60, in a dome-shaped nest of leaves. The eggs are eaten by jackals, otters, and monitor lizards, as well as mongooses. Man is the only enemy of the adult crocodile.

MARINE SNAKES AND LIZARDS [ORDER SQUAMATA]. About 16 genera of sea snakes are recognized, all of them confined to the Indo-Pacific. The most widely distributed species is the yellow belly *(Pelamis platurus)*, which ranges from Madagascar to Panama and from Japan to New Zealand. Although it is less than 1 m long it is able to cross wide ocean gaps, as its range also shows; but this fact seems to have escaped the notice of some paleontologists who argue that the presence of comparably sized fossil reptiles in South America and South Africa in Carboniferous times must mean that these two continents were joined since supposedly the reptile concerned *(Mesosaurus)* would be too small to cross the Atlantic. Sea snakes are often gregarious; Lowe (1932) records swarms numbering millions crossing the sea between Sumatra and Malaya, scattered over a region of sea some 60 miles across. Specimens that reach New Zealand (the extreme southern range) are solitary females, probably swept southward by currents; they penetrate several miles inland and have been known to enter houses and even get into beds, to the great surprise of New Zealanders who are taught at school that no snakes occur in that country. Most species are venomous; like all water snakes, sea snakes have

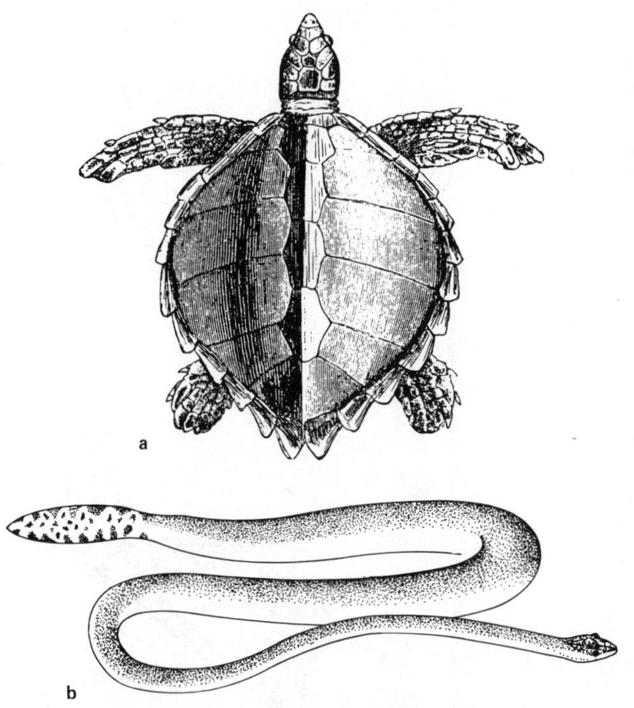

Figure 34. (a) *Caretta* (loggerhead), X 0.1, all warm seas, Chelonia (15, from Cuvier, *Règne Animal*); (b) *Pelamis platurus* (yellow-belly sea snake), X 0.2, warm Indo-Pacific south to Cook Strait (23).

a compressed, flattened tail, adapted to eel-like swimming movements. They feed on fishes. Some sea snakes have become adapted to life in freshwater lakes on oceanic islands.

In contrast to snakes, very few lizards have adopted a marine life. In the Galápagos archipelago there is a marine iguana. In the island of Komodo in Indonesia a giant monitor lizard *(Varanus komodoensis)* frequents the seashore and frequently has been observed swimming at sea. Neither of these species forsakes the coast for the open sea, however.

Marine Birds (Class Aves)

Seven orders of birds include members that are specially adapted for life in or near water, having, for example, webbed feet or long legs adapted to wading in shallow marginal water, and analogous modifications enabling them to find food in water environments. Adaptations of the kind indicated have enabled birds to occupy not only inland waters, but also marine habitats. In the following review representatives of most aquatic bird groups are noted, so that the material is applicable also to the section on freshwater environments (Chapter 9). The systematic characters of the orders of birds are set out in Table 19.

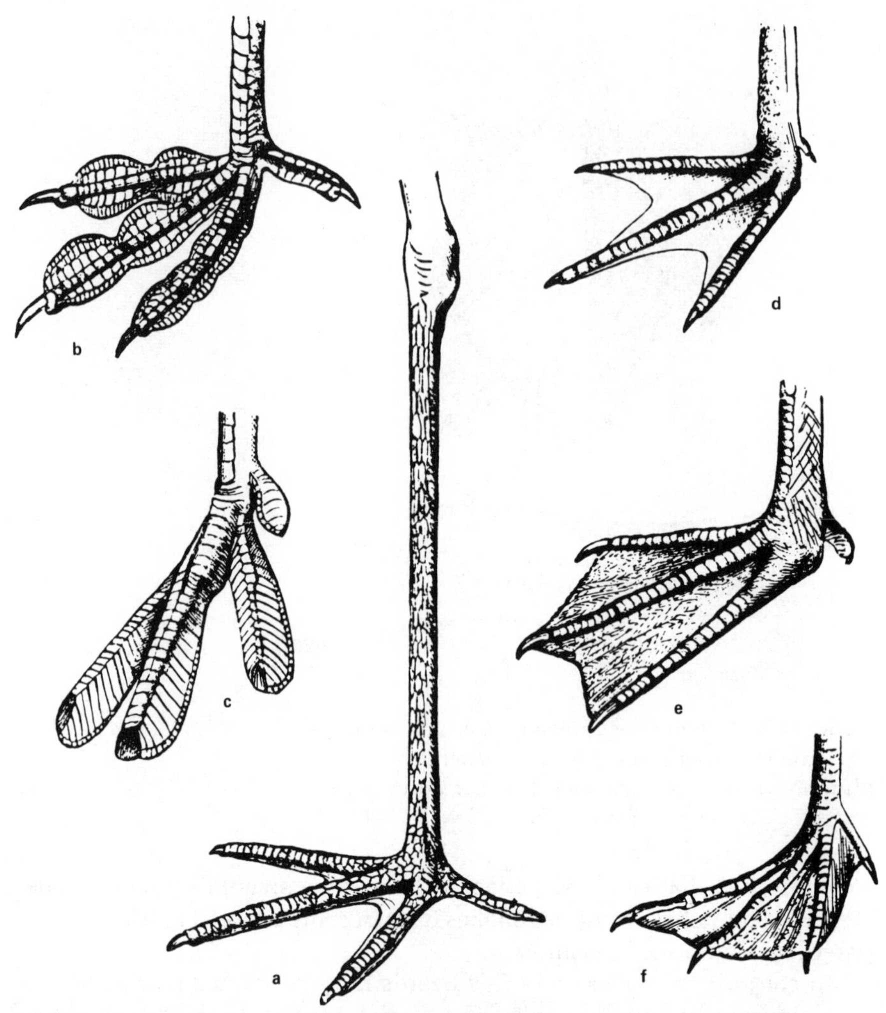

Figure 35. Key characters of oceanic and shore birds. (a) Order Ciconiiformes, family Ardeidae, herons; (b) lobed webs of some marsh birds occasionally found on coast; (c) fringed webs and flattened nails of grebes, order Podicipiformes (not in Table 19), essentially freshwater birds, but migrating across oceans; foot as in (d) wading birds, order Charadriiformes, and (e) duck, order Amseriformes; (f) order Pelecaniformes (15).

PENGUINS [ORDER SPHENISCIFORMES]. Penguins are exclusively marine birds of the far southern oceans, one species alone reaching the equator at the Galápagos archipelago. They occupy the southern islands and the Antarctic continent itself. They prey on fish and are themselves eaten by marine mammals such as sea lions and toothed whales.

The single surviving family of penguins (Spheniscidae) comprises the most completely aquatic of all living birds. The largest living species is the

Table 19. Key characters for recognizing the orders of marine birds (class Aves).

Class characters: warm-blooded vertebrates in which wings are normally developed, only the two hind limbs used for walking; feathers normally occur, and eggs laid. Seven orders, as follows:

Characters	Order	Common name
Wings converted to flippers used for swimming; Southern Hemisphere only	Sphenisciformes	Penguins
Wings normally developed, used for flying		
Hind toe joined by web to the front toes	Pelecaniformes	Pelicanlike birds
Only front toes connected by a web		
Nostrils open to exterior through raised tubes on bill	Procellariiformes	Albatrosses, petrels, and shearwaters
Nostrils open directly to exterior, not by way of tubes		
Ankle joint (tarsus) flattened from side to side; legs placed far back on body, diving birds	Gaviiformes	Loons
Ankle joint not obviously flattened; legs not set far back		
Hind toe very long; bill, neck, and legs long	Ciconiiformes	Herons
Hind toe small or absent		
Outer edge of bill serrated or fringed	Anseriformes	Ducks, geese, and swans
Outer edge of bill straight	Charadriiformes	Gulls, skuas, and waders

emperor penguin *(Aptenodytes forsteri)*, of Antarctica. However, a much larger fossil species *(Pachydyptes ponderosus)* lived in New Zealand during the Oligocene (about 35 million years ago) and reached a size of about 90 kg (200 lbs); for comparison, the emperor reaches 40 kg (90 lbs). Emperors and the other species that live on surface ice or nest on exposed sheets of Antarctic tundra or rock or the remote islands of the southern ocean all seem to have highly developed communal instincts and form very large rookeries, or penguin cities. In these communities there are regulated egg-laying territories and organized traffic routes for birds going to, or coming from, the fishing grounds. Newly emerging penguins coming fresh and clean from the sea show an evident distaste for penguins that have been ashore for long enough to become fouled with excrement; thus the two lines of outgoing and incoming traffic are kept apart. Penguins of the more temperate islands do not form such clearly organized communities and, although they may fish in groups, they tend to nest alone in burrows (or under beach houses of *Homo sapiens*) and the accent is on family rather than community ties. An example of the latter group is *Eudyptula minor*, the little blue penguin of New Zealand and Australia.

LOONS [ORDER GAVIIFORMES]. Loons are diving and fishing birds of the northern part of the Northern Hemisphere; they are about the size of a goose. The legs are placed far back on the trunk so as to favor propulsion in water and to make walking on land difficult; the birds can fly but take off only from water. Other characters are as given in Table 19. There is a single living family, containing four species. The best-known is the species distributed over the southernmost part of the range, namely, the great northern diver, or common loon *(Gavia immer)*. Certain diving birds of the Cretaceous period, the Hesperornithiformes, somewhat resembled loons, differing in having the wings vestigial and teeth in the jaws; whether there is any genetic relationship between those two stocks is doubtful, but it is clear that birds with a similar role in the ecosystem to that of loons today had already developed about 100 million years ago.

ALBATROSSES, PETRELS, AND SHEARWATERS [*Order Procellariiformes*]. The order encompasses oceanic fish-eating birds, readily recognizable by the peculiar tube-shaped nostrils. The group apparently is

Table 20. Key characters for recognizing the families of petrels (order Procellariformes).

Character	Family	Common name
Nasal tubes formed of horny material, placed one on either side of the upper ridge of the bill (culmen), so as to leave a clear median strip along bill	Diomedeidae	Albatrosses
Nasal tubes in contact, placed along the median part of the culmen of bill		
No hind toe; Southern Hemisphere only	Pelecanoididae	Diving petrels
Hind toe present		
Two distinct nasal tubes, fused together along culmen	Procellariidae	Prions and shear-waters
Nasal tubes fused together to make a single tube	Hydrobatidae	Storm petrels

ancient, for an albatross bone has been recognized in the Eocene sediments of New Zealand, of some 60 million years ago. The largest members are the existing albatrosses, especially the wandering albatross (*Diomedea exulans*) of the southern ocean (south of 30° south), where the wingspread reaches 3.7 m (12 feet). There is a great range of size within the order, the smallest members beings the petrels, which skim over the wave crests like sparrows; the smallest species is a petrel with the rather long name of *Holocyptena microsoma*. All members of the order are able to spend months at sea, sleeping afloat or on the wing. One southern albatross specimen is known to have circumnavigated the world at least 3½ times, for it was observed on successive occasions in Australia, New Zealand, and South America, each time recognized by a distinctive bird band placed on one leg. The larger species of the order feed on fishes and squid; the smaller species take plankton.

DUCKS, GEESE, AND SWANS [ORDER ANSERIFORMES]. Members of this order are mainly found on or near freshwater, but some species may be reckoned as shore birds. Best-known of these are the eider (*Somateria mollissima*), of Arctic shores. In the Auckland Islands group to the southeast of New Zealand there occurs a subantarctic species of very feeble flight, *Nesonetta aucklandica*; this duck frequents sheltered coves, finding food among the kelp beds. The black swan of Western Australia (*Chenopis*

Table 21. Key characters for recognizing the families of the pelicanlike sea birds (order Pelecaniformes).

Ordinal character: hind toe joined by a web to the front toes. Six families, as follows:

Character	Family	Common name
Bill hooked at the tip		
The horny plate forming the upper part of the bill (culmen) about the same length as the ankle joint (tarsus) of the leg	Phalacrocoracidae	Shags, or cormorants
The culmen is two or three times longer than tarsus		
Tail short, not forked	Pelecanidae	Pelicans
Tail long, deeply forked	Fregatidae	Frigate birds
Bill straight, not hooked at tip		
Chin region covered by feathers	Phaethontidae	Tropic birds
Chin region naked		
Tail rounded	Anhingidae	Darters
Tail pointed at tip, or wedge-shaped	Sulidae	Gannets and boobies

atrata) was introduced into New Zealand some decades ago and has become immensely numerous on salt-water lagoons and in sheltered harbors.

PELICANLIKE BIRDS [ORDER PELECANIFORMES]. Members of this order are aquatic birds, characterized by having all four toes joined by a web. The included families may be distinguished in Table 21. Most members of the order are found in tropical and subtropical waters, both fresh and salt; some members, particularly the gannets and the shags, occur in cool northern and southern seas. All are fish eaters. Group nesting is highly developed, particularly among the gannets, which have large colonies on various rocky headlands and scattered rocky islets in far northern and southern seas. Among the better-known genera may be noted *Sula* and *Moris* of the Sulidae; *Phalacrocorax* among the shags; *Fregata,* or frigate birds, tropical and subtropical long-winged species, which characteristically snatch food in flight from other birds; *Phaethon,* the tropic birds with long tail plumes, colored bright red in one pantropical species; and *Pelecanus,* the well-known pelicans, with a large gular pouch for holding fish below the large bill.

HERONS [ORDER CICONIIFORMES]. Seven families make up this order, but only two of them frequent marine environments. These are distinguished in Table 22. The order is made up of long-necked and long-legged birds. The herons, which include reef and coastal species, feed mainly on fishes and other smaller aquatic animals; the flamingoes have a strainer type of bill, adapted for taking small crustaceans and other animals from lagoon and lake floors; some species of flamingo inhabit salt estuaries and mangrove swamps.

GULLS, SKUAS, AND WADERS [ORDER CHARADRIIFORMES]. This is an order of about 16 families, of which 6 families are restricted to nonmarine habitats. The other families may be distinguished in Table 23.

The order is a very large one, and its members range widely. Remarkably long annual migrations are performed across the oceans by some species, such as the golden plover *(Pluvialis dominicus)* and Arctic tern

Table 22. Conspicuous marine families of the heronlike birds (order Ciconiiformes).

Ordinal characters: aquatic birds; the wings developed normally for flying; the front toes connected by a web; the hind toe well developed; the bill, neck, and legs all elongate. Seven families, but only the following are conspicuous:

Bill curved, front toes fully webbed	Phoenicopteridae	Flamingoes
Bill straight, front toes incompletely webbed	Ardeidae	Herons

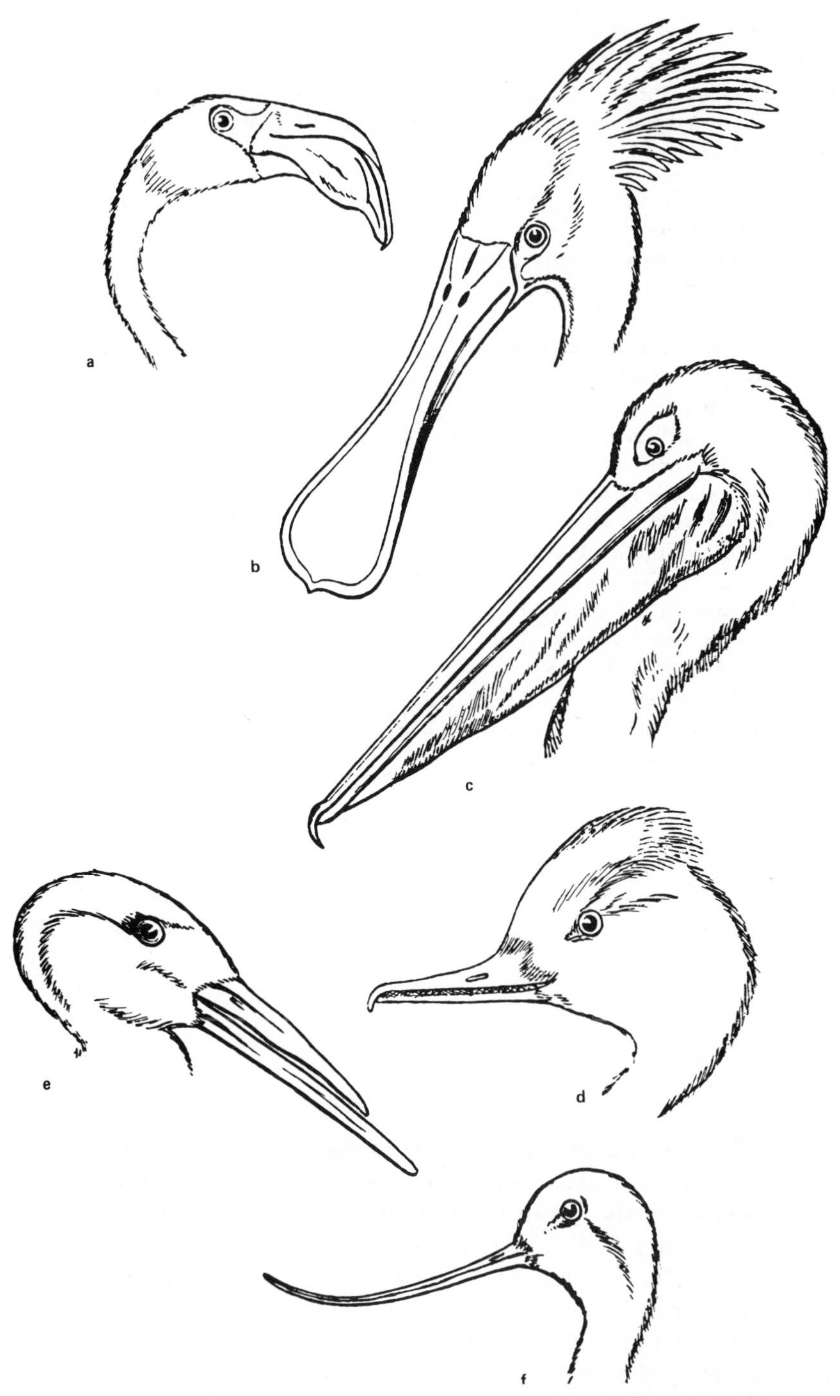

Figure 36. Key characters of aquatic birds (continued). (a) Phoenicopteridae, order Ciconiiformes; (b) spoonbill, Plataleidae, Ciconiiformes, but marsh rather than marine environments; (c) Pelecaniformes, Pelecanidae; (d) *Mergus*, filter-edged bill of Anseriformes; (e) skimmers, Rhynchopodidae, Charadriiformes; (f) avocet, Recurvistridae, Charadriiformes (15).

Table 23. Key characters for recognizing the families of wading birds (order Charadriiformes).

Ordinal characters: front toes alone connected by web; nostrils not enclosed in tubes; hind toes small or absent; outer edge of bill straight, not serrated. Ten families, as follows:

Character	Family	Common name
The bill flattened in vertical plane (compressed), bladelike	Rhynchopodidae	Skimmers
The bill is not obviously compressed		
Front toes completely webbed		
No hind toe	Alcidae	Auks
Hind toes normally present		
Upper bill formed of 3 horny plates	Stercorariidae	Skuas
Upper bill formed of a single horny element	Laridae	Gulls
Front toes webbed only at the base of toes, or without webs		
Claw of middle toe less than half as long as ankle joint	Recurvirostridae	Stilts and avocets
Claw of middle toe longer than half the length of ankle joint		
Toes with fleshy lobes along their sides	Phalaropodidae	Phalaropes
Toes without such lobes		
Bill long, not curved downward at tip		
Bill flattened in vertical plane at tip	Haematopodidae	Oystercatchers
Bill not flattened in vertical plane at tip		
Bill stout, not slender	Charadriidae	Plovers
Bill slender	Scolopacidae	Snipe, curlews, and sandpipers
Bill short, turned downward at tip	Glareolidae	Pratincoles

(Sterna paradisaea). Following are some of the better-known species and genera: gulls, genus *Larus*, range the shorelines of the world (and also occur on lakes and rivers); in recent years in many countries gulls have been changing their habitats to occupy, more or less permanently, city garbage dumps, where they feed on refuse; some flocks now occur far inland. Terns are related to gulls but have forked tails like swallows. Oystercatchers frequent rocky shores of the world; plovers, knots, and sanderlings favor sandy shores. Quiet lagoons are the main haunts of the avocet sandpipers. The Alcidae, comprising auks, puffins, and their relatives, are mainly found on far northern coasts and rocky islets. Skuas, or sea hawks, are predators of Arctic and Antarctic seas. One species, *Stercorarius parasiticus*, breeds in Greenland and other Arctic lands during the northern summer and appears in New Zealand during the southern summer, when

it is not a breeding species. Skuas prey on the eggs and young of other birds, the Antarctic species parasitizing the adelie penguin in this way. Skuas attack land birds when opportunity offers, and in the farmlands of the Southern Hemisphere they even attack lambs and sickly sheep, as gulls also occasionally do. Skuas are birds that even naturalists do not defend, for their depredations on other wild life are so severe as to rival those of man.

MARINE MAMMALS [*Class Mammalia*]. Three orders of mammals have become adapted in greater or lesser degree to life in the sea. All retain the characteristic features of mammals, having regulated body temperature, a viviparous habit, and the requirement to breathe air, so they are obliged to return to the surface of the sea at regular intervals. Table 24 sets out the diagnostic features.

DOLPHINS AND WHALES [ORDER CETACEA]. With the exception of one family of freshwater dolphins all Cetacea are marine. The included families may be distinguished in Table 25.

The most ancient known members of the order are known as the Archaeoceti, extinct carnivorous toothed forms occurring in Eocene sediments of about 60 million years ago. The last known survivors inhabited New Zealand seas in the Miocene, about 25 million years ago. The an-

Table 24. Key characters for recognizing the orders and suborders of marine mammals (class Mammalia).

Class characters: warm-blooded vertebrates in which hair is developed and, in the aquatic orders, the young are born after placental development in the uterus. Three orders, as follows:

Mammals in which a full dentition occurs, with prominent canine teeth for seizing prey; sharp chisellike incisors for cutting flesh; and large cusped carnassial molars for tearing prey animals apart	Carnivora	
Forelimbs not modified into swimming paddles	Fissipedia (suborder)	Polar bears and sea otters
Forelimbs paddle-shaped; hind limbs present	Pinnipedia (suborder)	Seals
Mammals in which canine teeth lacking and hind limbs lacking		
Teeth either all uniform or else lacking altogether	Cetacea	Dolphins and whales
Teeth of 2 kinds, incisors and molars; tropics only	Sirenia	Manatees and dugongs

Table 25. Key characters for recognizing the families of aquatic mammals (order Cetacea).

Ordinal characters: aquatic mammals in which the forelimbs are modified to assume the form of swimming paddles (flippers); the hind limbs are lacking; the teeth are either all similar or else lacking and replaced by horny baleen plates. Eight families, as follows:

Characters	Family	Common name
Teeth developed, at least in the lower jaw; no horn baleen plates (suborder Odontoceti)		
Numerous teeth in both jaws		
Teeth spade-shaped; snout short and blunt	Phocaenidae	Porpoises
Teeth conical; snout usually pointed in front	Delphinidae	Dolphins
One very long twisted tuck in the upper jaw of male	Monodontidae	Narwhals
No functional teeth in upper jaws		
Numerous teeth in the lower jaw	Physeteridae	Sperm whales
Only 2 or 4 teeth in the lower jaw	Ziphiidae	Beaked whales
No teeth developed at all; instead, a fringe-filter of fibrous horn plates (baleen) developed around the margin of the upper jaw (suborder Mysticeti)		
Throat region marked by conspicuous longitudinal grooves		
Numerous grooves developed on the throat; dorsal fin present	Balaenopteridae	Finback whales
Only 2 or 4 throat grooves; no dorsal fin	Eschrichtidae	Gray whales
Throat region without any longitudinal grooves	Balaenidae	Right whales

NOTE: A few genera are not identifiable without recourse to anatomical detail beyond the scope of this book.

cestor of these early whales is thought to have been some North American creodont carnivore, but this is uncertain. Two other suborders of whales are known, both still extant: the Odontoceti, or toothed whales, with a simplified dentition derived evidently from the Archaeoceti; and the Mysticeti, or baleen whales, which have lost the dentition. The Odontoceti are carnivorous, and they comprise most surviving cetaceans. They are active predators, feeding on large animals such as fishes and squid, including apparently giant squid. The common dolphin *(Delphinus delphis)* has been known in written records since the time of the Phoenicians. It grows to a length of 2.5 m (8½ feet). Dolphins travel in groups, called schools, and accompany ships for considerable distances, performing intricate leaps, entering the sea head first with scarcely a splash. Their interlocking teeth and crocodilelike jaws permit seizure of fishes, which are taken whole, tossed into the air to facilitate swallowing them head first. The harbor

Figure 37. Dolphin (*Delphinus delphis*), Cetacea (Cuvier, in 15).

porpoise of the north Atlantic (*Phocaena phocaena*) also travels in schools; favored prey include herring, mackerel, and squid. Similar species occur in the south Atlantic and in the south and north Pacific. Most porpoises have a relatively blunt snout. The name dolphin is applied to a fish, and some ichthyologists claim that all cetaceans so-called should be called porpoises; however, historical tradition, and the meaning of the ancient Greek word *delphis* clearly dictate the priority of the cetacean to the name. The grampus, or killer whale *(Grampus orca)*, is a member of the Delphinidae, ranging all seas from pole to pole almost. The sperm whale *(Physeter catodon)* is one of the largest whales (males to 18 m, or 60 feet, females half as long), and has a relatively large head, though small jaws; it ranges most seas, feeding mainly on squid; the pigmy sperm whale (*Kogia breviceps*) is a related cosmopolitan species, about 3 m long.

The rorqual, or finback, whales belong in the Mysticeti, having baleen plates and a planktonic diet, combined with a taste for cod and herring. Right whales are also baleen whales, but they lack the longitudinal throat grooves of the rorquals and gray whales. They were formerly very common but nearly extinguished by the whalers a century ago. Right whales are placed in the genus *Eubalaena*, also *Balaena*. The largest animal in the world is the blue whale *(Sibbaldus musculus)*, a member of the rorquals reaching a length of 30 m (100 feet); like other baleen whales it is a plankton feeder.

Whales are rapidly becoming extinct, owing to persistent overhunting, and the failure of international agencies to implement recommended conservation measures. The probable economic collapse of the industries based on whale products may occur early enough to save some stocks for future generations. The only effective conservation measures hitherto have been temporary circumstances dictated by accident or human perversity; for example, the almost total destruction by fire of the Union whaling vessels in the Bering Sea by the *Shenandoah* brought respite to the northern whales. In 1871 and again in 1877 the Arctic whaling fleets were destroyed by pack ice, once more delaying the final extinction by a few years. The New Zealand whaling industry, based on the southern humpback whale, collapsed abruptly a few years ago when the last remnants of the regional herd failed to put in the expected annual appearance on migration. This pattern of abrupt disappearance is likely to be seen elsewhere.

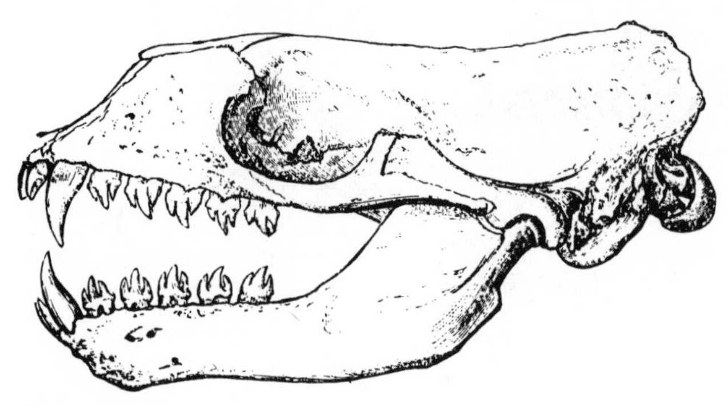

Figure 38. Skull of leopard seal showing full dentition and enlarged canine teeth, characters of order Carnivora (36).

MANATEES AND DUGONGS [ORDER SIRENIA]. The Sirenia are herbivorous aquatic mammals of tropical coasts and coastal seas and estuaries, having the characters given in Table 24. Two families are distinguished: (1) dugongs of the Indo-West-Pacific, tail fluke deeply notched (Dugongidae); (2) Atlantic manatees, tail fluke without any two-fold division (Trichechidae). The dugongs *(Dugong dugon)* occur along the coasts of northern Australia, Asia, and East Africa. The manatees *(Trichechus* spp.) range the Atlantic tropical coasts of America and West Africa. Fossils show that sirenians were already as modified for aquatic life in the Eocene, 60 million years ago, as are modern forms. The closest related mammals are believed to be the elephants. Some of the fossil sirenians preferred habitats such as coral reefs and areas of the continental shelf; the extant forms prefer sheltered estuaries and lagoons. *Hydrodamalis,* exterminated within recent times, inhabited the Bering Sea; it was discovered in 1741, when it occurred in large herds. By 1770 it had almost been extinguished by excessive hunting, the last specimens finally succumbing in 1854.

SEALS [ORDER CARNIVORA, SUBORDER PINNIPEDIA]. The Pinnipedia include the seals, walruses, and sea lions, all aquatic carnivores largely marine in habitat; they are strongly modified for aquatic life, but other features of their morphology imply a rather clear relationship with the terrestrial carnivores, such as cats, bears, dogs, so the pinnipedes are usually treated as a suborder of a larger order, Carnivora, to include all the types mentioned. The oldest known fossil seals date only from the Miocene, about 26 million years ago, and by that time they had already assumed virtually all the characters of extant pinnipedes. Three families of pinnipedes are recognized, as shown in Table 26.

The Otariidae comprise the sea lions, sea leopards, and sea bears. They occur in both the Northern and Southern hemispheres, usually near coasts. They spend a good deal of time on land, where they are able to move about rather freely, for the hind flippers can also be used as limbs for

Table 26. The families of seals (suborder Pinnipedia).

Subordinal characters: Carnivora with a complete dentition, including conspicuous canine teeth; the forelimbs modified to form paddles for swimming. Three families, as follows:

External ears present	Otariidae	Eared seals
No external ears visible		
Upper canines developed as long tusks	Odobaenidae	Walruses
Upper canines not tusklike	Phocidae	True seals

walking, galloping, and climbing rocks. The largest species are the sea lions, which reach a length of 4 m (13 feet). The males arrive at the breeding grounds in the spring and fight fiercely for territory near the seashore; defeated males are compelled to settle further inland, where the property values are depressed on account of the long journey needed to reach the fishing grounds offshore. Most of these settlements are on remote islands of the far northern and far southern oceans. The females arrive after the initial settlements have been established. They are then rounded up into harems of up to 15, the largest dominant bulls having the biggest harems. Excessively jealous of the attentions of unattached males, the harem bulls neither eat nor sleep during the breeding season; as this lasts several weeks, they are enfeebled at the end of the mating sessions. The cubs are born on land, and weigh a few pounds.

Walruses occur only in the Arctic Ocean. There are only two species, *Odobaenus rosmar* in the Arctic Atlantic and *O. divergens* in the Arctic Pacific. Large males grow to about 3.5 m long, with a weight of 1400 kg (1.5 tons). The long tusks are used for excavating the seabed, for they feed almost exclusively on infaunal invertebrates such as bivalve mollusks and some crustaceans. Walruses live in herds. They may gather on drifting icefloes, huddled together in vast biomasses. They are unaggressive animals but have close ties and defend their herds fiercely when attacked.

The true seals have a more extensive modification of the hind flippers for life in the water; therefore they are not active on land. They have thick, close fur and thick protective heat-insulating layers of fat (blubber) beneath the skin. This permits them to occupy extremely cold regions, such as Antarctica. The largest species are the elephant seals, also the largest of all Pinnipedia, growing to 6 m (20 feet). There are two species, *Mirounga leonina* of the southern ocean and *M. angustirostris* of southern California. The name is derived from a short, hollow inflatable trunk on the male, who uses it as an amplifier for his vocalizations. The well-known common seals of various species are much smaller members of this family. Some are easy to tame and have become attached to their owners, as terrestrial carnivores also do.

Some species of Pinnipedia are severely hunted, either for skins or oil. Of walruses alone the joint Russian-American hunting treaty permits 12,000 to be killed each year in the north Pacific. Far more are killed than can be used, and the floating carcasses of about 6,000 are abandoned every season. Two hundred such carcasses washed ashore at the town of Kotzebue in Alaska on August 12, 1971. About 50,000 seals are clubbed to death each year by hunters operating in the waters of the Bering Sea, under joint Russian-American agreements. Nearer at hand, Canadian, Norwegian, and other hunters operate in the eastern seaboards of Canada, their activities often involving intolerable brutality; the Canadian-Norwegian catch for harp seal for 1973 was 160,000.

POLAR BEARS AND SEA OTTERS [ORDER CARNIVORA, SUBORDER FISSIPEDIA]. Sea otters *(Enhydra lutris)* were once very common on north Pacific coasts; the species was nearly extinguished after its pelt was made fashionable women's wear by the empress Catherine the Great. Successful conservation measures are now bringing it back to its former habitats, and it has spread as far south as Monterey, California. Sea otters feed on sea urchins, which they eat while floating on their backs, sometimes cracking open the shell with a stone, at other times holding the urchin in the two forefeet like a hamburger and taking bites out of the uninviting object. The return of the sea otter to California will reduce the excessive sea urchin populations that developed after the otter's original destruction. This change will stop the constant wastage of kelp, which had been drifting to sea following urchin attack on its seabed holdfasts. The conservation of the kelp will restore the abalone populations, provided that man can contrive to stop inserting himself into the food chain.

POLAR BEARS [*Thalarctos maritimus*]. Polar bears are now approaching the status of an endangered species on account of excessive hunting for trophies and pelts. The species is thought to have originated in Eurasia during glacial conditions and to have spread to North America. It now has a general Arctic distribution. Occasional specimens drift as far south as Newfoundland each year, and as many as 50 per annum once used to drift on icefloes across the Denmark Straits from Greenland to Iceland. The numbers now, however, are rapidly being reduced. The species is the only aquatic bear. Modifications for life in cold seas include hair on the sole of the foot, permitting a firm grip on the ice, a method adopted by skiers, too, when fitting skins beneath the runners, the hair directed toward the rear. In winter polar bears eat seals and walruses, and in coastal areas a great deal of fish. In summer they are mainly herbivores. This variety in diet, according to season, characterizes some other species of bears. In fall the female retires to an ice cave to hibernate and to give birth to cubs in winter. The mothers care for the cubs assiduously, sharing out food if the supply is short. When alarmed, polar bears become extremely dangerous animals.

8

PLANETARY CIRCULATION: THE VECTOR FIELD

Sea winds and ocean currents are the vector components of the biosphere and constitute a special category of planetary influences on the biota, characterized by *directional* properties. Whereas the scalar components govern the diurnal and seasonal activities of feeding, sleeping, growth, and sex life, the vector components influence the *dispersion* of organisms. There are two dominant vector fields, that of the *winds* and that of the *ocean currents.* The two fields are causally related, but by no means identical.

The first scientific interpretation of the causes of winds and ocean currents was developed in 1735 by an English astronomer, George Hadley, who invoked two other discoveries by his contemporaries to explain the observed facts. The two principles were (1) the conservation of angular momentum, propounded by Isaac Newton, and (2) Boyle's law of gases. Hadley deduced, we believe correctly, the motions of the atmosphere produced by the operation of these laws. Before examining Hadley's findings, however, we may pause to glance over the history of man's ventures on the high seas: for, apart from their intrinsic interest, these experiences throw light on the probable effects of the global winds and currents on the dispersion of many kinds of living organisms, which, unlike man, do not keep written historical records of their travels.

SOME HISTORIC LANDMARKS IN OCEANOGRAPHY. A summary of a few of the outstanding episodes in oceanic exploration might list the following entries (many more could be added; countless more have surely gone unrecorded):

ca. 8000–7000 B.C.	First sea crossings by Neolithic man (to offshore islands)
ca. 2100 B.C.	Egyptians publish *Story of the Shipwrecked Sailor*
ca. 1500 B.C.	Minoans sail the Etesian winds
ca. 800 B.C.	Hesiod composes sailing instructions
140 B.C.	Wu emperor voyages to India
30 A.D.	Monsoon sailing is discovered; Hippalos first crosses the Indian Ocean from Aden
166 A.D.	Greeks enter the Pacific Ocean: Alexander's embassy to the Court at Hanoi
385–391 A.D.	Te Sasi leads 20 Polynesian ships from Java to Tahiti
ca. 450 A.D.	Polynesians discover and settle Easter Island
ca. 900 A.D.	Toi voyages to New Zealand
1000 A.D.	Leif voyages to North America
1425 A.D.	Cheng Ho completes voyage from China to Africa
1492 A.D.	Columbus discovers the trades and antitrades and measures the equatorial current, providing the foundations for study of the Atlantic Ocean
1735 A.D.	Hadley explains the planetary circulation

THE PRINCES OF THE ISLES. From the mountains of Sparta, when the air is clear, the eye may range southward and eastward to the nearer Cyclades, to Cerigo and the snow-capped peaks of western Crete. Seemingly some early watcher from Cape Malea conceived in Neolithic times the craft that somehow carried men to Crete; archeological remains attest that about 10,000 years ago men reached and settled the region that afterward was the site of Knossos. Other settlements were made on the coast of Crete, and the presence of Baltic amber in their ruins implies that the northern sea lanes were open. By early Minoan times the Cretans began to venture further.

Five hundred miles to the southeast, across the open Mediterranean, lay the kingdom of Egypt, site of another and earlier civilization. The Egyptians were bound by the needs of agriculture to pattern their life on the annual rise and fall of the Nile. Their rivercraft, which sailed the great inland waterway, do not seem to have sailed to sea beyond the sight of land, though they did later perform long coastwise voyages to Somalia (the Land of Punt), and later still around the coasts of Asia to India (Land of Irihya) and Ceylon (Rising-Sun Land), and eventually established a settlement in Java (East Land). But these long voyages do not seem to have been achieved until the first millennium B.C. Before the end of the second millennium B.C. the Egyptian ships were much more restricted as to their range. A novelette of the twelfth dynasty (about 2100 B.C.) tells of its hero who, with 150 other companions, set out on a voyage to Mine Land, suffering shipwreck en route. The sole survivor, clinging to a piece of the ship's timbers, is cast ashore on an island, and later rescued. In view of the supposed date of this document, it seems strange that Egyptian ships apparently did not sail out into the Mediterranean. At all events, the next episode in exploration seems to have taken the Egyptians totally by surprise.

"These are the Princes of the Isles in the midst of the great green sea"—so read the hieroglyphs on an Egyptian mural painted during the reign of the pharaoh Thothmes III, about 1475 B.C., and the people depicted are garbed in Cretan dress and carry Minoan vases. From this mural, and other evidences, we can date the discovery of the first planetary wind from the Minoan era.

Figure 39. Arrival of the Princes of the Isles at the court of Thothmes III. The costume and Cretan vases identify the visitors as Certans (2a).

THE ETESIAN WIND. In the summer, the desert lands of North Africa become heated, and the overlying air mass expands and rises, thereby leaving a low-pressure region into which cooler air flows from the Mediterranean. The gentle north wind so generated was called by the Greeks the etesian, or annual, wind. It carried the frail seacraft of Crete to Egypt, the return voyage to Crete apparently being made by coastwise sailing around the shores of Chaldaea and Phoenicia to Ionia, and thence by way of the Greek coast back to Sparta and Crete. The later Greeks and the Romans used this route regularly, the summer imperial post from Rome arriving therefore by sea in Alexandria, returning to Rome by overland chariot mail. In winter none but a fool ventured on the Mediterranean.

After the collapse of the Mycenaean civilization the Phoenicians inherited the sea roads of Minos. Their trade routes spanned the Mediterranean, Cadiz being colonized in 1100 B.C., Carthage in 950. Greek settlers founded Marseilles about 600 B.C. Throughout this period, and on into Roman times, coastwise sailing was the rule, and only in summer did any ship venture on the high seas. It is possible, however, that unrecorded voyages may have taken place. From Ezekiel's account, the Phoenician trade routes were far-flung and their ships sturdily constructed. Circumstantial evidence is now accruing that seems to imply that a semitic voyage to Central America may have occurred during the Carthaginian era.

With the return of civilization to Greece ca. 800 B.C. after recovery from an unexplained period of eclipse, Greek ships began to ply the Mediterranean and also the Red Sea. A body of maritime law was established, very similar to our own, and surviving briefs and prosecutions handled by the Athenian barrister Demosthenes throw an amusing light on malpractices in the fourth century B.C. From the first century A.D. we have a personal memoir of a traveler named Lukas, who embarked on a disastrous

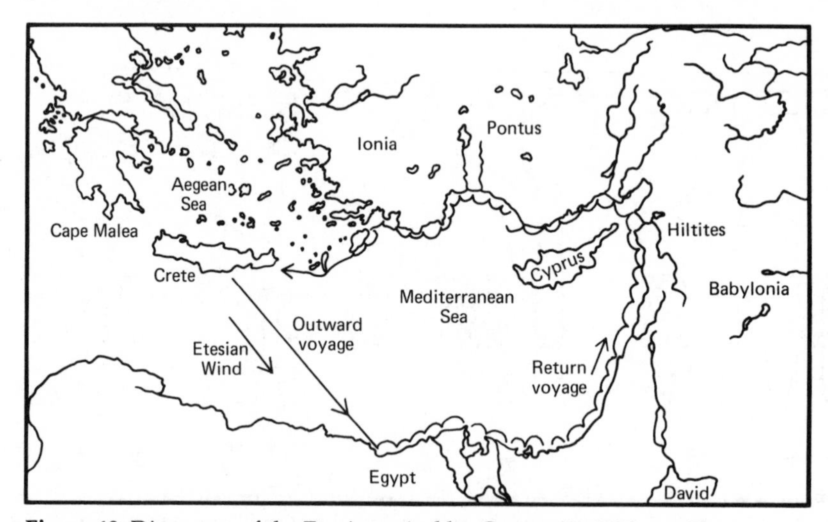

Figure 40. Discovery of the Etesian wind by Cretans in 1800 B.C. The return voyage from Egypt was made by coastwise rowing and sailing around the shores of Asia Minor and via the Aegean Islands (23).

voyage late in the season. Although the ships were large (one of those used by Lukas carried 276 souls) they lacked strength to withstand rough wind and heavy seas.

THE DISCOVERY OF THE MONSOON WIND. While these developments had been occurring in the Mediterranean world there had been a corresponding evolution of navigational skills in the northwest Pacific. Along the coasts of Asia and Indonesia the locally popular dhows and junks had established a network of coastwise sailing routes stretching from the Asiatic mainland southward to the independent kingdoms of Sumatra and Java. The peoples of Vietnam at that epoch spoke an Oceanic language, still surviving among the present-day Montagnards and destined to spread across the Indo-Pacific from Madagascar in the west to Easter Island in the east, and southward to Samoa and New Zealand. Probably the first Polynesian voyages began during the millennium before Christ, at first restricted to the Indonesian region, later to span half the world. During the Han dynasty (206 B.C.–A.D. 221) China assumed in the East a commanding role paralleling the contemporary rise of the Roman Republic in the West. By the year 140 B.C. when the Wu emperor ascended the dragon throne, Chinese navigation had reached the point at which the first thalassic voyage could be mounted. The emperor himself led the expedition, accompanied by most of his court officials. Leaving the then capital Ch'ang An, in northern China, the ships sailed 3000 miles southward to the Straits of Malacca, then, rounding the Malayan Peninsula, set their prows northwest to cross the 1200 mile open stretch of water of the great Bay of Bengal. Landing at "Huang-chih" (which modern Chinese scholars believe to have been Conjeveram, near Madras) the emperor marketed his cargo of gold and silk and returned to China with pearls, crystal, and precious gems. We are not told by the historian (Pan Ku, A.D. 2–92) how the voyage was accomplished, but it seems probable that somehow the Chinese utilized the seasonal winds called the monsoon. It was, in all events, the greatest voyage in the history of man. The Han court subsequently was disrupted by civil revolution, and China withdrew from the international scene for nearly 300 years.

HIPPALOS CROSSES THE INDIAN OCEAN. About A.D. 30 a certain sea captain named Hippalos was engaged in the spice trade between Egypt and northwestern India. His name suggests to scholars that he was an Alexandrian Greek. Like his predecessors for centuries since the time of Ezekiel he followed the coastwise route around Arabia, then northeast around the Persian Gulf, following the coasts of what is presently called Pakistan, to reach the delta of the Indus. In the course of his protracted voyages (the round trip lasted three years), Hippalos became aware of the general trend of the Asiatic coasts and realized that theoretically a direct route to India could be achieved by sailing northeast from Aden. The advantage of such a precarious venture would be to avoid paying the imposts levied by the Arab chieftains along the western Asian coasts. The practicality of such a voyage would depend on whether favorable (that is, following) winds might be encountered.

To this question Hippalos applied a novel answer, reinforcing his arguments by his long experience of the seasons of the Indian Ocean. He pointed out that in the summer months the wind blows with remarkable regularity from the southwest, whereas in the winter the wind reverses its direction to blow with equal regularity from the northeast. He had, in fact, detected the main features of the monsoon cycle, though its causes

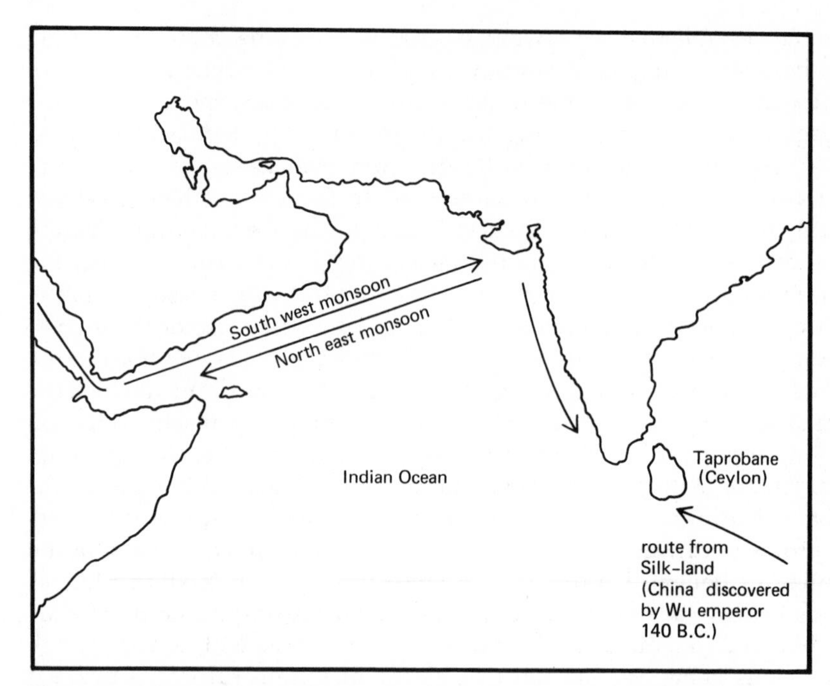

Figure 41. Discovery of the monsoon route to and from India by Hippalos in A.D. 30. The outward voyage from Aden was performed on the southwest monsoon in the summer, the return trip on the northeast monsoon in winter (23).

were to await elucidation by a later generation. Putting his theories and the courage of his crew to the test, Hippalos sailed one summer on the southwest monsoon, arriving in India in the unheard-of short space of about three months; the return voyage was performed on the northeast monsoon, and the ship returned to Egypt within the course of a year for the entire outward and homeward voyage. Almost overnight the trading routes were revolutionized, with up to 100 ships setting out for India in a single season. By the time of Claudius and Nero some Greek and Roman captains were sailing directly to Taprobane (Ceylon), and Roman currency flooded the east, as the innumerable coins found in India and Ceylon bear mute testimony. In Ceylon the Greek traders encountered a mysterious new fabric, silk, attributed by their Sinhalese hosts to a distant land to the east, whence traders brought it to India.

But before the Western visitors had time to contemplate their next step, the Chinese acted first. According to the contemporary historian Fan Yeh, the Chinese emperor An Ti in the year A.D. 120 had been entertained by a troupe of Roman musicians and circus performers, thoughtfully directed to him by the king of Burma. This visit, reinforced by favorable reports of the Roman world sent back from Antioch by an earlier overland embassy, must have prompted the Chinese to open the way for Western captains to receive friendly guidance across the Bay of Bengal, through the

Malaccan Straits, and so on to the Gulf of Tonking. An Egyptian Greek named Alexander is reputed to have performed the first passage, landing at Haiphong. Contact with the imperial officers was established at Cattigara, thought to be Hanoi. By A.D. 166, in the reign of the Han emperor Huan Ti, an embassy from Rome arrived bearing gifts from the Roman emperor An Tun, by which Western scholars recognize the name of Antoninus (Marcus Aurelius Antoninus was emperor in A.D. 166). No trace of these transactions remains in the scattered records of the Roman world, and we are wholly dependent on the Chinese historians.

The eventual collapse of civilized contacts between East and West as the two empires disintegrated all but obliterated the cultural results of the voyages. One lasting consequence, however, was the compilation of the *Geographies* of Strabo, and the *Almagest* of Claudius Ptolemy, a Greek astronomer of Alexandria and a contemporary of Marcus Aurelius. Latitudes and estimated longitudes were tabulated for numerous distant Eastern cities and coasts, and these provided the view of the world that the Moors and Hebrews brought to Spain and that Columbus used as the basis of his projected voyage westward to the Indies.

THE DISCOVERY OF THE PLANETARY WINDS OF THE ATLANTIC. Like Hippalos before him, Columbus was inspired by his previous experience of oceanic winds. His voyages on the west African coasts had taught him to expect with virtual certainty a steady wind from the east when sailing south from Spain and approaching the Canary Islands, in about 30° north latitude. Knowing that occasional reversals occur, he reckoned on being able to perform a westward passage on the east wind and to use fortuitous reversals of the wind direction to sneak back to Europe again. In fact he found no such thing. The east wind turned up, as expected, when he reached the Canaries, and it carried him right across the Atlantic in 70 days. But on attempting to return by the same route, Columbus found no west wind. Forced to tack, he gradually beat his way painfully northeastward until he neared the vicinity of Bermuda and crossed the thirtieth parallel. Almost at once he picked up the longed-for west wind and was swept safely back to Spain. As soon became plain from the voyages that followed, Columbus had established the main outlines of the planetary wind system—a great flow from east to west over the equatorial belt between 30° north and 30° south latitude; and two great belts of west-to-east winds, north of 30° north, and south of 30° south. The navigations that followed during the next two centuries disclosed that a similar pattern of winds exists in the Atlantic, the Indian, and the Pacific oceans. The tracks of the voyages show how effectively the sea captains learned to use these winds, planning their navigations to extract the best possible advantage of the air flow of the planet. Soon the equatorial east wind came to be called the *trade wind*, and the west winds to the north and the south the *antitrades*. By the eighteenth century scientific explanations were forthcoming as to the causes of the circulation.

THE CAUSES OF THE MONSOON CYCLE. In summer the dry desert and plateau region of Asia becomes heated, and the overlying air mass expands and rises and then flows outward at a higher altitude. The low-pressure region remaining below is therefore subject to the inflow of lower-level air, mainly from the cooler air mass over the Indian Ocean. This northward-moving air near the air-sea interface begins its journey from a part of the earth's surface near the equator where the air initially has a west-to-east velocity of 1000 miles per hour imparted to it by the earth's rotation. But as it flows northward it enters regions where the west-to-east motion of the earth's surface progressively decreases (the surface velocity of the earth falls off as the cosine of latitude, becoming zero at the poles). Since the initial angular momentum of the air cannot be destroyed, the equatorial air as it moves northward acquires a relatively increased west-to-east component of motion with respect to the surface. Thus, instead of the incoming cool air arriving from the south, it appears to be flowing from the southwest. This air, laden with oceanic moisture, reaches India as the southwest monsoon.

The converse is true of the winter months. The central Asian highlands become cold, high-pressure regions, whence air flows outward near the surface of the earth. The portion that flows southward toward the Indian Ocean enters regions of increasing angular velocity and so has insufficient velocity to keep pace. Lagging behind the rotating surface of the earth, its direction appears to change from being a north wind to become a wind from the northeast. This is the dry northeast monsoon.

A monsoon is a regional wind that exhibits properties of a planetary wind, that is, its directional qualities, or *vector components*, are related to the earth's rotation and are determined by the principle of conservation of angular momentum.

SOME PARAMETERS OF THE PLANETARY AIR FLOW. As already noted, the parallels of 30° north and 30° south serve to divide the oceanic air into three main zones, namely, an equatorial belt of east-to-west winds and a northern and southern zone each of west-to-east winds. Following are relevant clues as to the nature of this system of air-flow.

The system conserves angular momentum, because: (1) At any instant as much air is flowing from west to east as is simultaneously flowing from east to west. This proposition can be illustrated by considering the properties of a sphere. The area of the surface of a sphere is $4\pi R^2$. The area of that part of the surface between 90° and 30° is πR^2 for each hemisphere. The area between 30° and 0° is also πR^2 for each hemisphere. If we assume the atmosphere to have a uniform depth, then it follows that equal volumes of air move in the two opposed directions. Thus the total air flow neither adds nor subtracts from the total angular momentum of the rotating planet.

(2) The angular momentum of the moving air is a function not only of its velocity but also of its radial distance from the earth's axis. Thus, in order to balance, the velocity of the air of the equatorial belt (which lies furthest from the earth's axis) would have to be less than the velocity in the

Figure 42. Inferred surface currents of the oceans in late Cretaceous times, about 100 million years ago. The shaded areas mark transgressions of the sea over the continents. The Gulf Stream flows through Tethys to produce a northern Indian gyre (E), the other gyres being as today. The Panama seaway is open. The apparent curvature of the equator is due to the inference that the Cretaceous poles of rotation did not match those of today (23).

reverse direction of the air in the two west-wind belts. This is observed to be the case, for the trade winds are gentle and steady, whereas the winds of the antitrades are strong and often of near hurricane strength (as in the roaring forties).

Vertical air flow lies between zones of horizontal air flow, because: (1) The doldrums, or so-called windless region, form a narrow belt along the equator, where the barometric pressure is low. A barometric low is commonly produced when air is moisture-laden since water vapor weighs less than dry air. But over the ocean all air is moist (unless it has just arrived there from some dry region such as a desert). So moisture alone cannot be the cause of the barometric low pressure of the doldrums. The only other cause that can be invoked is that the low pressure is produced by air rising. This is highly plausible, because equatorial air receives most solar energy, and therefore becomes hotter than other air, decreases in density and floats upward.

(2) The analogous regions of so-called windless air, the horse latitudes, register high barometric pressure. In the same latitudes (30° north and 30° south) over land, dry desert air is present. It looks as if we have here descending dry air reaching the surface of the earth from a higher level of the atmosphere.

THE GLOBAL PATTERN EMERGES: HADLEY'S SYNTHESIS. The foregoing facts and inferences were synthesized into a simple cohesive theory by George Hadley in 1735. Hadley's theory states that: (1) Heated air in the equatorial region over the ocean expands and rises and flows outward to the north and to the south at a higher level in the atmosphere. After being cooled in the upper levels, and thereby losing its moisture content, it descends as dry air at latitude 30° north and south of the equator.

(2) Surface air flows toward the equator from the south and from the north to replace the air that rises. Conservation of angular momentum necessarily means that such air acquires an east-to-west component of motion, because it is entering a region of higher velocity of the earth's surface in the west-to-east direction. This effect produces the easterly winds called trades.

(3) The cold air descending at the horse latitudes spreads to the north in the Northern Hemisphere and to the south in the Southern Hemisphere. Since this air is entering regions of lower surface velocity of west-to-east rotation, the west-to-east component already present in the northward and southward spreading air exceeds that of regions it is entering. The conservation of its angular momentum causes it to assume the character of a west wind, since it is already rotating faster than the earth beneath it.

Hadley's theory not only makes sense of the atmospheric motions, but it also explains all the major features of the global circulation of the oceans.

THE OCEAN CURRENTS ARE DRIVEN BY THE WINDS. This proposition may be inferred from the observed fact that all major ocean currents flow in the directions we would expect if they are in fact wind-driven.

THE AIR–SEA COUPLE. In particular cases the proposition may easily be proved. For example, in the northern Indian Ocean, where the regional monsoon wind reverses its direction every six months, the underlying waters of the ocean similarly reverse their direction of flow, to keep in phase with the wind. Thus the Somali Current, which flows along the northeast coast of Africa reverses its direction shortly after each reversal of the monsoon. During the southwest monsoon, the Somali Current flows from the southwest. When the northeast monsoon begins, the Somali Current gradually slackens its flow, then comes to a halt, then slowly begins to pick up speed in the southwest direction, that is, begins to flow from the northeast. The frictional effect of moving air over the surface of the sea is called the *air-sea couple*. It is exactly analogous to the effect produced if one blows across a dish of water, so as to set the water in motion.

THE OCEANIC GYRES. The oceanic circulation can be simplified into a global pattern of five great rotating whirlpools, called *gyres* or gyrals. Two of the gyres lie in the Northern Hemisphere and rotate in the clockwise direction. They occupy the north Atlantic and north Pacific,

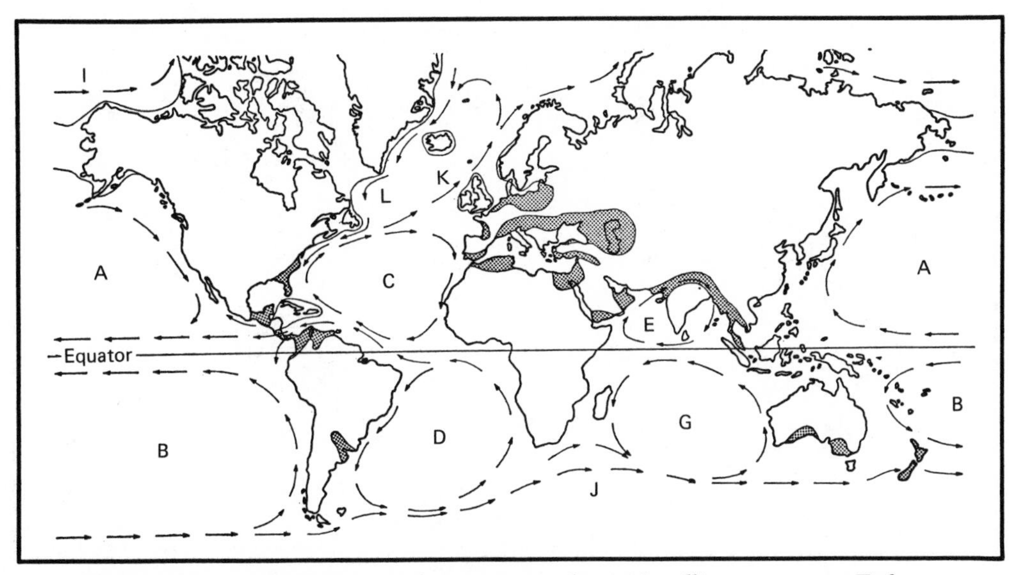

Figure 43. Inferred Miocene surface currents about 25 million years ago. Tethys is now bisected by the rise of Arabia and southwest Asia, so the Gulf Stream is deflected to produce the North Atlantic Drift (K). The Panama seaway remains open (23).

respectively. The remaining three gyres lie in the Southern Hemisphere and rotate in the counterclockwise direction. They occupy the south Atlantic, south Indian, and south Pacific oceans, respectively. The north Indian Ocean is of very limited extent at the present stage in earth history and, as already noted, carries a monsoon-driven current. During most of geological history the north Indian Ocean has been much larger than it is now, covering much of Asia; at such epochs it would carry a rotating gyre like other oceans.

Each ocean basin is bounded by eastern and western continental shorelines. Each has an overlying equatorial air mass that moves from east to west. Each has a poleward overlying air mass that moves from west to east. Under such conditions it is obvious that northern oceans must develop clockwise rotating gyres and southern oceans must develop counterclockwise rotating gyres. The topic can be treated mathematically, but it would serve no purpose to do so in this context. The separate parts of a gyre—the eastern side, or the equatorial part, and so on—are commonly designated by particular names, as, for example, Gulf Stream or Kuroshio. A table of the more important of these currents is given below. Near its center a gyre rotates slowly, so the currents have little velocity; nearer the edge of a gyre the velocities are usually highest. The speed commonly falls within the range of a quarter knot (about 400 m per hour) to 1 knot (1.85 km per hour).

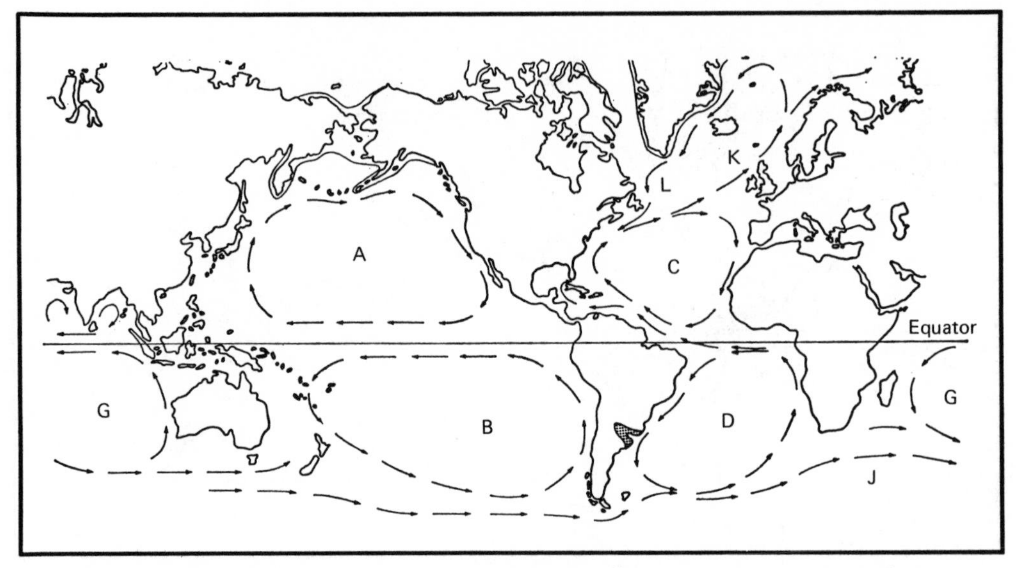

Figure 44. Late Pliocene surface currents about 3 million years ago, initiating the existing pattern. The Panama land bridge has been elevated and the principal gyres have their modern disposition (23).

THE TEMPERATURE OF AN OCEAN CURRENT. The water mass of a gyre gradually becomes warm as it passes through the equatorial part of the gyre. Water holds its temperature for a considerable period of time, so the western sides of oceans (which are supplied with water coming from the equatorial region) are also warm. The water loses its heat while passing through the poleward part of a gyre, and so the eastern sides of oceans are also cooler, since they are fed by waters arriving from the cold part of a gyre.

CORAL REEFS AND OCEAN CURRENTS. An important effect produced by the temperature distribution is that coral reefs tend to grow on

Table 27. Major ocean currents.

GYRE	DIRECTION OF ROTATION	NAMES OF PARTICULAR CURRENTS CONSTITUTING THE GYRE			
		SOUTHERN LIMB	WESTERN LIMB	NORTHERN LIMB	EASTERN LIMB
North Atlantic	Clockwise	North equatorial current	Gulf Stream	North Atlantic Drift	(Diffuse)
South Atlantic	Counterclockwise	West-wind drift	Brazil current	South equatorial current	Benguela current
North Pacific	Clockwise	North equatorial current	Kuroshio	North Pacific current	Californian current
South Pacific	Counterclockwise	West-wind drift	(Diffuse)	South equatorial current	Peru, or Humboldt, Current
South Indian	Counterclockwise	West-wind drift	Algulhas Current	South equatorial current	West Australian current

NOTE: The inferred evolution of the system is illustrated in Figures 42–44.

the eastern margins of continents, for these are warmed by the western water masses of the gyres. The western margins of continents have only impoverished reefs of very limited extent, for the waters are here too cool to promote coral growth. Table 27 shows the more important ocean currents distinguished by mariners and oceanographers.

Some Regional Currents

THE NORTH ATLANTIC DRIFT. This, the northern continuation of the Gulf Stream, is the north Atlantic equivalent of the west-wind drift of the Southern Hemisphere. The North Atlantic Drift bifurcates off western Europe, one part flowing northeast into the Arctic Ocean, the other part continuing southward along the north African coast, to rejoin the equatorial current.

THE LABRADOR AND EAST GREENLAND CURRENTS. Tongues of cold water emerge from the Arctic, presumably to compensate for the entry of North Atlantic Drift water into the Arctic. These cold currents follow the eastern seaboard of the Canadian and U.S. Atlantic states, excluding them from the moderating influence of the Gulf Stream, which latter therefore lies offshore.

EQUATORIAL COUNTERCURRENTS. Narrow eastward-flowing seasonal currents lie on or near the equator in all oceans and therefore divide the westward-flowing equatorial waters into a northern and a southern section. It has been supposed that the countercurrents are analogous to rivers, comprising water that is flowing downhill from the western sides of oceans, under the doldrums, where no air is flowing horizontally. The elevated western sea level of each equatorial oceanic region is attributed to the piling up of wind-driven water, carried westward by the trades. Such water can flow backward (that is, in an easterly direction) along the downslope produced by the lack of trades along the equator. This explanation has been disputed.

OCEAN CURRENTS DURING THE PAST 100 MILLION YEARS. For much of geological time the Indian Ocean has covered extensive areas of Asia and Europe by a shallow epicontinental sea known to geologists as Tethys. During the Cretaceous period a large part of North America was also covered by a shallow epicontinental sea known to geologists as the Sundance Sea. During the greater part of the past 100 million years a broad seaway has separated North America from South America. About 30 million years ago Tethys was bisected by the elevation of the Middle East region. The western part of Tethys then became the Mediterranean Sea, and other residual fragments are represented by the Caspian and Black seas. The eastern half of Tethys vanished to become dry land. Thus the

Indian Ocean lost its northern gyre. With the closure of Tethys, the northern gyre of the former northern Indian Ocean was disrupted. One of the residuals is considered (Fell, 1967) to be the present North Atlantic Drift complex, the Gulf Stream system having supposedly developed after the closure of the Mediterranean from the Indian Ocean. These inferred changes in oceanic distribution, and in the ocean currents, are illustrated in Figures 42–44.

THE DISPERSION OF MAN. As the foregoing passages show, important voyages, accidental and deliberate, were aided or made feasible through the effects of the oceanic winds and ocean currents. Many unrecorded voyages must have been made, for during the past 200 years records have been made of accidental drift voyages by Polynesians, blown off course when fishing, but subsequently making a safe landing on distant shores in the Pacific. Such drift voyages have always followed a path predictable from the known behavior of the planetary circulation. However, by the early centuries of the Christian era fast triremes and biremes had been invented, powered by oars, and capable of sailing in any direction, irrespective of the winds. These vessels, restricted mainly to naval squadrons on account of the expense of the manpower (200 men in 3 shifts), probably led Greek and Roman explorers further afield than extant Western records indicate. The hieroglyphic tablets of Easter Island, presently under translation, show that the Polynesian ships of the fourth century could sail into the teeth of the trade winds for distances up to 9000 miles and that the motive power was supplied by oars. Thus the story of early Pacific exploration is more complex than hitherto was suspected, and Greco-Roman influence in ship design and navigation must have been far-flung, as witness also the Chinese court records quoted above. In the National Museum of Athens there is exhibited an astronomical computer found by divers in the wreck of a sunken Greek vessel of the first century B.C. So complex is its mechanism that 50 years elapsed after its discovery before an engineer and a specialist in Ancient Greek were able to solve its structure and function as a navigational instrument. In this way recent studies have disclosed a degree of sophistication in navigational skills totally unexpected, and similarly Polynesian materials in the Easter Island archives are yielding evidence of a fundamental comprehension of latitude and longitude on the part of early voyagers in the Pacific, and their dependence on astronomical observations.

THE DISCOVERY OF CORAL ATOLLS. The contribution of Polynesians to the exploration of the Pacific is illustrated by the earliest known definition of a coral atoll, found in an Easter Island tablet presently under translation. It reads in part, ". . . a rampart of coral enclosing a deep lagoon, suitable therefore as an anchorage. The soil of the land is a salty limestone, yielding by way of vegetation some protection from the northerly winds. Showers of rain wash away the soil. There are good coconut trees. . . ." The tablet is undated, but the form of the hieroglyphic letters suggests a probable date of around 450 A.D.

9

LIFE IN INLAND WATERS

A body of freshwater in its natural state is nearly always an independent ecosystem, unless it happens to rest on a bedrock containing toxic minerals. As in the case of marine habitats, a freshwater lake or pond comprehends a nonliving environment and associated biota interacting on each other to produce an exchange of materials and energy between the living and nonliving components. The energy of freshwater ecosystems is derived mainly from photosynthesis performed by algae suspended as plankton and by the higher plants growing on the bottom. A variable amount of energy is often also derived from leaves, twigs, and pollen from terrestrial plant communities and blown or washed into rivers and lakes. Streams and rivers also constitute freshwater ecosystems, as do various types of wetlands. Most freshwater ecosystems are presently undergoing catastrophic deterioration and destruction as a result of the diversion of urban and industrial pollutants into natural drainage systems, most of which either lead to or actually constitute natural freshwater ecosystems.

The oldest known civilization had its origins in the valley of the Nile

River, whose annual floods determined the rhythm of agriculture of ancient Egypt. Thus it was that the earliest recorded scientific investigation of any ecosystem was carried out about 2400 B.C. on the fauna and flora of that river. This was in the time of the fifth dynasty of the old kingdom when arts and sciences had already blossomed, and a learned academy, presided over by Ptah-hotep, had not only developed clocks and sundials, cultivated astronomy and geometry, and determined a value for the constant π, but also set in motion a detailed biological survey of the freshwater fauna and flora. The surviving documents depict the species with such accuracy that generic and specific identification of the freshwater fishes and plants offers no difficulty. It is also worthy of note that the first experimental demonstration of an ecosystem occurred in London in 1841 when Nathaniel Ward constructed a balanced freshwater aquarium, following principles he had predicated in 1837. Ward's theories were based on earlier researches by Joseph Priestley, which seemed to forecast that it would be possible to associate selected plants and animals in an isolated environment, with access to air and sunlight, and that such an association would maintain itself indefinitely. Within a year after the demonstration Thynne established in London the first marine aquarium, again employing the methods developed by Ward for the freshwater model. Thus, it was these pioneer studies in freshwater biology that ultimately led to the whole ecosystem concept presently dominating our view of the natural world about us.

THE GEOMORPHIC CYCLE. Rain and its geological agents, the rivers and streams, very largely control the form that a landscape assumes, at least in regions where there is sufficient rainfall or other precipitation to assure the development of constantly flowing streams. When a part of the seafloor is elevated by the long-term buckling of the earth's crust, it comes under erosion attack by fresh water. The first streams form in whatever fortuitous depressions exist on the original surface, and thereafter valleys are cut, and the shapes of hills and mountains develop under the influence of the whole system of natural drainage. As a surface matures, the valleys become widened, and the rivers meander from side to side, producing flood plains, while the hilly parts become gradually subdued. Eventually, unless further uplift occurs, the landscape assumes the form of a *peneplain*, or nearly level surface. Throughout the (rather few) millions of years required for this *geomorphic cycle*, forests or grasslands have to adapt their dispersions and forms to that of the land surface on which they grow. Thus running water exerts a considerable influence on the nature of environments; for a habitat reflects the *accumulated effects* of earlier periods of the geomorphic cycle. In the same way, if climatic changes occur such as to modify the nature of the precipitation, or of the manner in which the precipitation is discharged from the land surface, corresponding changes in the land forms inevitably follow. Thus many temperate regions of the earth today still bear the scars of former glaciations, and much of the soil of North America, for instance, owes its character to events during an ice age that came to a close 12,000 years ago.

Figure 45. The widespread distribution of water plants from one continent to another, and in remote islands it may in part be attributed to migratory waterfowl. (a) *Typha angustifolia* (cosmopolitan cattail, or bulrush); (b) *Butomus* (flowering rush); both sketches from notebooks of Leonardo da Vinci; (c) the obliterative coloration of *Todorna* (sheldrake), sketched in a Nile marsh by a fourth dynasty painter, about 2800 B.C. (37, 2).

THE ORIGIN AND LIFE SPAN OF LAKES. Whereas rivers are likely to be very ancient features, always forming and persisting in response to the natural precipitation pattern of the regions where they flow, lakes on the other hand are necessarily transient though spectacular environmental features. This fact, surprising though it may seem at first sight, is attested by many pieces of evidence.

For example, most of the lakes, large and small, of the temperate Northern Hemisphere are found to be occupying depressions in the earth's surface that were produced by glaciers during the ice ages of the past 2 million years. The same is true of those parts of the Southern Hemisphere that were glaciated. Now it is a matter of simple observation to show that, over a period of time, a lake becomes gradually shallower, and therefore contains less water, because as the lake floor rises the supernatant water overflows and is lost. The cause of the relentless shoaling process is traceable to the constant delivery of sediment to a lake by its tributary streams; also a good deal of dead vegetation accumulates on the bottom. Most lakes are relatively shallow, and their life spans are to be estimated in terms of a few thousand years, after which time they will have filled up with solids and discharged their fluid content.

Several inferences can be drawn from these data. One is that in a very short span of time, geologically speaking, most of the lakes of the temperate regions will have vanished. By the same argument it follows that a few thousand years ago these lakes must have been larger, deeper, and more numerous—a fact that geologists have long since established on strictly geological evidence. Again, it follows that 5 or 10 or 100 million years ago, when no great ice ages had occurred on the earth for a long space of geological time, there must have been fewer lakes than there are today. It does not follow, however, that all lakes are related to glaciation. Some lakes occupy explosion craters produced by volcanoes, others occur in depressions of the earth's surface due to other causes, other lakes again occupy valleys that have been dammed by landslides or by moraines left by glaciers. A few lakes can be shown to be very old geological features. But on the whole lakes are short-lived phenomena.

THE ANTIQUITY OF FRESHWATER FAUNAS AND FLORAS. Some extremely ancient types of living creatures occur in lakes and rivers—for example, the fish called bowfin (*Amia calva*), in North America, and many others in various parts of the world. If lakes are short-lived features, how is it possible for a lake to contain an ancient fauna? Would not the ancient fishes of freshwater lakes have died out when their lakes dried up millions of years ago?

The answer to these questions evidently depends on the relationship between lakes and the streams that feed them. Merely because a lake dries up does not mean that its tributary streams vanish. On the contrary, the streams, unimportant though they may seem in comparison with a great lake, are really the long-lived features, and taken over the period of tens or hundreds of millions of years that a drainage system may exist, the total volume of water contained in the streams would far exceed that of the lakes they may have fed at one time or another; for most of geological time such streams have probably flowed more or less directly into the sea, without lakes along their routes. Now study of lake faunas nearly always discloses that the neighboring streams, or the tributary streams, also contain the same species. Further, if two streams are eroding on two opposed sides of a mountain chain, one or the other may eventually cut the watershed between the two, and in such case the stream that lies at the lower level captures the entire part of the drainage system of the other that lies at the higher level. This process, known as *headwater capture*, must inevitably result in the periodic transfer of a fauna from the river on the one side of a mountain chain to that on the other side. Young stages of fish are commonly passed in the upper headwaters of river systems; so it would be the juvenile stages that would actually be transferred across mountain ranges. The adult stages would form in the newly occupied river at appropriate elevations and in suitably quiet waters by the simple process of downstream migration of the maturing transfer species.

Table 28. The world's largest rivers.

NAME	DISCHARGE (IN CUBIC METERS PER SECOND)	LENGTH (IN KILOMETERS)	CONTINENT
Amazon	200,000	6,280	South America
La Plata-Parana	78,000	3,900	South America
Congo	56,000	4,600	Africa
Yangtze	22,000	5,000	Asia
Ganges-Brahmaputra	20,000	2,900	Asia
Mississippi-Missouri	17,000	6,200	North America
Mekong	17,000	4,200	Asia
Mackenzie	13,000	4,000	North America
Nile	12,000	6,500	Africa
St. Lawrence	11,000	3,400	North America
Volga	10,000	3,700	Europe

NOTE: The volume of the Amazon nearly equals that of the sum of all the other large rivers combined.

Thus it can be seen that however perilous it may seem, geological processes do ensure the long-term persistence of freshwater habitats in the form of streams and rivers. So it is evidently in these habitats that ancient fishes have tended to persist. When lakes form, as after any of the Pleistocene glaciations (see Chapter 16), naturally there will be an explosive expansion of the range of ancient and other fishes; and when the lakes dry up, there will be an equally natural contraction of the ranges. Such cyclic expansions and contractions must have occurred constantly throughout geological time.

In parts of the earth's surface where the crust is restless, as in the Andes of South America, lakes such as Lake Titicaca, with its ancient and

Figure 46. Headwater cataract region of Tambur River, Himalayas, 3500 m., lower limit of *Pinus* (see also Chapter 16), sketched by Sir Joseph Hooker (31).

unique fauna, probably persisted as lakes for much longer, because the constant heaving of the crust will tend to create depressions as fast as they fill up with sediment. Indeed, recent measures of the rate of elevation of various mountain systems bordering the Pacific suggest values like 5000 feet of elevation per million years; the elevation is nearly always tilted, so lakes would inevitably form in the depressed parts. Titicaca may well have moved about as the tilts occurred, and could we see a time-lapse film of the Andes taken from outer space at the rate of, say, one frame every 100 years, we might well expect that such a speeded up film would depict Titicaca as rolling and sliding about from one bed to another, but always maintaining its integrity and a continuity of faunal and floral populations.

Table 29. The world's largest, highest, and deepest lakes.

NAME	VOLUME (IN BILLIONS OF CUBIC METERS)	AREA (IN SQUARE KILOMETERS)	MAXIMUM DEPTH		HEIGHT ABOVE SEA LEVEL		CONTINENT
			FATHOMS	KILOMETERS	FEET	KILOMETERS	
Caspian	88.0	440,000	536	0.98	−85	−0.03	Eurasia
Baikal	23.0	34,700	900	1.62	–	–	Asia
Superior	12.0	80,700	222	0.41	602	0.18	North America
Tanganyika	10.0	31,600	780	1.43	2,534	0.77	Africa
Nyassa	8.0	28,500	370	0.68	1,650	0.48	Africa
Michigan	5.7	58,000	153	0.28	580	0.17	North America
Huron	4.6	59,600	125	0.22	581	0.17	North America
Victoria	2.6	67,800	–	–	3,720	1.13	Africa
Ontario	1.7	19,500	130	0.24	247	0.07	North America
Aral	0.95	67,900	40	0.07	–	–	Asia
Titicaca	0.70	8,500	167	0.30	12,506	3.81	South America
Maracaibo	–	10,400	–	–	–	–	South America
Eyre	–	25,500	–	–	–	–	Australia

ISOLATED POND FAUNAS. Occasionally, especially in clay country with considerable topographic variation, a landslip occurs after heavy rain and produces a raised barrier along the side of a hill. Rainwater accumulates in the new hollow, and a pond or tarn results. At first the new water body is devoid of any fauna or flora; the accumulating rainwater will stagnate for a while, for the grasses will die and decay, and insects and earthworms formerly inhabiting the drowned region will also die unless they made their escape before the water accumulated. But this lifeless phase is followed within a few months by a total change. Waterweeds make their appearance in ever increasing variety, and by the end of the first year there is also a fauna of water snails and various aquatic crustaceans. How can such aquatic animals and plants appear, as it were, spontaneously, where none were before?

This question was studied by Hutchinson, a sheep-farmer naturalist whose land was especially subject to earthquake deformation and to the sudden appearance of new tarns on mountain sides. He found that almost certainly the water snails arrive on the feet of migrating birds such as duck and teal, which habitually seek out small bodies of water. Other aquatic animals are probably also transferred in this way, as Darwin speculated; and seeds and spores of aquatic (and nonaquatic) plants are apparently

Figure 47. Immature stream valley in tropical forest near Petropolis, Brazil; (*Mazama*) forest deer, or brocket (see Chapter 12) (9).

transferred when adhering to plumage, or by evacuation in feces after having been swallowed by the migrating birds at another location.

Eventually an isolated pond of this type acquires a vertebrate fauna. Tadpoles appear, evidently as the progeny of adult frogs that migrate through wet grass at night from other water bodies. Eels and catfishes of certain species leave their ponds if the food resources are exhausted and migrate by night over land, seeking new waters. In this way the first vertebrates occupy the new water body. After a much longer lapse of time a natural drainage pattern of tributary streams could develop, eventually leading to other fish-fauna transfers by headwater capture—though, unless the new pond is a large one, it is more likely to silt up before such an event could occur. Another immigration route can develop if the tarn overflows and forms a stream of egress; such a stream would inevitably flow into some preexisting stream, thereby providing, in wet weather at least, an aqueous route to the pond; crustaceans such as crayfish make use of shallow trickles of that kind in their migrations overland. So it is apparent that seemingly isolated bodies of fresh water are often much less isolated than appears at first sight.

THE EFFECT OF CROP DUSTING. A major disaster struck these natural regenerative processes with the advent of crop dusting. Insecticides killed not only the insects to which they were directed, but almost everything else. The rainwash from dusted hills flowed into streams and by them was delivered into lakes. Whole ecosystems were devastated and an unnatural overgrowth of plants was precipitated by the temporary accumulation of nitrogenous materials of animal origin delivered by bacteria acting on the dead invertebrates and fishes. To this was added in settled regions near cities the pollution of industrial waste chemicals and household wastes. These pollutants, if toxic, killed the surviving organisms and converted the wetlands to unpleasant stagnant wastes; or, if nutrient, promoted a gross overgrowth of plant life called *eutrophication*; the excessive

Figure 48. Mature lower valley with flood plain and backwaters, rapids in foreground, River Dee near Abergeldie, Scotland; *Pinus sylvestris* to left, with mixed birch (*Betula*) and willow (*Salix*) forest (see Chapter 16) (10).

plant life exhausted the oxygen at night (when plants respire in the same manner as animals) and led to further fish kills through lack of oxygen, or for lack of sufficient invertebrate grazers to maintain the ecosystem cycle. To what extent these changes are irreversible is not presently clear. But an ominous sign is the fact that Francis Hutchinson's collection of snail specimens, when taken to a leading museum some years after his death, proved to contain numerous species no longer found in the areas where he collected them. And for other invertebrates the story is similar; wholesale exterminations have occurred within the space of a few decades.

If problems of this kind are to be solved one tool needed in the study is careful documentation of what was present in particular places before and after twentieth-century pollution began to destroy the environment. In the unpublished notebooks of Victorian naturalists such as Hutchinson, and in the collections of plants and animals they made, we have such data. Though considered local and trivial at the time they were made, their observations can tell us much. By the same token, seemingly trivial notebooks kept by present-day observers of nature, recording what is seen and when and where it is seen, may not improbably come to acquire presently unforeseen importance to our successors. The rather numerous names of plants and animals that occur in this text reflect a view that the best field biology is always accompanied by a reasonably close attention to the actual species that occur in given environments; a more generalized approach leads to inaccurate assessments of ecosystems, and especially tends to overlook changes in the species or generic content of particular elements of an ecosystem.

FRESHWATER HABITATS. Two main types of habitat are obviously distinguishable, that related to streams and rivers, where water flows more or less constantly in one direction, and that related to lakes and ponds, where the water is more nearly stationary and such currents as occur tend to be wind-generated. The habitats are defined as follows:

1. *Lotic series*, also called *channel series*: those habitats characterized mainly by flowing water in rivers and streams. The subdivisions are (a) the *rapids*, or *headwater zone*, where the flow is swift, with intervening waterfalls, and confined to narrow, steep valleys V-shaped in section; (b) the *midvalley zone*, where pools and rapids occur at intervals; (c) the *flood-plain* zone, where the flow is slower and dispersed across broad mature valley floors; (d) *estuarine zone*, near the point of entry into the sea or lake.
2. *Lentic series*, also called *basin series*: those habitats characterized mainly by standing waters as in ponds and lakes, also swamps. The subdivisions are (a) the *littoral zone*, shallow marginal waters sufficiently illuminated for the growth of rooted plants; (b) the *epilimnion*, or *limnetic zone*, the open area of water beyond the littoral zone, occupied by plankton and nekton; (c) the *hypolimnion*, or *profundal zone*, the deeper cold region underlying the epilimnion and

commonly separated from it by a layer of water in which the temperature changes rapidly (the *thermoclime*). This zone may be lacking from shallow lakes and ponds.

CIRCULATION. In rivers and all lotic systems the water flows in one direction, passing successively through the various zones. So the biota change in an orderly sequence from the headwaters to the outlet, and particular communities characterize the successive parts of the circulation. Narrow-leaved and resistant plants, and strong-swimming, or demersal, animals are in the swift-flowing zones.

In lakes the water is circulated by the action of wind, and when a wind-generated flow in the epilimnion strikes the littoral boundary a vertical downflow is generated, with a consequent upflow on the other side of the lake. Strong temperature differences between the epilimnion and the hypolimnion tend to oppose the wind-driven circulation. Thus, lakes tend to be more markedly stratified in summer (when the epilimnion is heated by the sun) and in the winter (when the epilimnion may freeze over). Circulation is most marked in the spring and fall, and at these seasons shallow lakes may assume a uniform temperature *(homothermal state).*

Swamps may be viewed as part of the lentic series, often representing one-time lakes or ponds on which so much decayed vegetable matter has built up over the ages that most of the water has been lost through overflow. The circulation is here reduced to almost zero.

Phytoplankton

In lakes an analogous situation exists to that in the sea. The epilimnion carries a population of phytoplankton, which is one of the two major sources of energy for the lake ecosystem. In summer the phytoplankton is actively growing and producing organic materials. A shower of plankton debris falls to the bottom where the degraders convert it back to mineral nutrients. Thus during the summer plankton bloom, the surface waters become depleted of nutrients, and the bottom waters enriched. The spring and fall thermal adjustments permit the wind to perform an overturn of the waters, thereby renewing the capacity for production.

ALGAE. The groups that occur in the sea occur also in fresh water, except that Phaeophyta are lacking, and Rhodophyta are very rare. One additional phylum may be noted: phylum Cyanophyta (blue-green algae). These algae occur mainly in lakes and streams, though a few marine forms are known. A few also grow in hot springs (thermal algae) at temperatures up to 77° C. Some participate in the form of symbiosis with fungi that leads to the formation of lichens, though this occurs mainly in terrestrial locations. All Cyanophyta are of microscopic size, though these unicellular organisms may form colonies of considerable size on lake surfaces. The distinctive feature of these organisms is that the genetic material is scattered through the middle part of the cell, instead of being organized into a cell nucleus. Also the chlorophyll is scattered through the outer part of the cell, instead of being concentrated in plastids. The cells are often enclosed in gelatinous material. In addition to chlorophyll other pigments are present, notably the blue-green *phycocyanin*, and some species produce a red pigment. Food material is stored in the form of glycogen, a carbohydrate similar to starch. Some species have a fecal odor and can render fresh water undrinkable, either through its unpleasant smell or from actual toxins. Some species grow in soil and increase its fertility. Some species are collected and dried for food by man, especially in tropical countries, for example, *Spirulina platensis* in Africa.

Cyanophyta reproduce by simple fission, without sex cells. As might be expected for so simple an organization, Cyanophyta are extremely ancient organisms, known to occur in Proterozoic rocks as fossils up to 2.3 billion years old.

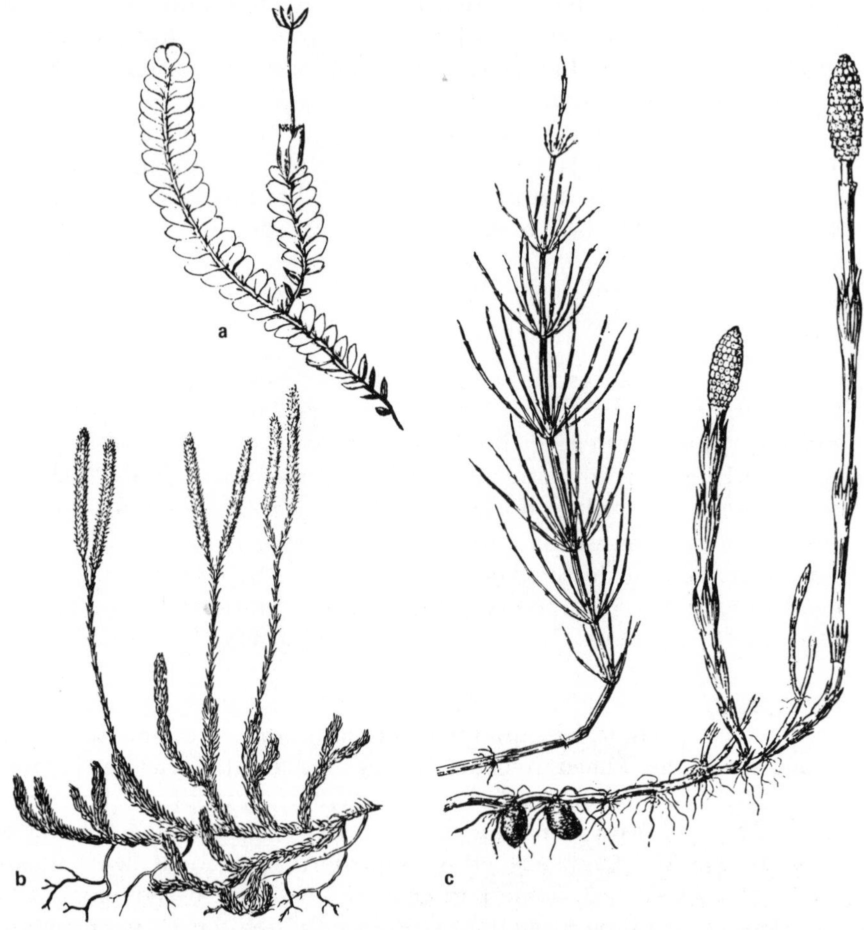

Figure 49. (a) *Plagiochila* (liverwort), X 0.6, Hepaticae; (b) *Lycopodium clavatum* (club moss), X 0.5, Lycopodineae; (c) *Equisetum* (horsetail rush), X 0.5, Equisetineae (41).

FLOATING PLANTS. Some members of the phytobenthos break away from their rooting substrate, float to the surface, and there continue to live as floating phytoplankton. In addition, some embryophytes live as permanent floating denizens of the epilimnion. These include the water ferns (order Hydropteridiales), small relatives of the true ferns, from which they differ in forming two types of spore instead of one type only. There are two families of water ferns: Salviniaceae, examples being *Azolla*, scaled plants, often tinted red, of American origin but now widely distributed in the warmer regions of the world, and forming communities at the surface resembling a mat; and *Salvinia*, the water shamrocks. There is also the family Lemnaceae, the duckweeds, scalelike monocotyledons a few millimeters across, forming dense floating mats in water bodies of all continents. In summer these floating plants grow so close together that the communities they form resemble the vegetation of the soil, so much so that unwary animals (such as frog-hunting carnivores) may leap onto a duckweed-covered pond and disappear beneath the green blanket, emerging some seconds later in obvious high dudgeon.

Zooplankton

In most rivers the water flows too fast for any significant amount of zooplankton to survive. In ponds and lakes a limited zooplankton develops, much less varied in systematic content than in the sea, largely because the marine benthic invertebrate phyla that have planktonic larvae are poorly represented in freshwater habitats. A detailed classification is not feasible here, but following are some of the commoner freshwater planktonic animals. Protozoa are well represented, though the forams are very rare, and radiolaria are lacking. Floating minute clusters of colorless or brown or green cells occur as resting stages of bottom-dwelling freshwater sponges, phylum Porifera. Sexual stages of the Coelenterata occur rarely in plankton, resembling microscopic jellyfishes; see the section on zoobenthos (p. 171). Small organisms a millimeter or so in length, with one or two rings of cilia at one end, and a multicellular body, are the so-called rotifers, forming a separate phylum Rotatoria; they are remarkably sensitive to the temperature range and acidity (*p*H value) of water, factors that therefore determine their distribution. The phylum Platyhelminthes, or flatworms, is commonly represented in freshwater plankton by larval stages that resemble microscopic skates, with a flattened anterior region and a lashing tail; these are called *cercaria* larvae and are infective stages of parasitic trematodes allied to the liverfluke. They are referred to under zoobenthos, to which some of their hosts belong. Other larvae of flatworms are the *coracidium* larvae, spherical minute balls of cells, ciliated, with three pairs of spicules at one pole; the adults of these are tapeworms, parasitic on fishes and on water birds. The young free-swimming stages of nematodes may also occur as plankton, minute S-shaped worms, also referred to under zoobenthos. The youngest stages of freshwater clams are free-swimming; they resemble miniature clams but have a tuft of cilia by which they can move in the water; they seek out the gills of freshwater fishes, on which they are temporarily parasitic before assuming life at the bottom as benthic organisms.

Of the zooplankton more easily seen by the naked eye, various members of the phylum Arthropoda may be noted. The larvae of insects are sometimes aquatic and free-swimming—for example, mosquito larvae (see zoobenthos, below). Among the crustacean arthropods of lakes, freshwater shrimps, water fleas, copepods, and mysids may be noted; the water fleas, which very much resemble fleas, are members of the crustacean order Cladocera; the others mentioned have already been noted as marine plankton.

Phytobenthos of the Supralittoral Zone

The plants that grow on the bottom, or that live on the soil forming the margins of lakes and rivers, constitute a rich and varied flora, with members of most of the main groups of higher plants taking part. Table 30 gives a summary of the systematic classification of the plants to be mentioned in the following sections.

Plants of the supralittoral zone include all the trees, shrubs, and herbs that line the banks of lakes and of rivers, plants that favor soils that are permanently damp, and where there is a high water table.

Bryophytes commonly occur on the ground, both liverworts and mosses, particularly the bog moss *Sphagnum*. Lycopods are occasionally present, such as *Selaginella* in humid climates, forming a thin creeping mat. The equisetes are commonly represented in Northern Hemisphere floras by species of *Equisetum*, which favor the outer, more exposed parts of a lake or pond where the water level is subject to seasonal fluctuation. These plants usually form a belt or clumps of vegetation about 30–40 cm high. They are lacking entirely from the Southern Hemisphere, save where they have been accidentally introduced by man. Ferns of the genus *Athyrium*, spleenwort, and the chain fern *Woodwardia* occur in damp situations in northern lands. In the Southern Hemisphere a fern commonly found lining the banks of bodies of water is *Blechnum fluviatile*, the fronds simply pinnate, that is to say, resembling small fronds of a palm.

Among the flowering plants, monocotyledons are usually conspicuous in bogs and swamps, others grow with part of the plant immersed in the water, and some grow around the margins of the littoral and in the supralittoral zone. Notable here are the rushes (family Juncaceae), type genus *Juncus*. These favor damp soil rather than water. They resemble grasses but have a cylindrical stem at the tip of which small clusters of florets form in spring and early summer. In tropical countries members of the family Araceae, for example, Talo or Taro, of Indonesia and Polynesia *(Colocasia esculentum)* favor damp ground in the vicinity of water.

The dicotyledons characteristic of these situations in the Northern Hemisphere include a number of genera of trees and shrubs, notably: (1)

Table 30. The classification of aquatic plants.

Subkingdom Thallophyta: Plants without leaves, roots, or vessels and in which sex cells are released when they are formed. Freshwater examples include diatoms, green algae, blue-green algae.

Subkingdom Embryophyta: plants with leaves or leaflike organs of nutrition, and in which the sex cells are retained within the sex organ of the female after fertilization until the embryo is developed. There are two main divisions, or phyla:

Phylum Bryophyta: small plants, usually less than 10 cm high, of creeping or tufted habit, with leaflike scales but no woody stem, no flowers, no roots. Threadlike *rhizoids* attach the plant to the substrate, but do not contain vessels, so do not function as roots. There are 2 common freshwater classes:

(1) Class Musci: mosses, tufted mat-forming bryophytes, mostly in damp places.

(2) Class Hepaticae: liverworts, creeping flat expansions of seaweedlike tissue (thallus) found on wet or damp ground, or near waterfalls in the spray, and in caves, sometimes bearing erect portions of the thallus.

Phylum Tracheophyta: small herbs to large trees, having well-differentiated leaves, stems, and roots; internal transporting vessels for sap and water (hence the term *vascular plants*). There are 5 main classes:

Classes in which minute unicellular *spores* serve as the dispersal stages. These are:

(1) Class Lycopodineae: lycopods, or club mosses, mosslike herbs (ancient types were like trees)

(2) Class Equisetineae: horsetail rushes, herbs with side stems arranged in rings at intervals along the main erect stems (ancient types were as large as trees).

(3) Class Filicineae: ferns and tree ferns, ranging from herbs to tall palmlike trees.

Classes in which large *seeds* with food reserves serve as the dispersal stages. These are:

(4) Class Gymnospermae: the needle-leaved conifers and related trees.

(5) Class Angiospermae: flowering plants, ranging from herbs to tall trees. There are two subclasses: (a) Monocotyledoneae, with parallel leaf veins (for example, rushes, sedges, grasses), and (b) Dicotyledoneae, with net-veined leaves, aquatic examples being water lilies, watercress, willows, elders, bog oaks, swamp mahoganies, mangroves.

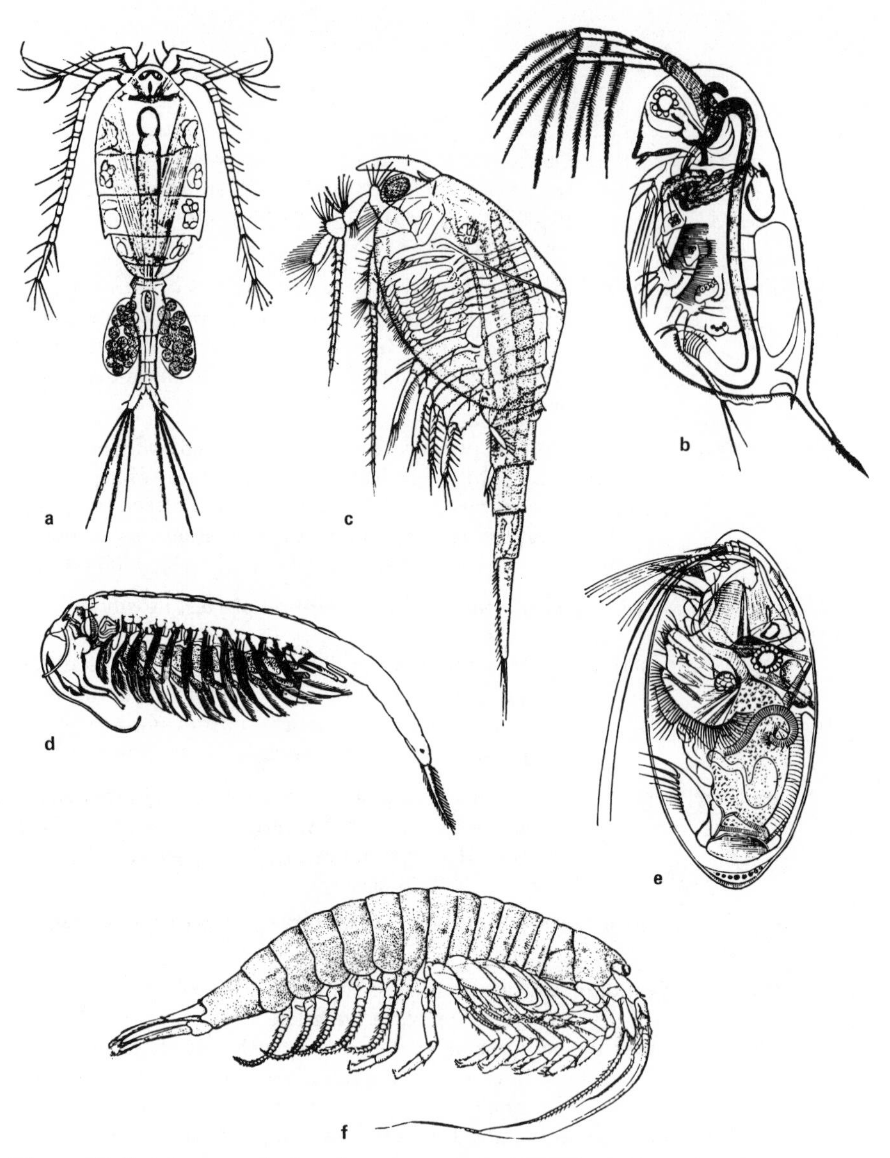

Figure 50. Freshwater planktonic crustaceans. (a) *Cyclops*, X 20, female with egg sacs, Copepoda (15); (b) *Daphnia* (waterflea), X 10, Branchiopoda (15); (c) *Nebalia* and (d) *Branchipus*, X 20, Phyllocarida; (e) *Cypris*, X 10, Ostracoda; *c–e* (15); (f) *Anaspides*, X 2, ancient form surviving in Tasmanian streams, Anaspidacea (29).

willows (genus *Salix*) and cottonwoods (genus *Populus*) both of the family Salicaceae; these trees frequently have their roots in the water, where they help to prevent erosion. (2) Birches (genus *Betula*, family Betulaceae), especially in northern North America the canoe, or paper, birch *(B. papyrifera)*, and in the southern states river birch *(B. nigra)*; birches are favored food and building materials for the beaver. (3) Alders (genus *Alnus*); These are mainly shrubs; though in Arizona and Mexico the alders are trees; they are closely related to birches and are classified in the same family Betulaceae. (4) Among the maples the red maple (*Acer rubrum*, family Aceraceae) is a water-loving species. (5) Elders (genus *Sambucus*, family Caprifoliaceae) include shrubs and small trees that favor damp soil and lake or river banks; some species live in seasonally dry water courses, in this respect recalling the Southern Hemisphere trees next noted.

In the Southern Hemisphere the foregoing groups of trees are absent (except as introductions by man). Ecological equivalents include trees such as river she-oak (*Casuarina cunninghamiana*, family Casuarinaceae), the graceful drooping foliage of which somewhat recalls the weeping willow. The genus occurs in Africa and Indonesia, but is mainly Australian. She-oaks are among the trees that line the seasonally dry water courses called billabongs. Coolibah, or river gum *(Eucalyptus coolibah)*, and also a giant river gum *(E. camaldulensis)* reaching 24–30 m high (80–100 feet) of the family Myrtaceae, have the same habit. *Callistemon*, or bottle-brush, is a moisture-loving shrub of the same family. *Tristania suaveolens*, the so-called swamp mahogany of the Queensland coasts, occurs in swamps and in the neighborhood of fresh water. It is not a true mahogany, but a member of the Myrtaceae, like many other Australian trees.

In the Northern Hemisphere gymnosperms are represented by members of the family Pinaceae such as tamarack or larch (genus *Larix*), black spruce (*Picea mariana*), and arborvitae (*Thuja occidentalis* a member of the Cypress family, Cupressaceae); all tend to grow in cold, moist locations such as bogs in the southern part of their range, whereas they occupy higher ground in the northern parts of their range. A somewhat analogous distribution occurs in the Podocarpaceae of the Southern Hemisphere, with some genera such as *Dacrydium* favoring lowland swamps and the sides of water courses as well as open ground in higher or colder regions, the individual species varying as to habitat.

Phytobenthos of the Littoral Zone

Unlike the trees of the supralittoral zone, the phytobenthos shows much less regional differentiation, the plants of the freshwater lakes and streams having a much more uniform character throughout the world. The plants are more abundant and more luxuriant in the tropics but tend to belong to the same families and genera as in cooler regions. Ecologists distinguish three main subzones:

1. EMERGENTS. Emergents are plants of the phytobenthos that are rooted under water and that grow so tall as to have much or most of the plant body exposed to the air. These include the best-known aquatic plants because they are so conspicuous beside lakes and rivers, and the graceful lines and reflections that they offer have made them favorite subjects for painters and poets since the days of the Pharaohs. They include

equisetes such as *Equisetum fluviatile* and *E. littoralis*, and also some well-marked families of monocotyledons, especially the following:

(a) Cattail reeds, or bulrushes (family Typhaceae), with a virtually cosmopolitan species, *Typha angustifolia*, known in North America as cattail reed, in British countries as bulrush, and in New Zealand as kaupo. It is easily recognizable by its tall cylindrical flowering stem up to 2 m high, surmounted by a cylindrical spike of densely crowded florets near the tip, the seeds subsequently dispersing with the aid of silky threads. The slender leaves arise from the base and are as tall as the stem. These plants form dense communities in shallow swampy areas around the margins of lakes. Their pollen and buds are eaten by man in some countries.

(b) Sedges (family Cyperaceae). Sedges are grasslike plants but the stems are solid, rather triangular in cross section, and they lack the joints (nodes) of grasses. The flowers are inconspicuous, at the tips of stems where an umbrellalike cluster of leaves may form. The most famous sedge is the papyrus plant of the Nile *(Cyperus papyrus)*, which grows 2 m high. Smaller similar species include *Cyperus alternifolius* from Madagascar, commonly grown as a house plant. In North America the name bulrush has been transferred to the sedges of the genus *Scirpus*.

(c) Reeds (family Gramineae, genus *Phragmites*). Reeds are tall grasses up to 2–3 m high, with conspicuous tufted flower heads on hollow stems rounded in cross section, suitable therefore for making primitive Panpipes. More terrestrial relatives, also called reeds, are the genera *Arundo* and *Cordaiteria*, variously known as prairie or pampas grasses, often in swamps. Reeds, like other grasses, have jointed stems. In North America reeds are commonly called bamboo in some districts, but this usage is incorrect. Rice *(Oryza sativa)*, like the reeds, is also a member of the grass

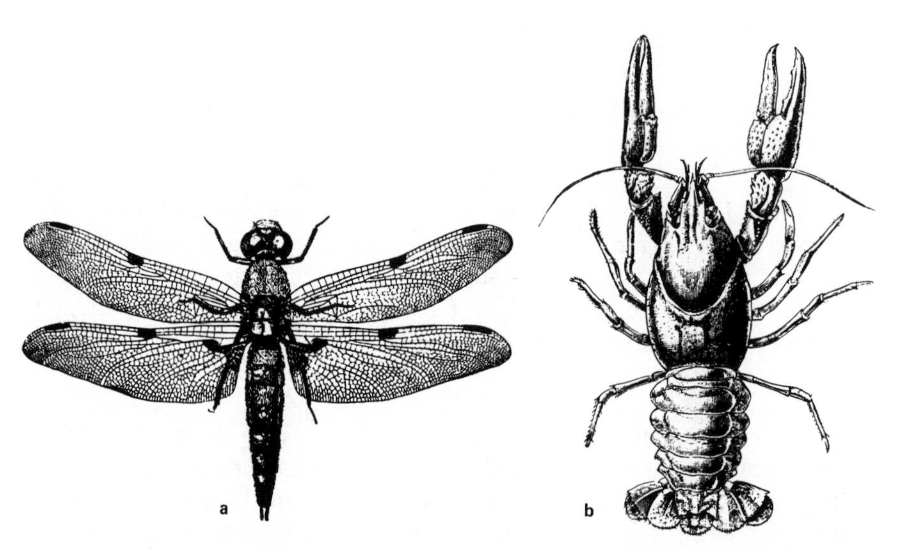

Figure 51. (a) *Libellula* (dragonfly), X 1, Odonata; (b) *Astacus* (freshwater crawfish), X 0.5, Decapoda (25).

family Gramineae. Believed originally to have been a native to southern India, it is now grown throughout the tropical and warm temperate regions of the world. Only the forms known as padi rice require the water to cover the roots.

(d) Arums (family Araceae). These are aquatic and marsh plants. The arrow arum of North America (*Peltandra virginica*) has arrow-head shaped leaves and a spike of small densely clustered flowers enclosed in a narrow green cup called a spathe.

(e) Arrowheads (family Alismataceae, *Sagittaria* spp.), also having arrowhead-shaped leaves, but with an ascending stem bearing white whorls of flowers. A giant species in the Southern Hemisphere, *Sagittaria montevidensis*, grows to 2 m high.

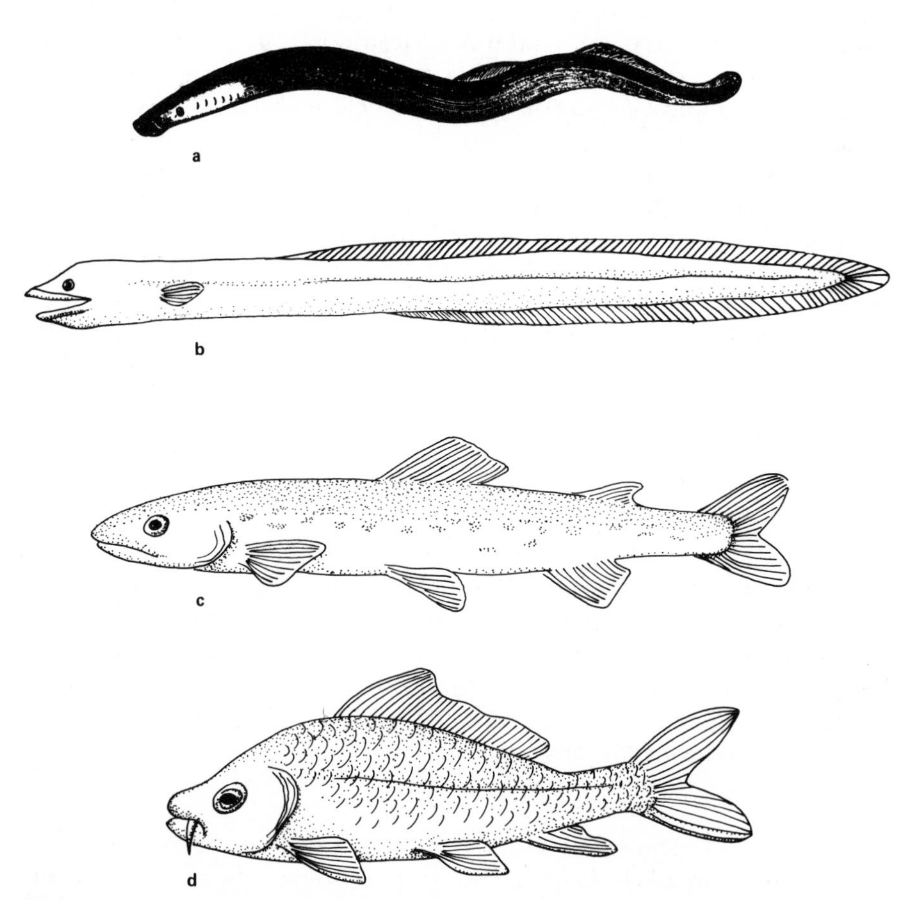

Figure 52. (a) *Petromyzon fluviatilis* (freshwater lamprey), X 0.1, Europe, Cyclostomata (15); (b) *Anguilla dieffenbachii* (eel, or Maori tuna), X 0.1, New Zealand, Anguilliformes (23); (c) *Salmo* (trout), X 0.1, Northern Hemisphere, Salmonidae (23); (d) *Cyprinus* (carp), X 0.5, Northern Hemisphere, Cypriniformes (23).

2. THE FLOATING LEAF SUBZONE. Phytobenthos are rooted on the bottom and have the entire plant body submerged, the leaves alone reaching the surface, where they lie either in the air-water interface or floating on the interface with a more or less upturned margin to prevent flooding of the leaf when water birds walk across them. Conspicuous members of this community are the water lilies and water hyacinths. As in the case of the emergents, the plants of this subzone also have a rare beauty in the aqueous setting, and they too have evoked the wonder of poets and musicians. To the Greeks the water nymph personified a water lily; on the somber tarns of the Norwegian highlands *Nymphaea* elicited the muse of Henrik Ibsen and Edvard Grieg. The name lotus was applied by various classical writers to different plants, one of which was an Egyptian water lily.

(a) Family Nymphaeaceae, principal genus *Nymphaea*, with about 40 species. The tropical species are more brilliantly colored, and although perennials (like all water lilies) they can be cultivated as annuals in temperate countries. The giant South American water lily *(Victoria regia)* develops leaves up to 2 m in diameter. A large Asian species is *Euryale ferox*, occurring in tropical rivers. The genus *Nelumbo* (subclass Dicotyledoneae) yields edible tubers.

(b) Family Potamogetonaceae. Other water plants in this ecological grouping belong to the families Ruppiaceae, Potamogetonaceae, and Zanichelliaceae, all of which are combined as the one family Potamogetonaceae by some botanists. Their members are widely distributed. The water hyacinth (*Eichhornia*) is native to South America but has been introduced into Florida, where it is spreading widely and obstructing the passage of some rivers; the plant bears flotation bladders that support the heavy clusters of floating leaves. *Cabomba* is a feathery-leaved relative of the water lilies.

3. THE SUBMERGED SUBZONE. This is the deepest part of the littoral zone, where the plants are wholly submerged. Among the groups represented are green algae (Chlorophyta), with the genera *Chara* and *Nitella*. Mosses include the genus *Fontinalis*, which occurs on the bottom in the deepest part of the zone. Various monocotyledons occur, notably the eelgrasses (family Hydrocharitaceae), with genera such as *Hydrocharis*, *Elodea (Anacharis)*, *Stratiotes*, *Valisneria* (eelgrass). These plants are commonly seen in freshwater aquaria.

LOTIC PHYTOBENTHOS. Most of the plant formations mentioned above are likely to occur in the appropriate circumstances in connection with both lentic and lotic systems. Where a river has extensive—or even restricted—backwaters, conditions simulate those of ponds and the plants are similar. The littoral zone of a sluggish river is essentially like that of a lake or pond. Where, however, a stream is swift-flowing it is obvious that there can be no permanent floating plankton, and the turbu-

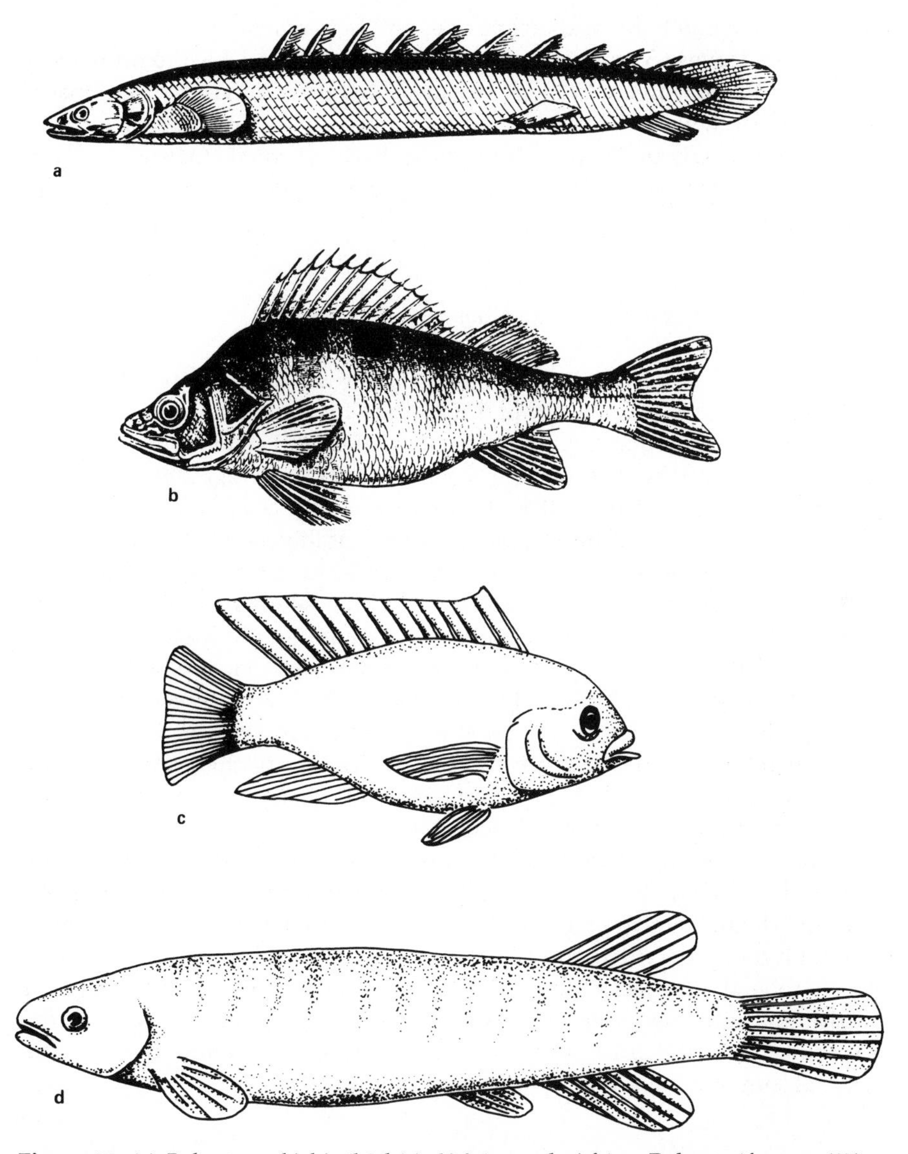

Figure 53. (a) *Polypterus bichir* (bichir), X 0.1, north Africa, Polypteriformes (15); (b) *Perca fluviatilis*, X 0.5, Eurasia, Perciformes (15, from Cuvier); (b) *Tilapia* nilotica (African cichlid), X 0.5, as painted by a Nile artist, 2700 B.C., Perciformes (2); (c) *Galaxias*, X 0.5, New Zealand (also Australia, South America), Galaxiidae (23).

lence will also tend to destroy the fragile leaves of many of the phytobenthos species that occupy standing waters. In turbulent situations the plants tend to be resistant, wiry forms, or encrusting algae, or filamentous forms whose leaves will not be broken by the current. In the Chilean Andes *Nierembergia rivularis* is an aquatic member of the characteristic South American potato family Solanaceae; in response to its unusual habitat it has adopted a mat habit and creeps over and between stones; it has narrow dark green leaves. Some related species have adapted for life on the Argentinian Pampa. Even in swift streams, however, inevitable slow backwaters occur where the stream takes a bend, or where it has recently abandoned a part of its bed; here grow beds of musk *(Mimulus)*, mint *(Mentha)*, watercress *(Nasturtium)*, and other genera encountered in lentic series.

Freshwater Zoobenthos

Most benthic animals in freshwater environments are invertebrates. The phyla shown in Table 4 for marine environments occur, with certain exceptions, also in freshwater lakes and rivers. The phyla that *do not* range into fresh water are the Echinodermata together with some minor phyla. In the cases of certain phyla, some classes are absent, although other classes of the phyla are present, notable absentees from fresh water including the anemones and jellyfishes among Coelenterata, and the octopuses and squids among Mollusca. Since lakes and ponds tend to accumulate silt, brought in by tributary streams, the bottom is generally soft and silty; this means that many of the hard-bottom sessile (attached) animals of marine environments cannot inhabit freshwater environments because the habitat is unsuitable. Marine organisms lacking any physiological mechanism for maintaining the salt content of the body fluids are also excluded, and probably echinoderms are absent for this reason, no kidneylike organs occurring in those animals.

Protozoa similar to those of the plankton occur in the benthos. One family, the Euglenidae, is of interest in that some of its members possess chlorophyll and thus are autotrophs (and often are classified as unicellular Chlorophyta by botanists therefore), whereas certain other members, very closely related, may lack chlorophyll and have therefore a holozoic mode of nutrition, capturing other organisms or feeding on their dead remains (saprophytic nutrition). Animals such as these may stand close to the original common ancestry of plants and animals.

PORIFERA. Freshwater sponges are few in kinds and not usually conspicuous elements of any fauna. They occur in rivers as well as lakes. *Spongilla* is a widely distributed example. Freshwater sponges form resting stages in unfavorable seasons, namely, winter in temperate regions, and also in monsoon regions when the rivers dry up during the dry winter. The spicules of freshwater sponges resemble the spool of a roll of film and are siliceous. Symbiotic algae occur in the tissues.

COELENTERATA. The only freshwater representatives are solitary hydrozoans. These animals either have a simple life history, with only one stage (the polyp) as in *Hydra*, or they may have a complex life

Figure 54. Oldest known record of salmon (*Salmo*). A herd of deer (*Cervus elaphus*) apparently crossing a stream, engraved on antler by a Palaeolithic artist of Lorthet, Hautes-Pyrenees, about 8000 B.C. (1). Compare also Figure 141.

history, as in *Microhydra*, where the polyp stage inhabits the benthos, and a free-swimming medusoid occurs as a planktonic phase (originally described as a distinct genus *Craspedacusta*).

PLATYHELMINTHES. These are the flatworms, members of which occur also in the sea and on land. The free-living members are placed in the class Turbellaria, often spoken of as planarians, and used in growth-gradient experiments. The other members mentioned, Cestoda and Trematoda, are parasites in other animals. Their larval stages are free-living and therefore turn up in freshwater samples. The coracidium larva grows into a tapeworm, inside its host. The cercaria stage of the trematodes is a temporary free-swimming stage that seeks out a suitable host animal and then bores into its skin. The infection called swimmer's itch arises when a cercaria attacks a human being instead of its correct host (often a fish); finding the physical conditions in man unsuitable, the larva remains in the skin and travels a sinuous tunnel, marked by a red line in the skin of the wrong host. Some trematodes have man as a natural host, and such species can be dangerous to life if they enter vital organs. The disease schistosomiasis is an example.

NEMATODA. These are the roundworms, many of which are free-living in water or soil, others parasitize plants, and others parasitize animals. All can be recognized by their characteristic movements, which consist of rapidly flexing the body into an S-shape, and then abruptly reversing the S-bends into a mirror image of the first position, these two positions alternating rapidly. Hookworm disease and elephantiasis due to filariasis (Filaria infection) are examples of nematode parasitism of man.

ANNELIDA. Two classes are encountered here that are common in fresh water but less so in marine habitats. The leeches (class Hirudinea) spend much time as either free-swimming or bottom-dwelling animals, attaching themselves periodically by suckers to hosts whose blood they suck. *Hirudo* is a common freshwater genus, formerly used for blood-letting by old-time physicians, who were called leeches in Saxon times. The other class, Oligochaeta, comprises the earthworms and their smaller freshwater relatives, often found in the mud of lakes. It is believed that the Hirudinea evolved from Oligochaeta.

MOLLUSCA. The freshwater members are water snails and clams. The numerous kinds of freshwater snails are mainly herbivorous. Well-known genera include *Limnaea*, the pond snail, which is also an intermediate host for trematodes parasitic on vertebrates; *Planorbis*, the ram's horn snail used in aquaria; *Ampullaria*, the large apple snail of tropical rivers and lakes. The freshwater bivalves are mainly infaunal mud-dwellers. *Unio* is an example.

CRUSTACEA. These are common and important members of fresh waters. *Mysis relicta* occurs in brackish seas such as the Baltic, and also as a relict of the ice ages in various freshwater lakes of Europe and North America. Its marine relatives were noted earlier. Of the other groups, *Daphnia*, the water flea, exemplifies the Cladocera; it is preyed on by *Hydra*. Ostracods, which also occur in the sea, resemble miniature clams when their bivalved carapace is closed. The Anostraca include the fairy shrimps and brine shrimps. These are very small forms, which not only endure, but actually require, a period of desiccation, and when rain comes, or ice melts to form a temporary pond, the eggs hatch out and produce mature individuals in a few weeks. For example, *Branchionecta* was collected from high-altitude pools in Montana on September 6, only seven days after rain filled the dry glacial hole on August 30, at 8000 feet. The freshwater copepods are exemplified by *Cyclops*, which swims in short jerks, the female trailing paired egg sacs. The freshwater Decapoda include crabs, for example, *Hymenosoma*, and various lobster-shaped crayfishes, such as *Astacus* in the Northern Hemisphere and *Paranephrops* in the Southern Hemisphere.

INSECTA. These are primarily terrestrial anthropods, but many have become adapted to spending part of the life history in water.

The larval stages on the bottom of lakes and rivers, and the adults flying above the water surface, form a significant part of the diet of numerous freshwater fishes and other vertebrates such as frogs. Orders that have aquatic representatives include: (1) *Hemiptera*, bugs with suctorial mouthparts; these are predators on other aquatic invertebrates; *Corixa*, the water boatman, is an example. (2) *Odonata*, dragonflies, large glider-shaped insects with 2 pairs of long similar net-veined wings, and with predatory aquatic larvae; *Libellula* is an example. (3) *Ephemerida*, May flies, with 2 pairs of dissimilar wings and 2 or 3 threadlike cerci on the tail; they are eaten by fishes. (4) *Trichoptera*, caddis flies, with aquatic larvae that construct cases from sand or twigs; the adults have 2 pairs of wings covered by silky hairs. (5) *Coleoptera*, water beetles; *Dytiscus* is an example. (6) *Diptera*, flies and mosquitos with aquatic larvae; *Anopheles* transmits the protozoan parasite *Plasmodium*, the causal agent of malaria.

Freshwater Nekton

Vertebrates make up the nekton of freshwater environments, and the same classes of fishes and air-breathing vertebrates occur as in marine habitats, with the addition of the Amphibia, the frogs and salamanders that are lacking from saltwater faunas. On account of the reduced benthos in freshwater habitats as compared with marine, the nektonic predators are obliged to prey more on one another.

Of the groups other than fishes, the birds of freshwater environments are for the most part the same groups as have been referred to in Chapter 7, in the marine context, so they are not further discussed here. The mammals, though relatively more important in freshwater than in marine contexts, belong for the most part to groups essentially related to terrestrial environments, and so they are mentioned in later chapters. The principal ordinal groups of fishes and their diagnostic characters are set out in Table 31.

LAMPREYS [CLASS CYCLOSTOMATA]. These eel-like animals are distinguished from true fishes by their lack of paired fins and lack of jaws, the mouth being suctorial. They are related to the marine hagfishes, from which they differ in having no barbels around the mouth and in having an indirect mode of development in which a larval stage called *ammocoete* occurs. The ammocoete much resembles the primitive protochordate called *Branchiostoma* and develops from the eggs that are laid in the upper tributaries of streams. As the ammocoete slowly develops into a lamprey it makes its way downstream toward the sea, where the adult life is commonly passed. However, large lakes such as the Great Lakes of North America may serve as the adult habitat instead of the sea. The adults are active predators, attacking trout and other fishes. In the Northern Hemisphere *Petromyzon* is a representative genus, and *Geotrea* occurs in all the

Table 31. Diagnostic characters for identifying the principal orders of freshwater fishes.

DIAGNOSTIC CHARACTERS	COMMON NAME	CLASS OR ORDER
Body elongate, eellike, lacking paired fins, and having a circular suctorial mouth	Lampreys	Class Cyclostomata
Body shape that of a typical fish, with paired fins and upper and lower jaws; skeleton cartilaginous, without bone	Sharks and related forms	Class Chondrichthyes
Body shape that of a typical fish, or eel-like, with upper and lower jaws, and usually with paired fins well developed; skeleton of bone	Bony fishes	Class Osteichithyes
Pectoral and pelvic fins widely separated and set low on body		
The plane of the paired fins horizontal; rows of bony plates along sides of the body; tail with upper lobe longest	Sturgeons	Order Acipenseriformes
Not so		
Dorsal fin subdivided into numerous finlets	Bichirs	Order Polypteriformes
Dorsal fin not subidvided into numerous finelts		
Dorsal fin entire, reaching along most of back region	Bowfins	Order Amiiformes
Dorsal fin short, located about midway along back		
Weberian ossicles present	Carps	Order Cypriniformes
No Weberian ossicles	Herringlike fishes	Order Clupeiformes
Dorsal fin placed far back, near tail fin	Garpikes	Order Lepisosteiformes
Pelvic fins well developed; numerous bone fin spines in fins	Perches	Order Perciformes
Pelvic fins small; 3–4 erect spines precede the dorsal fin	Sticklebacks	Order Gasterosteiformes
Pelvic fins absent, caudal, dorsal, and ventral fins all united	Eels	Order Anguilliformes
Pectoral fins attached vertically to body, widely separated from pelvics	Minnows	Order Cyprinodontiformes
Functional lungs present; nostrils with internal connexion to mouth	Lungfish	Superorder Dipnoi

southern continents. The discontinuous distribution of *Geotrea* was employed as a biogeographical argument in favor of the hypothesis that the southern continents were formerly united; but when the marine adult stages were found to roam the southern ocean, even as far south as Antarctica, it was realized that the supposed discontinuous distribution was a fallacy. Before the cutting of the Niagara canal, *Petromyzon* was lacking from the Great Lakes. After the lamprey gained access to the lakes, a catastrophic collapse of the freshwater fisheries of the Great Lakes followed, on account of the severe predation exerted by the lampreys. Lampreys are very ancient types of animal, with an ancestry apparently going back to early Paleozoic times, about 400 million years ago. Their astonishing effect on the Great Lakes fishery is a demonstration of how successful these animals can be and helps us to understand their remarkable persistence through time.

SHARKS [CLASS CHONDRICHTHYES]. A few sharks occur in lakes and rivers, for example, in India, *Carcharias gangeticus*, and in Lake Nicaragua, *Carcharias nicaraguensis*. Radio-tagging has recently established the fact that the Nicaraguan species is able to return to the Caribbean Sea, apparently by means of the river that flows from the lake. The related sawfish *(Pristis pristiurus)* may also enter fresh water. In Australia marine sharks occasionally swim far up rivers to attack people and animals on river banks hundreds of miles from the sea.

BICHIRS [ORDER POLYPTERIFORMES]. The order is confined to tropical Africa and comprises the genera *Polypterus* and *Calamoichthys*, readily recognizable by the subdivision of the dorsal fin into numerous finlets. *Polypterus bichir* ranges the Nile and Lake Rudolf; *Calamoichthys* with other species of *Polypterus* ranges the Niger and Congo rivers. These are ancient fishes and have the thickened enamel-covered scales (ganoid scales) of Paleozoic fishes.

STURGEONS [ORDER ACIPENSERIFORMES]. The young stages of these fishes are passed in freshwater inland localities, the adult stages in the sea; see page 114.

GARPIKES [ORDER LEPISOSTEIFORMES]. With the one surviving genus *Lepisosteus*, garpikes are confined to North and Central America. Distinguished by the alligatorlike snout, ganoid scales, and the dorsal fin set very far back, near the tail. These are predators that tolerate a variety of environmental conditions and are very tenacious of life under poor conditions. As living fossils of great scientific interest, it is to be hoped their survival resources include the means to overcome man-made toxicity and impoundment in the waterways of the only continent where they now exist.

Figure 55. Preening swan (*Cygnus*) observed by Theodotos of Klazomenae in the wetlands of Ephesus and engraved on silver coin, fourth century B.C.

BOWFINS [ORDER AMIIFORMES]. With the single surviving species *Amia calva*, bowfins are found only in the Mississippi Basin and in Lakes Huron and Erie, the sole remnant of formerly widespread Mesozoic fishes of the same order. Under recent conditions of environmental deterioration due to pollution and dam building, the future of the species has become very uncertain.

EELS [ORDER ANGUILLIFORMES]. As already noted (Chapter 7), eels breed in the deep sea, and the newly hatched young slowly make their way back to the coast, where they enter rivers. The immature and near-adult stages are passed far from the coast therefore. Eels, like catfishes of some genera, have the power to breathe on land for a limited period of time; at night, if they have exhausted the available prey in their pond or stream, eels will migrate overland to other water bodies. Thus newly formed ponds, once an invertebrate fauna has been established by the visits of transporting agencies such as migratory waterfowl, are apt next to acquire an eel fauna. The mature eels migrate downstream to the coast and thence out to the deep regions of the sea where they reproduce.

HERRINGLIKE FISHES [ORDER CLUPEIFORMES]. Though mainly marine, they have contributed some families to freshwater habitats.

The true herrings (family Clupeidae) are represented by such species as the alewife *(Pomolobus)*. Smelt (family Osmeridae, genus *Osmerus*) are marine fishes that often occur in freshwater habitats.

Salmon and trout (genus *Salmo*, family Salmonidae) are typically Northern Hemisphere fishes; the diagnostic structural feature is a second or posterior dorsal fin that lacks supporting rays, the so-called adipose fin. Fishes of this group commonly spend the adult years either at sea or in some large body of fresh water, and then migrate up rivers in the breeding year; the sexual products are shed in the water in the headwaters, and then the parents die. The young slowly make their way downstream and ultimately return to the sea or enter a lake. Related species have similar life histories.

Some Southern Hemisphere fishes of the family Galaxiidae seem to represent the Salmonidae. They have a similar life history, and consequently the genera and species have very wide distributions, with young stages occurring in South America, Australia, and New Zealand. This dispersal pattern was formerly used to support the theory of a onetime united southern continent, or Gondwanaland, but this was before the marine phases in the life history were known. The present-day proponent of the continental drift hypothesis bases his deductions on strictly geophysical evidence and, if he takes account of biological data, will adopt the position that the unification of the southern continents was an event so far back in time as to have no influence on the dispersal of freshwater fishes on the present epoch.

The family Osteoglossidae, with representatives such as the giant *Arapaima gigas* in South America, *Scleropages* in Australia, and *Heterotis* in Africa, seems to be ancient. The members have robust bony scales like Paleozoic fishes. This family, too, has been cited in support of continental drift theories, but this must be viewed with skepticism.

CARPLIKE FISHES [ORDER CYPRINIFORMES]. Members have similar external characters to those of the herring order, as shown on Table 31. They have, however, a distinctive internal feature, namely, the presence of a chain of bones called Weberian ossicles, between the inner ear and the swim bladder. It is believed that sound vibrations in the water are picked up and amplified by the taut bladder (serving as a resonant chamber) and so transmitted to the ear. The order is exclusively freshwater in distribution, having no marine members. There are very many families, of which the following are some examples.

Carp in the strict sense, family Cyprinidae, include such well-known members as the goldfish *(Carassius auratus)*, bred as an aquarium fish in China at least since A.D. 970. Carp are exclusively Northern Hemisphere forms, though man has introduced them into the southern continents, where they have become feral.

Several groups have developed electric organs, that is to say, modified muscular tissue capable of storing an electric charge in much the same way as in a battery, and then releasing the charge suddenly for the capture of prey, or the discouragement of enemies. The electric catfish of Africa *(Malapterurus electricus)* is a voracious night-hunting fish found in the northern rivers of that continent, placed in a separate family Malapteruridae. Ancient Egyptian tomb paintings from about 2000 B.C. depict the species. The family Gymnotidae of South America, with the species *Gymnotus electricus*, is another where electric organs are well developed. The family includes also the banded knife fish *(G. carapo)*, a fish that hides by day and hunts by night, having feebly developed electric organs.

The catfishes comprise a number of families of the order in which sensitive feelers or barbels hang from the lips, used as sensory organs to detect the presence of prey at the bottom. In the Great Lakes region of North America occurs the family Ameiuridae, with a body shaped like a tadpole, an example being the bullhead catfish *(Ameiura nebulosus)*. The family Clariidae, ranging from India to Indonesia, includes the walking catfish *(Clarias batrachus)*. As the name suggests, the fish can migrate overland by night from one pond to another. An albino form of the species has been accidentally introduced into Florida. The gill filaments of these fishes can remain damp and functional in humid air for periods of several hours. The family Callichthyidae comprises heavily armored catfishes with bony plates developed on the flanks, overlapping like tiles on a roof. The barbels occur on the upper jaw. The bronze catfish *(Corydoras aeneus)*, of South American streams from Venezuela to La Plata, is an example sometimes seen in aquaria.

The piranhas of the family Characidae are notorious South American fishes of a highly aggressive temperament, carnivores hunting in schools and sometimes attacking land mammals, including man, if they enter fresh water. *Rooseveltiella nattereri*, a red and black fish of oval outline about 10 inches in length is an example from the Amazon and Parana rivers. The family also has representatives in Africa, a fact sometimes cited as evidence of continental drift.

TOP MINNOWS [ORDER CYPRINODONTIFORMES]. These fishes are characterized by having the pectoral fins attached vertically to the body and widely separated from the pelvic fins. The family Cyprinodontidae includes *Fundulus* of North and Central America, sometimes called killifish. These fishes are useful in controlling mosquitos, which they eat in the aquatic larval stage. The family is restricted to North and South America, Africa, and Asia, lacking therefore from Australia and New Zealand. Its distribution vitiates evidence that freshwater fishes owe their dispersion to continental drift.

PERCHES [ORDER PERCIFORMES]. These fishes are easily identified by the conspicuous and numerous bony fin spines and the strong development of the pelvic fins (see Table 31). The order has already been mentioned under marine nekton in Chapter 7. A number of freshwater families occur, of which the following may be mentioned as examples.

Perches in the strict sense, that is to say, members of the family Percidae, have two well-developed dorsal fins. They are exclusively Northern Hemisphere temperate forms, with representatives in the rivers and lakes of eastern North America, Europe, and western and central Asia. The yellow perch *(Perca flavescens)* is a well-known American example.

The sunfishes, or freshwater bass, make up the family Centrarchidae, found only in eastern and central North America, where they are very common. These are perches with a strongly compressed body (that is, a body flattened in the vertical plane), and with an oval or rounded outline. Among well-known examples are the blue-gill sunfish *(Lepomis macrochirus)* and the large-mouth bass *(Micropterus salmoides).*

Two Oriental genera are *Betta,* the Siamese fighting fish (family Anabantidae), and *Toxotes,* the archer fish of Indonesia and Australia (family Toxotidae), fishes that catch insects by squirting drops of water at them.

Members of the family Cichlidae differ from other perches in having only one pair of nostrils. Most fishes have two nostrils on either side of the head, one member of each pair being used to take water into the nasal capsule for testing its quality, the other used for ejecting the water sample; the cichlids use the one aperture alternately for ingress and egress. Cichlids are distributed through Africa and South America, a fact sometimes cited as evidence of the former union of these two southern continents. However, the evidence is equivocal to say the least, for cichlids also occur in the Middle East and in southern North America, whereas they are absent from Australia, a continent supposed to have been united with the other two southern continents. The female of some cichlids protects the eggs by holding them in the mouth until such time as the young hatch out; the term mouth-brooder is used by aquarists for species of this type. An example is *Tilapia,* with representatives in Africa and Palestine. *T. galilaea* inhabits the Sea of Galilee and has been the object of fishery since classical times. This would appear to be the species that figures in various traditions in the New Testament, such as the feeding of the 5000 on 5 loaves and 2 small *Tilapia;* and the fish that regurgitated a denarius of the emperor Tiberius at an opportune moment would appear to have been a *Tilapia* mouth-brooder. The skeptical would do well to take account of the fact that coins and jewelry are not infrequently swallowed by bottom-dwelling fishes, if such objects are dropped to the bottom, and many unusual objects have been found in the stomachs of cod, for example. An attractive cichlid not unusual in aquarium exhibits is the jewel fish *(Hemichromis),* the name relating however to the beautiful aspect of the fish, which is native to the Nile and Congo basins.

Hot Springs

Boiling springs emerge at the surface in solfataras and other areas of subterranean volcanic heating of the water table. As such a spring flows down a slope it heats the surrounding rocks and soil and itself becomes cooler in the process. The result is the creation of a localized warm climate. The isotherms bend upstream and therefore uphill, bringing lower-level plants into higher levels. Thus, on the slopes of the volcanoes of the Tongariro region in New Zealand the subalpine scrub and even the *Nothofagus* (southern beech) trees can be seen from miles away occupying unusually high levels in shallow gorges cut by boiling streams, as at Ketetahi. Within the water itself, there is no life other than bacteria in the immediate vicinity of the outlet from below. But when the temperature has fallen to about 170° F or 77° C thermal algae make their appearance; these are members of the Cyanophyta. Further downstream the flora is successively enriched, and many animals also take advantage of the favorable microclimate. When hot springs emerge on the floors of lakes or of rivers similar local enrichments of the ecosystem are observed. Trout, for example, favor warm water in cold lakes. In estuaries an emergent hot spring will attract many marine fishes, and also animals such as the Sirenia. Although there seems to be little doubt that the trout and the sirenians really enjoy the warm water, or act as if they do, it must also be pointed out that warm water is more soluble than cold and hence provides more dissolved mineral nutrients, thereby enriching the growth of autotrophs, and the heterotrophs dependent on them. Some emergent hot mineral springs lose their powers of solution so rapidly on being cooled at the surface as to precipitate mineral matter on the spot. Limestone formed from such precipitation is called *tufa* and often forms tiers of horizontal terraces, as in Turkey and at Yellowstone. If silica is the dissolved material, then the material deposited at the outlet may penetrate (in solution) the tissues of plants and animals before it is deposited, and in such cases fossils form that retain the internal cellular detail of tissues with amazing fidelity. The Triassic petrified forest trees of the western United States are examples of such action.

10

THE ORIGIN OF LAND ANIMALS

During the Devonian period, from 400 to 350 million years ago, a group of freshwater fishes evolved functional lungs, with special nasal passages to permit breathing air. These animals are classified as a group of Osteichthyes in a separate extinct order, the Osteolepidiformes, of which a typical genus is *Osteolepis*. The lung was derived from the swim bladder, which opened into the floor of the gullet; once the nasal passages had evolved, air could be taken in through the nostril or the snout if the fish rose to the surface of the water, and the air then passed into the mouth and so into the lung. There are three known surviving genera of freshwater fishes that retain these features. They are called lungfishes. The living genera are *Neoceratodus*, found in the Mary and Burnett rivers of Queensland, Australia; *Lepidosiren*, in rivers of South America; and *Protopterus*, in rivers of Africa. The rivers they frequent exhibit marked seasonal variation in flow of water. In summer they may be reduced to a row of disconnected deep water holes, often crowded with crocodiles and fishes, with stagnant vegetation and a low oxygen content. In Africa such watercourses are called *wadi* by the

Arabs. In Australia they are called *billabongs*. The lungfishes are, of course, well adapted to survive such conditions, since they are not dependent on their gills and the poor oxygen content of the stagnant water. In other respects their habits vary. The Australian *Neoceratodus* is a quiet fish feeding on aquatic vegetation. *Lepidosiren* is a carnivore in marshes where it captures fishes and snails and can also inflict a dangerous bite on human intruders. *Protopterus* in Africa feeds on frogs, fishes, and invertebrates. Both *Lepidosiren* and *Protopterus* estivate in the dry season by excavating a burrow in the mud before it dries out.

THE ORIGIN OF AMPHIBIANS. The Devonian period was dry climatically, and extensive deserts are attested by the red desert sandstones that survive. Watercourses must have resembled the wadi and billabongs of today in dry regions. Toward the close of the Devonian a new group of vertebrates appears in the fossil record, called the Ichthyostegalia. These resembled the Osteolepidiformes, but their bones show that a characteristic fish character had been lost, namely, the lateral-line system. This is taken to imply that the adult stages were no longer aquatic. Thus the

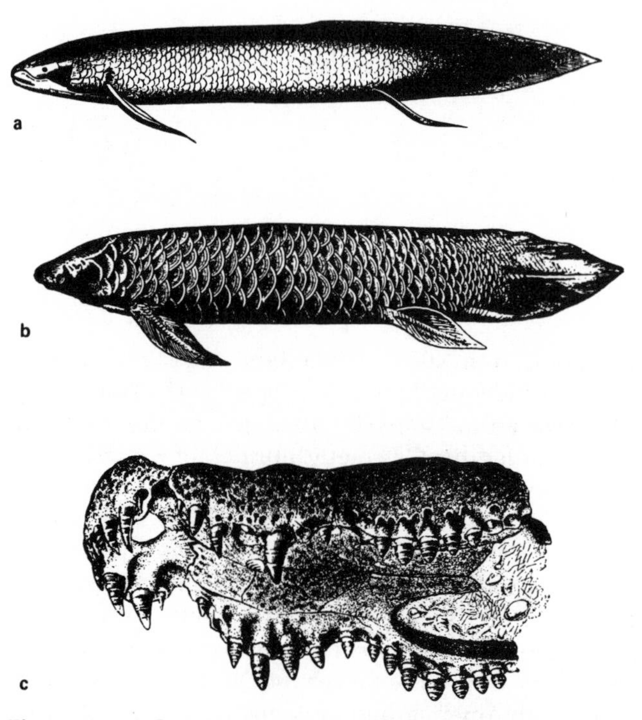

Figure 56. (a) *Protopterus annectens* (African lungfish), X 0.2, Dipnoi (15); (b) *Neoceratodus forsteri* (Queensland lungfish), X 0.1, Dipnoi (15); (c) *Alligator*, X 0.2, Eocene deposits of Isle of Wight, England, evidence of climatic change cited by Hooke (Chapter 4), Crocodilia (39).

gradual transition from lungfish to the nearly related first land animals seems to have occurred at that time. Soon afterward we find extinct Amphibia named Labyrinthodonts, early Carboniferous derivatives of stock like the Ichthyostegalia. Amphibians are dependent on water for reproduction, for the eggs and sperm are normally shed in water, where fertilization occurs, and the young stage (tadpoles) live until the gills are replaced by functional lungs. The types of Amphibia that arose during the Paleozoic are now all extinct, and modern amphibians are for the most part much smaller animals, with lighter skeletons, specialized for particular damp habitats not much frequented by other animals.

Only two major orders of Amphibia occur in present-day environments, and in many parts of the world only one of the orders is found, and then often only as minor or relict elements of the fauna. The two orders are (1) Urodela, the tailed amphibians or salamanders and newts, lacking from most parts of the Southern Hemisphere, and (2) Salientia, or frogs and toads.

THE ORIGIN OF REPTILES. Toward the close of the Carboniferous period, about 300 million years ago, the air-breathing amphibians began to develop characters that led to their complete emancipation from the need to return to water in order to reproduce. These new features involved an enlargement of the lungs, so that respiration no longer required the accessory exchange of gases at the surface of the body, by way of the damp skin; so now the skin could develop thick protective scales, thereby permitting the animal to wander into dry environments without danger of desiccation. Another new feature was the evolution of the penis, permitting internal fertilization, so that it was no longer necessary to shed the eggs and sperm in water. Also the eggs developed large reserves of nutritive yolk material and a protective shell. Development could not occur on land. All these new characters led to animals capable of spending much or all of their life on land. They were still dependent on climate, for there was no temperature-regulating mechanism in the body. Vertebrates with the characters listed are called reptiles and placed in the class Reptilia by systematists.

There are only four surviving groups of reptiles, each regarded as a separate order; in former times many other orders evolved, but later they became extinct. The four surviving orders may be distinguished at sight by the external features of the body. Tortoises and turtles from the order Chelonia, in which the body is covered by a shell, or *carapace*. The carapace is lacking from the other three orders. In reptiles, as in amphibians and fishes, the openings of the alimentary canal and the genital ducts occur together, the common aperture being termed the *cloaca*, of a slitlike shape. Reptiles that lack the carapace and have a cloaca directed longitudinally may readily be distinguished from those where the cloaca is transversely directed. The former have an additional internal character of free ribs imbedded in the abdominal wall; this assemblage comprises the large and well-known crocodiles (order Crocodylia), and also a separate order comprising the unique lizardlike New Zealand Tuatara (order Rhynchocephalia). The reptiles with a transverse cloaca make up the single order Squamata. Two suborders of the Squamata are readily recognizable, the lizards (suborder Lacertilia), in which legs, eyelids, and external ear openings are normally well developed; and the snakes (suborder Ophidia), in which the structures named are lacking.

Two orders of reptiles are commonly found in freshwater faunas, especially in the tropics; these are the Chelonia, represented by a variety of freshwater turtles and tortoises; and the crocodiles. Some snakes, as the anaconda of South America and the watersnake of North America (Natrix), habitually frequent the banks of rivers, and spend much time immersed in water.

CROCODILES. There are eight surviving genera of crocodiles, all confined to the tropics and associated with lakes and rivers, where they are the dominant carnivores. In the rivers of India and Burma occurs a species so different from all other kinds that it is placed in a separate family. This is the Gharial *(Gavialis gangeticus)*, with very long narrow jaws and numerous teeth (more than 27 teeth in the upper jaw). All other crocodiles have fewer than 22 teeth in the upper jaw, and the jaws are shorter and broader than in the Gharial; these forms are combined in classification to form the single family Crocodylidae. Two subfamilies are recognized: (1) the alligators, in which the upper jaw is relatively wider than the lower, so that certain teeth of the lower jaw close into sockets in the upper jaw; and (2) true crocodiles, in which all the teeth of the lower jaw lie outside the margin of the upper jaw when the jaws are closed. The alligators include the genera *Alligator* (southern United States and the Yangtsekiang River of China), *Caiman* of Central and South America, and two other South American genera. The true crocodiles include the pantropical genus *Crocodylus*, one other genus in central Africa, and a third genus with narrow jaws resembling a Gharial and restricted to Indonesia. On account of ruthless hunting for the leather market, all crocodiles are now under threat of extinction.

CHELONIANS. The freshwater turtles and tortoises occupy varying roles in the ecosystem. Some, such as the snapping turtles, are active predators and feed on fishes and waterfowl; others are herbivorous. Some, although related to freshwater forms, spend most or all of their life on land, such as the box turtles and tortoises, some of which are adapted for such dry locations as grasslands and even deserts. The main groups, exclusive of the marine forms mentioned in an earlier chapter, are as follows.

Two families have the carapace covered by skin, in which no obvious scales or horny plates are developed. Of these, the family Trionychidae, or soft-shell tortoises, have separate digits on the limbs. The other family, the Carettochelyididae, have paddlelike limbs and occur only in New Guinea.

Of the others, in which the carapace is covered by horny plates, there are six families, and most have limbs adapted for walking rather than swimming; that is to say, the limbs are not paddle-shaped. Two of the families are restricted to the Southern Hemisphere, constituting the side-neck tortoises so-called because of their habit of folding the head under the carapace by turning the neck to the side; these are the Chelydidae of South America, Australia, and New Guinea; and the Pelomedusidae of South America, South Africa, and Madagascar. The distribution of these two families has been used as an argument to support the theory that the southern continents were once all united. In the remaining four families the head is retracted under the carapace by the neck folding into a vertical S-shape bend. The snapping turtles and their relatives are recognizable by the long tail and constitute the family Chelydridae. Of the short-tailed forms, those with webs between the digits form the family Emydidae, and those in which the toes are separate, the family Testudinidae. The sixth family is distinguished by characters too technical to require mention here. The tortoises and turtles, like the crocodiles, are very ancient forms of land vertebrates, more or less adapted for life in or near rivers and lakes. Their long evolutionary history lends special interest to them as objects of scientific study, and their bizarre forms lend added interest to the ecosystems of which they form a natural part. Therefore it is now widely recognized by thinking people that these animals should not be hunted or merely destroyed because they take some fishes and waterfowl as food. They have lived on the earth for over 200 million years without causing damage to the ecosystem, and they should be protected from molestation by man, especially by supposed protectors of wildlife who use specious arguments, claiming that aquatic reptiles endanger the other aquatic fauna. Were such arguments valid, the other aquatic fauna would have disappeared millions of years ago.

SALT MEADOWS AND ESTUARIES. *Salt meadows* can be regarded as transitional regions, or *ecotones*, between the sea, the fresh water, and the land. Since it seems very probable that all the tracheophytes of the aquatic environment are descended from terrestrial ancestors, it is to be expected that these ecotones would carry vegetation of intermediate types, and this proves to be the case. True rushes *(Juncus maritimus)* are represented, and sedges, such as *Carex litorosa*. One of the bulrushes, *Scirpus lacustris*, occurs both in saline and freshwater areas. The yellow-flowered musk, called monkey-flower in America (*Mimulus* spp.), penetrates the salt meadow where freshwater streams intertwine across an outlet plain. In the Southern Hemisphere a scrubby shrub *Plagianthus* occupies salt meadows in Australia and New Zealand, each country having its own series of species; the genus belongs in the mallow family, Malvaceae. A member of the primrose family (Primulaceae) is the genus *Samolus*, which includes a cosmopolitan species and others that range the salt marshes and estuaries of the Southern Hemisphere; coastal rocks are also colonized. Plants of the kind mentioned in this paragraph are important in preventing subaerial and storm erosion of temperate coasts in both the Northern and Southern

Hemispheres and to this extent are analogous to the tropical mangroves, to which otherwise the similarity is slight.

MANGROVE SLOUGHS. On tropical shores where wave action is not severe there develops a fringe forest of trees and shrubs, all of which are tolerant of saline or brackish conditions and develop stiltlike prop roots or ascending roots (pneumatophores), used to prevent the silt from being washed away. These trees belong to various families, but are all spoken of as *mangroves*. One family, the Rhizophoraceae, comprises about 20 genera, including the black mangrove *(Rhizophora mangle)*, which ranges the coasts of both the Atlantic and Pacific tropical Americas. Another family, the Combretaceae, includes *Laguncularia*, ranging the tropics of both sides of the Atlantic; *Conocarpus*, with a similar distribution; and a third family, the Verbenaceae, is represented by the mangrove genus *Avicennia* in both the New and Old worlds. Other genera that participate are *Bruguiera*, *Sonneratia*, *Carapa*, and *Ceriops*. All show convergent adaptation to life in or near salt water and to a rising and falling tide at regular intervals. In Florida the mangroves are zoned such that *Rhizophora*, with its stilt roots, stands next to the sea; then on the landward side comes *Avicennia*, with pneumatophores; then *Conocarpus*, which is transitional to the tropical forest itself. The seeds of mangrove trees are adapted for dispersion over salt water, and some species are viviparous, so that the seed has already sprouted a stout anchoring root that penetrates the substrate when the seed falls off the tree. The mangrove belt is somewhat wider than the coral reef belt, for it reaches to northern New Zealand in latitude 35° south, whereas the coral reef belt ceases at about 30° south. Mangroves may extend along the banks of rivers, when their stilt roots form a kind of picket fence along the river bank, as in New Guinea.

Mangrove sloughs gradually are converted into dry land as the mangrove belt advances into shallow marginal seas, the accumulated silt and organic debris causing a natural reclamation of the seabed into dry land. On many coasts the mangroves are therefore not merely protective, but actively extending the real estate—a matter not appreciated by Floridian estate agents and others who have mercilessly destroyed the mangrove fringe, with resultant erosion of the coastline. Botanists regard the mangrove association as *climax formation*, for it does not evolve into a forest of different type and regenerates itself with the same species after destruction or injury. This is a rare circumstance in forests, where a *succession* of replacement plant associations usually follows destruction of a forest. Mangroves in the Indo-West-Pacific comprise some 10 species of trees, whereas on Atlantic coasts only 4 species are predominant. There is no equivalent in cool temperate and cold latitudes.

BOGS AND FENS. The term *swamp* is generally applied to formations where large plants such as trees occur, whereas *bog* is applied to situations that lack the firm foundation for trees, or where the climate or other factors inhibits the growth of large plants, as in peat bogs of subpolar or mountainous regions. *Quaking bogs* exist where a thick sheet of vegetable matter is floating on a water mass below. Bogs tend to be deficient in nitrogen, and so insectivorous plants such as sundews *(Drosera)* are favored. The moss *Sphagnum* is commonly found in bogs, and small sedges. Liverworts and *Lycopodium* may occur, also ground orchids (Monocotyledoneae); and in tropical or humid lands umbrella ferns *(Gleichenia)* and other ferns. In northern lands at high altitudes cranberries and blueberries (*Vaccinium* spp.) may occupy large boggy areas. Bogs over a long period of time accumulate the bones of various large animals that become bogged when traversing unstable ground; if the subsoil and bedrock are alkaline, as over limestone, the bones may persist for thousands of years, accumulating in such numbers as to give the impression of a catastrophic destruction of large herds of animals (whereas in fact the deaths occurred at intervals). If the subsoil is acid, then the bones dissolve. Thus in parts of New Zealand where the bedrock comprises noncalcareous Mesozoic sandstones the overlying bogs become acid, for there is nothing to buffer the humic and other acids that develop in bogs, and so the only remains of the former population of moas (giant birds that antedated European settlement) occurring in such bogs are the accumulation of gizzard stones; whereas over the Tertiary limestones the bogs yield thousands of skeletons. The skeletons are found standing upright, but complete only to the base of the long neck. Associated bones imply that the trapped moas were attacked by large extinct eagles, which destroyed the exposed portions. In Irish bogs the skeletons of a giant elk are found; and in Danish bogs human burials of the Bronze Age have remained in partial preservation on account of the presence of tannic acid derived from oak trees, the tannin having tanned the hides of the buried persons, preserving their external appearance, features, and clothing. When a bog occurs on low-lying land near the sea it is subject to periodic inundation by saline water and is then termed a *salt marsh* in North America, the commoner term in Britain being *fen* or *marsh*. The fens district of eastern England is an example. Salt marshes have their own special populations of species that are not adversely affected by salinity of the soil. In tropical sands the salt marsh is replaced by mangroves.

SWAMPS. A swamp may be viewed as a former lake basin now nearly filled with the accumulated silt brought in by contributory streams and the accumulated organic debris of former water vegetation; on this basis larger and larger vascular plants take root; eventually forests arising as trees occupy the ground. As in the more aquatic habitats the Monocotyledoneae are conspicuous, but a much higher proportion of Dicotylodoneae is usually present. Depending on how much water is present, a varying predominance of aquatics such as the bulrushes, reeds, and sedges will occur. The large grasses and reeds *(Arundo, Cordaiteria)* may be dominant in open situations; in New Zealand the swamps also include a palmlike giant agave (*Cordyline*), locally called cabbage tree, or ti, by the

Maori; and a shrub-sized agave *(Phormium tenax)* is highly characteristic. Swamp trees may include any of the supralittoral species listed on page 164. Where water is extensive the swamp cypress (*Taxodium distichum*) occurs in the southeastern United States; this is a deciduous gymnosperm, which formerly lived in Eurasia, but which died out there at the end of the Tertiary. In Asia *Metasequoia*, with a very different ecology, is a formerly widespread relative of *Taxodium*, now regaining its former distribution through the agency of botanists and interested gardeners. Dicotyledons are numerous and include herbs such as mint *(Mentha)*, forget-me-not *(Myosotis)*, swamp loosestrife *(Decodon)*, musk *(Mimulus)*, and many others.

EARLY SWAMP FORESTS. The oldest fossil forests appear to have been swamp forests, a circumstance that seems to imply that the large vascular plants evolved on swamps in response to the ecological niches that developed there, permitting a gradual transition from small plant modules to progressively larger ones. Going further back in time again, we find that the earliest known land floras developed on bogs; fossil peats from the Silurian of Australia and the Devonian of Scotland carry the oldest known land plant associations (though lichens must surely have existed on much more ancient land surfaces). It is significant that the earliest land faunas appear in the same contexts as the earliest land floras, lagging by some tens of millions of years, but apparently coordinated. This suggests that entire ecosystems are what really evolved, for any step or advance made by any component element of an ecosystem would have to influence the other dependent parts of the ecosystem. This, if it is true, suggests that a useful plan to adopt in surveying the terrestrial environments and their biota would be to begin with swamp forests dominated by ancient terrestrial trees, such as tree ferns and the oldest surviving types of gymnosperm forest, and inhabited by faunas of the oldest surviving kinds of animals. To realize this project, we turn now to the remotest parts of the Pacific, where premammalian forests of Cretaceous type still have recognizable remnants.

11

AUSTRAL FORESTS

Archaic Life Forms of Far Southern Lands

FOREST BIOMES. When an uninhabited potential habitat is created by some physical accident or catastrophe it is not immediately occupied by the species that ultimately become the resident community. Instead, there is usually a series of invasions, each being superseded by gradual changes, as more species enter and others disappear. This phenomenon is termed by ecologists a *succession*. For example, after a great forest fire in a forest of maple and beech, the bare land is first colonized by various grasses and herbs; then follow consolidation stages in which a denser turf of grasses like Kentucky bluegrass appear and various larger herbs such as goldenrod; then come blackberry, hawthorn, and New England aster; then follows a *subclimax* of trees such as hickory, oak and elm; lastly a *climax* stage of oak, sugar maple, and beech. The subclimax and climax stages commonly constitute a *forest* in those habitats suitable for the growth of trees.

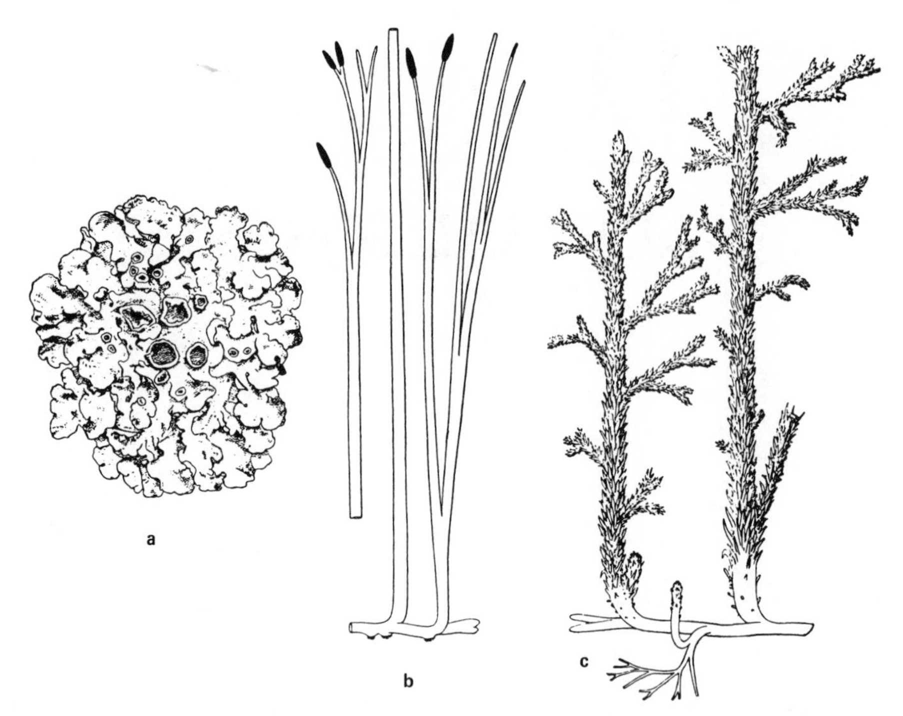

Figure 57. Ancient types of land plants. (a) *Parmelia*, X 0.5, a lichen that grows on trees and rock, Thallophyta (41); (b) *Rhynia*, a Devonian bog plant with leafless stems and spore-producing organs at the tips; (c) *Asteroxylon*, a leaf-bearing relative of *Rhynia*, Psilophyta (33).

A *tree* is any woody vascular plant that has a main growth axis, or *trunk*; a *shrub* is a woody plant that has no principal growth axis. At the present geological epoch trees are growth forms of gymnosperms and flowering plants (angiosperms); but in earlier geological epochs many spore-producing plants developed tree species, such as the lycopods, the equisetes, and the ferns, all of which were noted as herbs associated with ponds and lakes. A few ferns still have the character of trees (the tree ferns), but the giant lycopods and equisetes are extinct.

A successional sequence such as that outlined leading from grasses to oaks and maples is termed a *sere*. The nature of a sere depends on the substrate. Thus a *hydrosere* leads from the initial communities of a lake or pond to the later marsh subclimax or climax stages where forest trees have invaded, and bog stages formed intermediate successional forms of the hydrosere. On dry, sterile ground or rock a different sere occurs, which may be called a *xerosere*. The term *subsere* is often employed to denote a set of successional stages following the natural or artificial destruction of a preexisting sere. The term *prisere* is sometimes used for a primary sere, starting with virgin territory never before occupied, as when a seafloor is elevated to become land and is first settled by terrestrial organisms.

Forests may readily be classified into broad categories such as *coniferous forest* or *taiga* (when the dominant climax species are conifers such as spruce, pine, larch, hemlock), and *deciduous broadleaf forest* (such as the monsoon forests of the tropics and the temperate summergreen forests of the Northern Hemisphere), and so on. Such broad classifications involve great regional communities, often made up of species that differ from place to place, yet all of which share the property implicit in the name. Such great regional forests are called *forest biomes.* The term *biome* may also be applied to other comparable regional classifications of similar though different communities, for example, *grassland biomes* (including prairie, steppe, pampa, veldt), or desert biomes, and so forth. Biomes are characteristically dominated by plant species; since plants are not very conspicuous in most marine habitats, the application of the idea of the biome in marine biology has not been very consistent or profitable, and the concept has been omitted from the earlier chapters of the present text for that reason. However, if reference is made now to Table 1 a certain broad correspondence will be seen between the biomes named in the column headed "Nature of Land Surface," and the oceanographic features listed in the next column, as also the climatic features listed in the third column. Coral reefs would seem to constitute a natural marine biome, and so might the west-wind-drift regions of the Southern Hemisphere seas.

According to whether or not this viewpoint is taken under review, biologists and others have reached strangely discordant interpretations

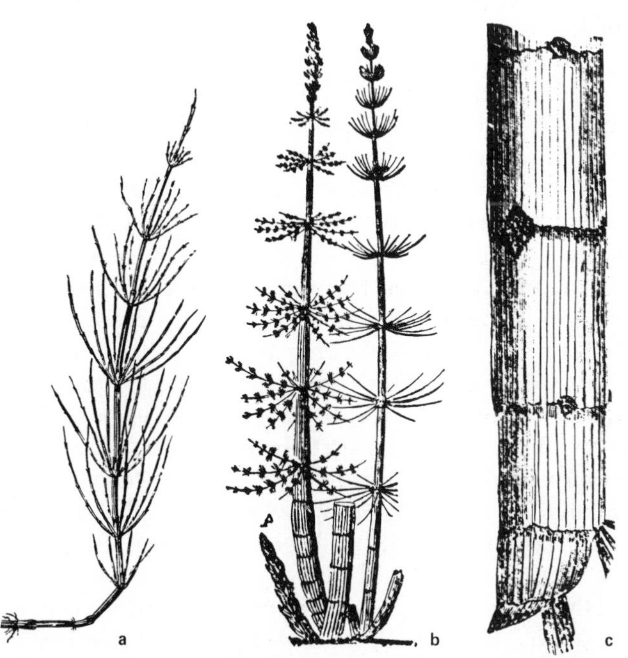

Figure 58. Equisetes. (a) Extant genus *Equisetum* (horsetail rush) (41); (b) fossil Carboniferous genus *Calamites,* greatly reduced; (c) stem of same, X 0.2 (19).

of the fact that the southern continents share similar faunal and floristic elements in much the same manner as do the northern continents. But this is a matter that is better considered after a review of the nature and content of the biota of southern continents. Biogeographers who have taken note of the shared elements of the faunas and floras of the far southern lands often use the term *austral* to describe such biota; thus an austral genus or family of plants or animals is one that is likely to have species in Australia, New Zealand, Tasmania, South America, and on the various oceanic islets between these widely separated lands. South Africa, which lies outside the west-wind-drift region shares little with the austral lands, though its flora has close affinities with that of the northern part of Australia, which also lies outside the west-wind drift; and, as might be expected, both South Africa and northern Australia share much with the intervening Indonesian and Asian lands of the tropical belt.

WHAT WERE THE FIRST LAND PLANTS? The earth is 4.5 billion years old, and life is known to have existed on the earth for most of that interval. Yet we do not encounter fossil land plants until the close of the Silurian period, about 400 million years ago. Are we to infer that no land floras occupied the continental surfaces for all the 4 billion years before that? Such a proposition seems unlikely when one glances at the extraordinary erosion shown in aerial photographs of the defoliated regions of Vietnam. Besides, the earliest microfossils from very ancient pre-Cambrian rocks suggest forms like Cyanophyta, and later pre-Cambrian rocks yield fossils that seem to be Chlorophyta, and fungi also seem to be present in early fossil deposits. Now fungi are Thallophyta that lack chlorophyll, and feed as heterotrophs on dead organisms. Fungi also have the property of forming intimate symbiotic unions with Chlorophyta and Cyanophyta, such unions being the peculiar plant growths called lichens. A chance meeting of a wind-blown fungus spore and a single celled chlorophyte or cyanophyte has often been observed to result in the production of a lichen.

Lichens have the power of occupying quite dry surfaces, such as rock and dry ground. They are very common features of habitats of nearly all types, except in the regions of cities; the toxic pollution of the atmosphere has destroyed nearly all lichens out to the outer margins of the suburban areas during the past few decades; and the trees whose bark was once covered by shaggy growths of lichen, as recently as about 1950, now stand bare-trunked as their own bark cracks and shatters. And the community of fascinating lichen mimics among the insects is no longer seen save in country districts. Similarly the rock-colonizing lichens that still cover the coastal cliffs and exposed reefs far from city pollution were once universally found.

Knowing this, and the antiquity of the symbionts that produce lichens, one is tempted to suspect that in the early days of the earth the continents must surely have been clothed in rock lichens. Rocks on exposed continental surfaces do not provide the conditions favoring the preservation of their biota as fossils. Perhaps that is why we do not know of ancient lichen bi-

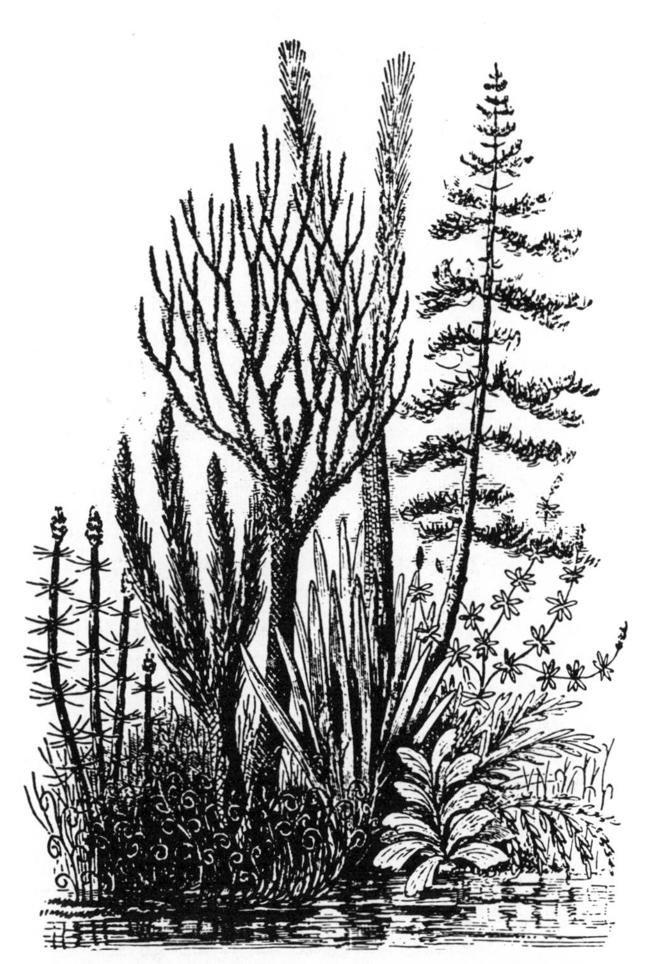

Figure 59. Late Paleozoic swamp forest restored; left to right: *Calamites*, carpet of psilopsids, two club moss trees, *Leptophleum* and *Lepidodendron*; *Cordaites* and *Dadoxylon*, early gymnosperms; and ferns or seed ferns (*Platyphyllum*) at water's edge (19).

omes. To test the feasibility of these ideas, a few years ago I arranged for a set of samples of impactite to be collected from the Henley meteorite craters of central Australia. It was possible, it seemed, that such a substrate (produced by meteorite explosions and the resultant deformation and ejection of parent sandstone, converted into a half-fused chert) would simulate the conditions on the early earth; and, being in the heart of a great desert, the lack of moisture would further simulate the most rigorous circumstances under which early plant colonists might have invaded the continents from the sea or fresh waters. When the samples were examined in the laboratory, minute pinhead-sized black dots, apparently tiny lichens, were found on two of the pieces of impactite. It is my view, perhaps not generally shared, that lichens probably did cover much of the earth's dry land before the Silurian period.

THE EVOLUTION OF FORESTS. The geological record becomes more explicit after the onset of the Silurian period, and from it we learn that *bog communities* of plants resembling rushes, but related to very primitive spore-producing plants called Psilotales (they have no common name), developed in Australia at the end of the Silurian. In the early Devonian similar plants occurred in Scotland, again associated with a bog. The fossils are referred to genera *Asteroxylon* and *Rhynia* and other genera, and they are preserved in great detail in chert. Chert is a siliceous rock often formed by hot springs. It is of interest to note that the two surviving genera of plants believed to be most nearly related to these ancient bog plants are found in New Zealand, sometimes in association with the warm soil around hot springs, whose silica is deposited in the vicinity. These plants, *Psilotum* (which ranges the tropics) and *Tmesipteris* (found in New Zealand, Tasmania, and the southwest Pacific islands), also frequent other habitats, particularly tree fern forests.

By the onset of Mississippian (lower Carboniferous) times, about 350 million years ago, swamp forests appeared, with equisetes and lycopods forming trees. These, and tree ferns, made up the lowland forests for the next 80 million years. Then came forests of conifers and, probably, the colonization of high upland country by evergreen forests of gymnosperms. Meantime insects had evolved as important forest inhabitants, and land vertebrates included amphibians and the first of the reptiles. These, and some of the succeeding stages in the geological evolution of the forest biomes, are summarized in Table 32.

EARLY AUSTRALIAN SWAMP ECOSYSTEMS. The scene illustrated in Figure 5 is a careful restoration of a Triassic swamp ecosystem from the district of Sydney in southeastern Australia. It has already been mentioned in Chapter 1 as yielding clear evidence that the environment was tropical or subtropical. The crocodilelike animal represents a labyrinthodont, an early type of amphibian of carnivorous habit; among genera known from Sydney Triassic rocks are *Cyclotosaurus*, reaching a length of 11 feet and having a skull 3 feet in length; also *Mastodonosaurus*. In addition fossils show that large reptiles roamed the dry lands. The prey of the giant amphibians were fishes, of which remains of the lungfish *Ceratodus* have been found in the Triassic sediments of Australia. Several other fish genera are known, including one shark. None of these animals could have lived in an environment subject to a long polar night, so that the supposed geomagnetic evidence that Sydney lay inside the Triassic polar circle is unacceptable.

The luxuriant broadleaf flora similarly points to a warm and humid climate. The largest of the broadleaf forms is *Macrotaeniopteris*, a fern, or seed fern, with several species. In the left foreground is *Equisetites*, a large horsetail rush. Other ferns or fernlike plants seen are *Phyllotheca* and *Thinnfeldia*. A very rich flora is implied by the long lists of fossils reported in the literature. Very large insects, including cicadas and bugs with a 20-inch wingspread *(Mesotitan giganteus)*, further emphasize tropicality.

Table 32. Geological history of forest biomes.

PERIOD	MILLIONS OF YEARS BEFORE PRESENT	VEGETATION	LAND VERTEBRATES	LAND INVERTEBRATES
Paleozoic				
Silurian				
	400	Bog Psilopsida	Lungfishes appear	First land scorpions
Devonian			Ichthyostegalia	
	350			
Mississippian		Swamp forests of equisetes (Calamites), lycopods (Lepidodendron), and tree ferns	Labyrinthodont amphibians	
Pennsylvanian			First reptiles	Extinct giant insects
	270			
Permian				
	225	Conifers become forest trees		Modern Odonata
Mesozoic				
Triassic		Araucarian forests appear	Amphibians reach and pass maximum	Coleoptera, Hemiptera, Ephemeroptera
	180			
Jurassic		Podocarp forests appear		
	135	Flowering plants appear	Dinosaurs; first birds and mammals	Microlepidoptera and hepialid moths
Cretaceous		First deciduous forests	Dinosaurs wane	
	70		First snakes (boas and pythons)	
Cenozoic				
Paleocene		Grasslands appear	Placental mammals become conspicuous	Social insects appear; Macrolepidoptera (butterflies, large moths)
Late Tertiary	5	Forests peak and decline		
Recent (Holocene)	Now		Mammals peak and decline	

NOTE: On each continent the Holocene decline in forests and mammalian faunas coincides with the appearance of man.

New Zealand: The Most Isolated Continent

On the map New Zealand appears as an archipelago of two larger islands and about 30 lesser offshore islands, with a total land area of only about 100,000 square miles, or roughly the same size as Great Britain, or the state of Colorado. In spite of its relatively small area, its geological structure is that of an independent continent.

Apparently always rather isolated (the dinosaurs seem never to have reached there, though they were present in Australia), New Zealand seems to have lost any land link it ever had with Indonesia about the middle of the late Cretaceous, say about 100 million years ago. The fossil record, and the present nature of the fauna and flora, shows that New Zealand never had a mammalian fauna of any kind save only for two species of bats, apparently wind-borne across the ocean. Polynesian man settled the country during the first millennium after Christ and brought the Pacific rat (*Rattus exulans)* as a stowaway and a domesticated barkless dog as a deliberate introduction. The human settlers found a land in which dense rain forests clothed the hills and grasslands and scrub covered parts of the low country, with swamp forests in the north.

In this habitat ranged about 20 species of gigantic flightless birds called by the new settlers moas (moa is the general Polynesian word for domestic poultry). The great birds were defenseless and all were exterminated before the first Europeans visited the country (there is a possibility that one small moa species was living about 1790, and some have speculated that it might still survive in the remotest parts of the south island). The rest of the vertebrate land fauna comprises various birds, mostly related to or identical with Australian and Indonesian birds, evidently transoceanic immigrants, a few reptiles, the largest being *Sphenodon*, the tuatara, the only known survivor in the world of a Triassic group of reptiles called Rhynchocephalia, some lizards, and a primitive genus of frogs, *Leiopelma*.

These facts point to prolonged isolation, and the most likely date of final loss of terrestrial connection with the Old World is given by the date of origin of snakes in the late Cretaceous, for no land snake has ever reached New Zealand, though the fauna includes two species of sea snakes, of which about 10 individuals seem to come ashore on the average each year. The isolation evident in the faunal content shows also in the flora. Although deciduous trees introduced from the Northern Hemisphere have flourished in the country, and trees such as the Monterey pine have become New Zealand's most valuable forestry species (maturing in half the time required in California), none of the characteristic deciduous trees of the northern lands, nor any of the needle-leaf gymnosperms, had reached the country by natural means.

The flora is rich in archaic types of plants and also shows structural affinities with the flora of adjacent tropical lands. Ferns, which average 3.5 percent of the species of vascular plants on other continents, in New Zealand comprise about 10 percent and on some of the offshore islands the ferns make up 15 percent of the vascular flora. Their size is as impressive as their abundance in actual individuals as much as in variety. To this highly differentiated archaic flora is added an austral element, of which the best-known and most thought-provoking is the *Nothofagus*, or southern beech, forest—shared by New Zealand, Australia, Tasmania, and Chile.

These discrete floristic elements, and the associated faunistic data, are discussed separately. However, before doing this, we must pause to glance at the regional aspect of the New Zealand forests.

NEW ZEALAND FORESTS: WHAT BIOME? To the visiting botanist from abroad, the New Zealand forest presents a baffling aspect of contradictory features. Setting aside the many genera and species that are restricted to the country and that obviously are to be explained as the result of 100 million years of separate evolution in remote isolation, there remain problematic features related to the physical structure of the forest. One of these incongruities is the fact that the lowland forests resemble tropical rain forest.

Figure 60. Fossils and living fossils, the fernlike plants. (a) *Neuropteris*, Carboniferous seed fern; (b) *Alsophila*, modern tropical tree fern (41); (c) *Alethopteris* and (d) *Odontopteris*, fernlike plants from the Carboniferous (19); (e) *Asplenium*; (f) *Cyathea*, a tree fern that occasionally reaches a height of 65 feet in New Zealand forests today; (g) *Blechnum*, a cosmopolitan form mentioned in Chapter 9; (h) a dicotyledonous tree, lancewood (*Pseudopanax*), of which the juvenile form with downward hanging leaves grows through the fern cover to become an ecodominant (23).

Figure 61. Late Cretaceous swamp forest including palms, gymnosperms (with *Araucaria* or related form to left), and some early broadleaf trees. (19).

Those forest areas, which were the most widespread before the country was cleared for settlement by the Pakeha during the nineteenth century, are composed of mixtures of species of gymnosperms belonging to the family Podocarpaceae and various families of broadleaf trees of the Dicotyledoneae: in various associations of species they are known collectively as the *mixed broadleaf-podocarp forest*. The English botanist Richards (1952) has made a comparative study of the characteristics of lowland forests of humid regions of the tropical zone (the belt enclosed by the Tropics of Cancer and Capricorn), and he lists the architectural features that distinguish this type of forest and the morphological characteristics of the trees comprising it.

In the *architectural category*, conspicuous features of tropical rain forest include *stratification*: the component plants form several distinct

Figure 62. Tiered structure of New Zealand forests. In the examples shown, (a) the emergent ecodominant is *Agathis australis*, reaching a maximum height of 150 feet; (b) ecodominant layer; (c) intermediate tiers; (d) field layer. In the other example to the right, mixed podocarp forest, the emergent is (e) rimu, (*Dacryocarpus cupressinus*), carrying (f) *Astelia* as tank epiphytes in the forks (23).

layers, depending on the size module of the plant, ranging from the uppermost canopy layers to lower layers of small trees and shrubs, and a field layer of mainly ferns on the forest floor. The uppermost canopy surface (ecodominant layer) is uneven, on account of the presence of very tall *emergent trees* that stand much higher than the others; these emergents are species of dicotyledons. In addition to the plants that are self-supporting, there is an abundance of climbing or suspended creeper plants, the

Figure 63. Plank buttresses form on a number of species of various tree genera, gymnosperm, and broadleaf, depending on the degree of wetness of the forest subsoil, *Beilschmiedia tawa* and *Podocarpus* being examples; the festoons of liane stems, *Rubus australis,* have collapsed from above by their increasing weight, and have now come to form a more or less intertwined mesh. These features of humid New Zealand forest simulate humid tropical forest (23).

lianes, and also numerous epiphytes (plants that grow perched on other plants), the epiphytic plants including woody shrubs as well as herbs without wood tissue.

Without going into detail at this juncture, it is sufficient to say that the description of the architecture of the tropical lowland rain forest given by Richards is also in fact a precise description of the New Zealand lowland mixed forest—with the sole difference that the tall emergents of the New Zealand forest include podocarp gymnosperms as well as broadleaf dicotyledonous trees, whereas in the tropical forest only the latter are present.

Turning now to the *morphological category* of features that distinguish tropical lowland rain forest, Richards lists: most forest trees have tall slender trunks with relatively small crowns, not greatly branched; the bark is smooth and thin, not rough and cracked or flaky; the tall columnar trunks are supported often by lateral plank buttresses arranged and shaped like the fins that stabilize a rocket; sometimes instead the trunk is propped up by stilt roots of the mangrove type. Flowers sometimes grow out of the trunk, or on old branches *(cauliflory* and *ramiflory),* fruits therefore forming in the same situations. Epiphytes that begin life as small perched seedlings may grow into great trees that kill their host (banyan stranglers).

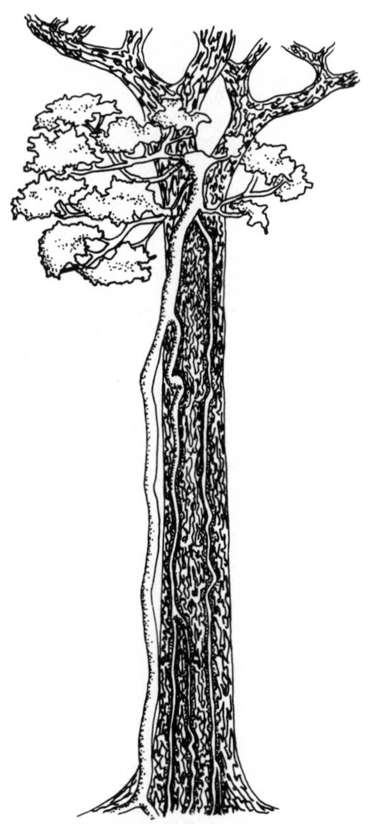

Figure 64. New Zealand forest architecture (continued). *Griselinia littoralis* exemplifies a group of genera that initially grow as epiphytes in a tree fork, later sending stemlike roots down the trunk of the host to reach the ground. The host tree in the case illustrated is a gymnosperm (*Podocarpus totara*), a tall ecodominant or emergent (23). Compare with Figure 86.

Aerial roots, pneumatophores (like those of *Avicennia*, already mentioned), may rise out of the soil or hang down from branches. Lateral secondary shoots may arise from the base of a tree (coppice shoots). The leaves tend to be of middle size, with smooth margins, and prolonged at the tip into acuminate *drip tips*. The pollination of the flowers is thought usually to be performed by insects.

Again without going into detail, it can be said that the foregoing is an accurate description of typical life forms in the New Zealand mixed forest, save only that drip tips do not occur on the leaves and stilt roots are found only in the mangroves of the far north of the country. Pollination is performed by birds and by the wind as well as by insects.

Other tropical rain forest characters in the New Zealand forests are the abundance of tree ferns and of epiphytic and ground-dwelling ferns; the occurrence of red-flowered parasitic mistletoes of the family Loranthaceae;

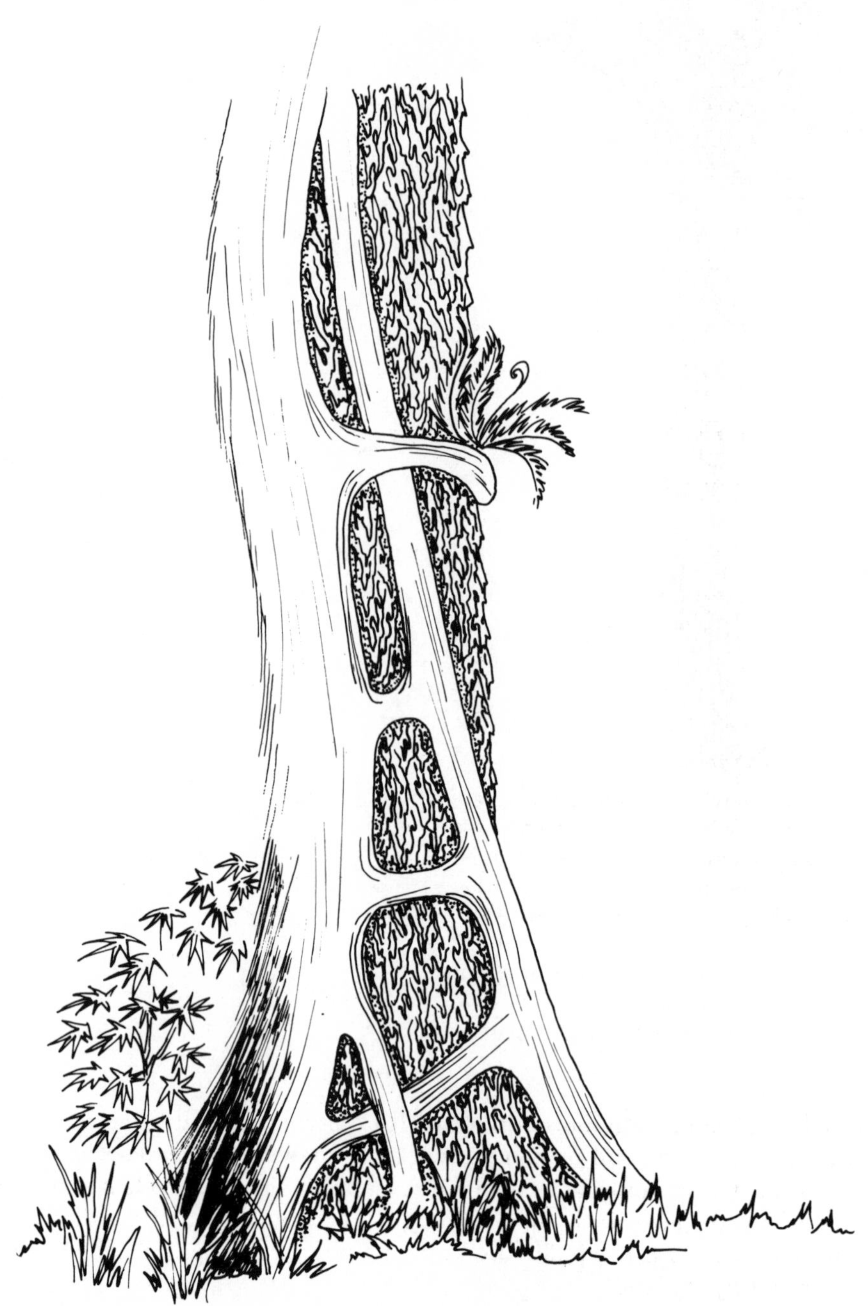

Figure 65. New Zealand forest architecture (continued). Some epiphytes, such as *Metrosideros lucida* (rata, a myrtle), produce such massive descending roots that when these reach the soil they supplant the rooting system of the host trees (*Podocarpus* and *Dacryocarpus*). Eventually the host is overtopped by the parasite and dies for lack of light. If rata seedlings germinate in soil on the ground they never grow larger than a small shrub. Strangler epiphyte is a term sometimes applied to these forms, but New Zealand botanists dispute the notion that any real strangulation is involved (23).

and the occurrence of trees whose juvenile stages have a single slender unbranched main trunk with large terminal leaves, until a height of several meters is attained, after which the main stem suddenly branches into a crown bearing small erect leaves.

The foregoing similarities to tropical rain forest are inconsistent with two categories of data—namely, *climatic* and *floristic*—for these point up the aspects in which the New Zealand lowland forests differ from tropical rain forest.

CLIMATIC DIFFERENCES. New Zealand is a sprawling archipelago spread across about 20 degrees of latitude. Restricting ourselves to the main islands plus those which appear to stand on the same continental block, the range of latitude can be reduced to about 14 degrees, from 34° south to 48° south. Translating this latitudinal spread into comparable Northern Hemisphere terms, for coastal regions lying between 34° north and 48° north latitude, we might list: (1) Crete and intervening lands as far north as Brittany, France; (2) Tokyo and intervening lands north to Khabarovsk, East Siberia; (3) Los Angeles and intervening regions north to Vancouver, B.C.; and (4) North Carolina and intervening states north to Newfoundland. No one cares to argue that such latitudinal stretches of territory could be considered as tropical, or even subtropical. Of the spans listed, the first, or western European comparison, comes closest climate-wise to the New Zealand situation.

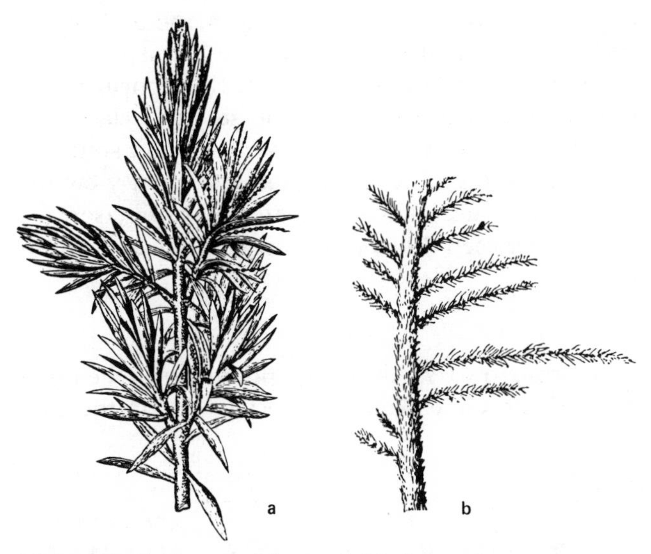

Figure 66. (a) *Dorycordaites* and (b) *Araucarites*, two tree types of the late Paleozoic and early Mesozoic that simulate New Zealand forest trees *Podocarpus* and *Dacryocarpus*; beyond the fact that all are gymnosperms, the degree of affinity is presently obscure (19).

TEMPERATURE AND RAINFALL. In an "average" lowland tropical rain forest, for example, in Ceylon, the range of mean temperature for the coldest month and hottest month is 20–25° C. For Wellington (41° south latitude) in mid-New Zealand the corresponding range is 8–15° C, and for Boston (42° north latitude) the range is −1°–+23° C. Thus the middle latitudes of New Zealand never reach even winter tropical temperatures and are never so cold nor so warm as the temperate climate of Boston. The annual precipitation for lowland tropical rain forest is usually within the range 200 to 500 cm per annum. For Wellington the rainfall is 120 cm per annum; for Boston the figure is 100 cm per annum.

Thus New Zealand's climate is temperate, without extremes, more uniform than the corresponding northern latitudes, and never tropical.

FLORISTIC DIFFERENCES. The New Zealand flora comprises genera that are not conspicuous (or not present at all) in tropical floras. The lianes do not belong to the genera that are conspicuous lianes in the tropics. The families of plants, however, do correspond with those conspicuous in tropical rather than temperate climates. The characteristic temperate northern families, such as oaks, willows, birches, maples, and elms are totally absent.

On the basis of these data New Zealand botanists have regarded the lowland broadleaf-podocarp forest as subtropical (Cockayne, 1926; Dawson, 1962), whereas overseas botanists prefer to classify it as temperate (Polunin, 1960; Dansereau, 1957). However, it has gradually been recognized that the temperate forests of the Southern Hemisphere are "radically different from those of the same latitudes of the northern hemisphere, probably due to the continental climate of the northern hemisphere and the oceanic climate of the southern" (Morse, 1969). If this conclusion is accepted, as is here proposed, it becomes feasible to set out a classification of forest types equally useful to botanists of northern and southern lands and decidedly more meaningful to zoologists who must view the faunas in these settings. It will be useful here to denote more precisely the differences in continentality between the northern, equatorial, and southern regions of the earth.

RELATIVE CONTINENTALITY. As already pointed out, the three great belts of circulation of the atmosphere and ocean divide the sur-

Table 33. Relative continentality.

REGION BETWEEN LATITUDES GIVEN	AREA OF LAND IN SQUARE MILES	TOTAL AREA IN SQUARE MILES	CONTINENTALITY (RATIO OF LAND TO TOTAL AREA)
90° north and 30° north	25,000,000	50,000,000	50%
30° north and 30° south	26,000,000	100,000,000	26%
30° south and 90° south	6,900,000	50,000,000	14%

face of the earth into four equal parts, one from the North Pole to the parallel for 30° north; one between 30° north and the equator; one between the equator and 30° south; and one between 30° south and the South Pole. Since the two zones on either side of the equator constitute a single trade-wind belt, and the other two zones each constitute west-wind belts, the ratios become 1:2:1. The ratios of land to total area of each of these three belts is given in Table 33. Continental climates, as is well known, are colder in winter and warmer in summer, and the presence of continents tends to disrupt the oceanic pattern of winds and ocean currents. The much greater continentality of the northern belt of the Northern Hemisphere leads to less even precipitation of rain throughout the year, for much of it falls as snow, or is locked up as ice and unavailable to plants, in winter; and the summer rains evaporate more rapidly owing to the higher summer temperatures. In the Southern Hemisphere these restrictions on the availability of water to plants are absent or much less severe. Hence the forests of the Southern Hemisphere south of the parallel of 30° south are much more humid than those of corresponding northern latitudes, and the growing season is not sharply interrupted by a severe winter, save only in the neighborhood of Antarctica in southernmost South America. Thus it is only in the far southern extremity of South America that any semblance of deciduous summergreen forest occurs. The similarity of New Zealand lowland rain forest to the tropical rain forest is evidently due to the comparable humidity plus the lack of pronounced seasonal variation.

If the proposition is accepted that the essentially oceanic character of the austral forests prohibits any close comparison with the essentially continental northern forests, we are no longer plagued by anomalies and need only define the southern forests in terms of their own intrinsic features. Three major biomes are here proposed, all of them evergreen. These are set out in Table 34.

Table 34. Austral forest biomes.

BIOME	CLIMATE	EXAMPLES
Austral rain forest	Humid-temperate	Tree-fern forest *Dacrycarpus* swamp forest Kauri forest Mixed broadleaf-podocarp forest
Austral beech forest	Humid-cool-temperate	*Nothofagus* subalpine forest
Austral sclerophyll forest	Semiarid	Snow-gum forest (subalpine) Montane gum forest Jarrah-Karri forest (Westralian)

THE DISTRIBUTION OF AUSTRAL FORESTS. Austral rain forest occurs in New Zealand and in parts of Chile. Austral beech forest occurs in the mountainous parts of New Guinea and New Zealand, in uplands and

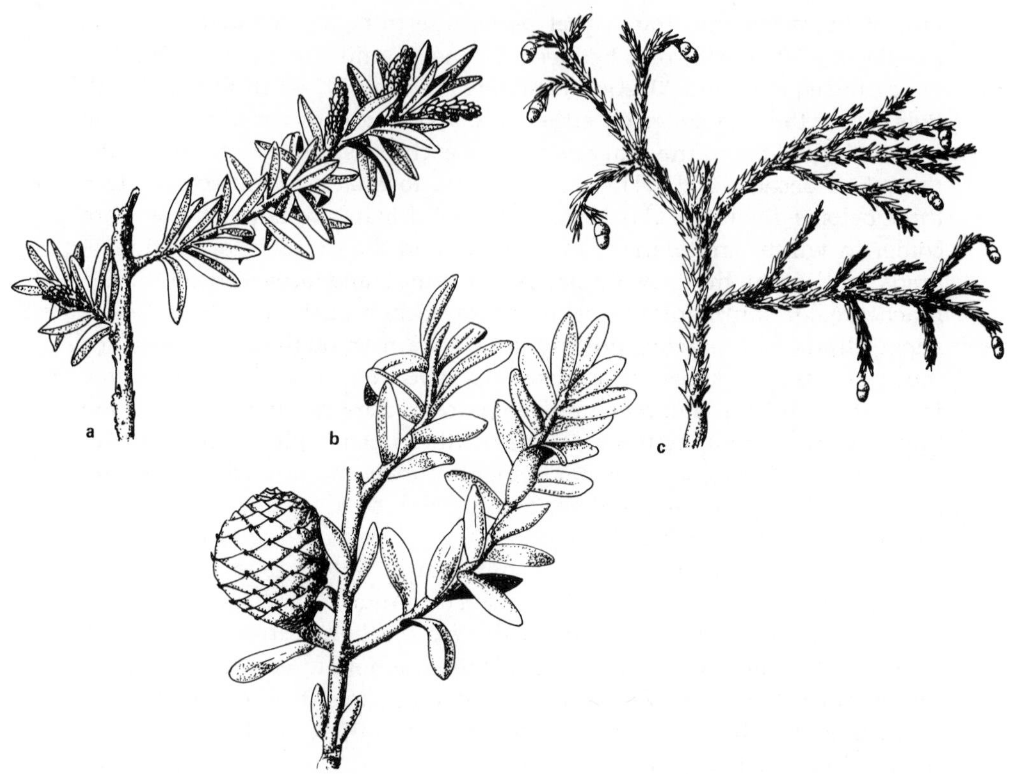

Figure 67. Austral gymnosperm trees. (a) *Podocarpus totara* (totara), Podocarpaceae; (b) *Agathis australis* (kauri), Araucariaceae; (c) *Dacryocarpus cupressinum* (rimu), Podocarpaceae; all New Zealand. Drawn by Desmond FitzGerald.

cool southern coastlands of New Zealand, Tasmania, eastern Australia, and southern Chile. The Austral sclerophyll forests are essentially restricted to the coastal lands of Australia and the southeastern high country of Australia. Sclerophyll forest occurs also in the northern monsoon regions of Australia and in the central desert and grasslands, but these biomes are described elsewhere in this text. Examples of these forests will be discussed briefly in the sequence in which they have been listed in Table 34.

The Rhythm of Life in a New Zealand Rain Forest

In most parts of the world where forests still occur there are two classes of denizens: those, such as the fruit-eating and seed-eating birds, the insectivorous birds, the omnivorous squirrels, the forest-grazing animals, and the arboreal creatures of the high canopy, which are active by day; and the animals of the dark, the predatory owls, the rats and mice, roaming skunks and raccoons, and many of the larger carnivores. When night falls, this second world of living creatures comes to life, and for about eight hours or so the air is filled with the sound of strange calls, mysterious footfalls, and obscure thumps, the snapping of twigs or the rustle of foliage brushed aside by unseen moving bodies. It is still this way in the tropics and in some of the northern woods of America, as it once was in the lost forests of Gaul and Britannia when the Romans first described them. It is still so in the eucalypt forests of Australia; and if the moon is shining, the night creatures call to one another with an intenser fervor, the more so if clouds are passing overhead, for then a whole chorus of cries swells every time the intermittent moonlight illuminates the forest clearings.

But it is wholly different in the remote forests of New Zealand. Here there are no native forest mammals, only a small silent forest rat, which came as a stowaway on the ocean-going canoes of the Polynesian voyagers. There is one native owl, the ruru, so-named by the Maori for its rather

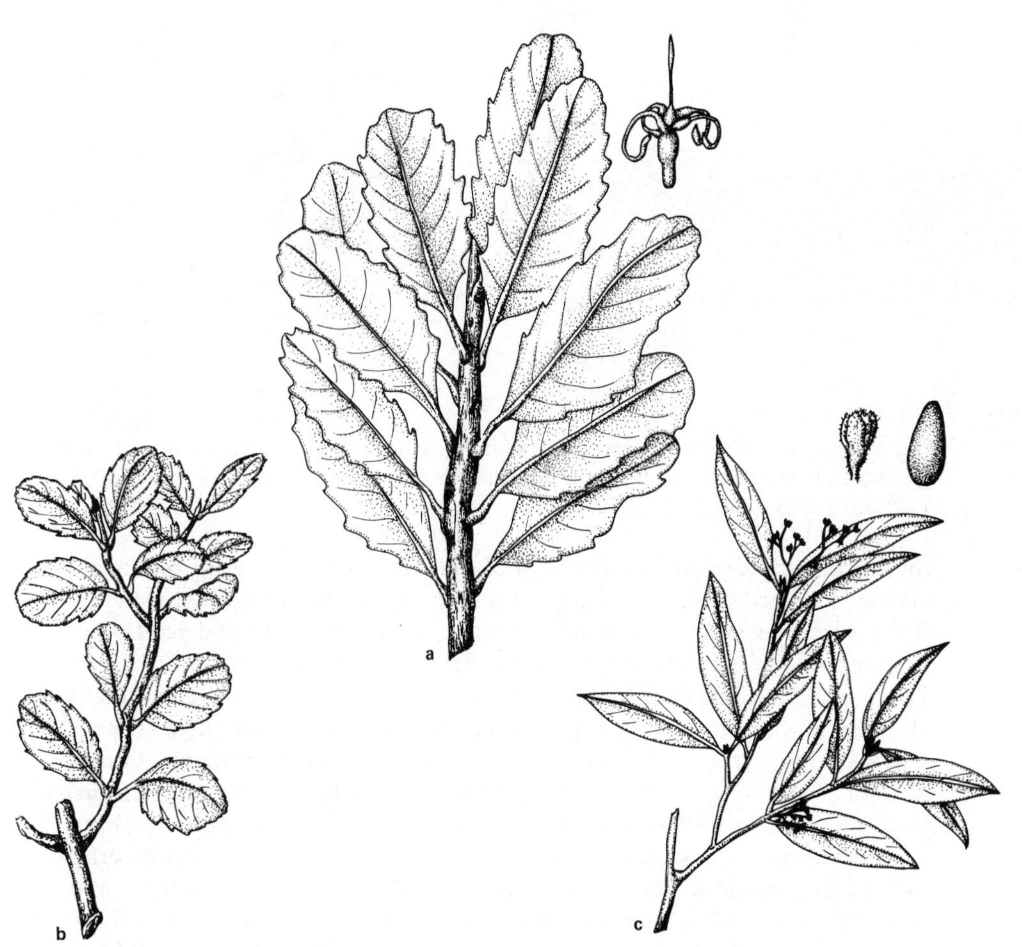

Figure 68. Austral broadleaf trees. (a) *Knightia excelsa* (rewarewa), Proteaceae; (b) *Laurelia novaezelandiae*, Monimaceae; (c) *Beilschmiedia tawa* (tawa), Lauraceae; all New Zealand. Drawn by Desmond FitzGerald.

Figure 69. (a) *Aristotelia racemosa* (wineberry), Tiliaceae; (b) *Metrosideros robusta* (rata), Myrtaceae; (c) *Eugenia maire* (maire), Myrtaceae; all New Zealand. Drawn by Desmond FitzGerald.

mournful and reiterated night call; and the kiwi stalks by night, sometimes crying in a shrill voice. And these calls serve mainly to emphasize the silence, or rather not so much silence but a continuous subdued rustling of the canopy leaves in the wind that almost always blows in these latitudes. There is also the constant drip of the rain forest after the rather frequent showers (about one day in every two is wet), and the murmur of streams and waterfalls. The practiced ear will learn to detect another curious undertone of stridulations, analogous to trilling of night crickets in northern lands, but much fainter and less musical; these are the wetas, roaming night insects of considerable size and possessed of large and powerful mandibles, wingless, with large thorny legs, striped like a faded tiger, and thoroughly repulsive. At night they make the forest floor their own. There is one islet south of the mainland, called Antipodes, where wetas have somehow crossed some 300 miles of sea, landed, and taken possession of the whole islet, for there are no other land animals there; their roving bands

Figure 70. (a) *Fuchsia excorticata*, Onagraceae; (b) *Griselinia littoralis* (broadleaf), Cornaceae; (c) *Pseudopanax crassifolia* (lancewood), Araliaceae; all New Zealand. Drawn by Desmond FitzGerald.

swarm over the beaches and feed on everything the sea casts ashore. On the northern islets another species, the weta-punga, is found and occasionally reaches a length of over 14 inches.

By day the forests are not much less silent than by night. The birds are active, but the noisier species prefer the high forest canopy, where their calls are somewhat subdued by the intervening foliage and the wind. The fly-catching species that frequent the lower strata are small creatures with high-pitched cries for the most part. However, in the more open stretches of forest the honey eaters come down for flowers at the lower levels, and their song at the right season is melodious, especially the voice of the bellbird and the related tui. In more upland areas the harsh cries of the parrots and one of the cuckoos punctuate the quietness. Where a stream winds through the forest the waterfowl send the decibel level soaring; but in general these must be the quietest and most peaceful forests anywhere on earth. In summer the loudest sound is often the chorus of cicadas high above the forest floor in the tall treetops.

Much of this character must have applied to the vanished forests of the late Paleozoic and also those of the Triassic and Jurassic periods. Reptiles are sun-dependent animals, active and rather silent by day, inactive in the cool night. I imagine that these forests, composed largely of conifers not greatly different from the kinds that still live in New Zealand, must have been the home of mainly day-active denizens. When night fell a near-silence would ensue, broken by much the same sounds as those of the wetas, only generated by the great cockroaches of those ancient woodlands, or by the distant chorus of frogs in some swampland by the forest edge. But there are even older forest ecosystems to be found in New Zealand, very limited in extent it is true, but nonetheless recognizable though pale reflections of the Pennsylvanian period, when spore-bearing trees grew in massed luxuriance and insects were the almost undisputed owners of the foraging rights. These ancient relics comprise what may be recognized as the following.

THE TREE-FERN FOREST. The earliest botanist to recognize the tree-fern forest as a distinct type of rain forest was William Colenso. In 1886 he wrote an account of one of his explorations in the North Island in which he notes:

> On a flat in the heart of the forest. . .I found a grove of very tall tree-ferns. I suppose they occupied about three roods of ground (ca. 0.3 hectare) and I estimated their number to be from 800 to 1,000. They were all lofty, from 25 to 35 feet high, and in many places growing so close together that it was impossible to force one's way through them. Their trunks were most profusely covered with the usual epiphytal ferns. . . . Familiar as I long have been with our New Zealand forests, I gazed with astonishment in this deep and secluded grove . . . for I had never before witnessed such a grand display.

Perhaps the most striking tree-fern forest ever discovered is that described by another New Zealand naturalist, the late W. R. B. Oliver, on Sunday Island, in the Kermadec group. Here he found veritable giants. One species, which he named *Cyathea kermadecensis*, has trunks 20 m (65 feet) high and up to 2 m thick at the base (that is, a girth of 20 feet!), and with individual fronds 4 m long (13 feet) and 0.9 m wide. His photographs confirm the dimensions cited, though it is evident from them that nearly all of the girth is made up of massed adventitious aerial roots, structures that commonly occur on the outer surface of the trunks of more moderately scaled tree ferns. Together with his giant ferns Oliver found another species, which he named *Cyathea milnei*. It grows to about half the height of the first-named but redeems its modesty in that dimension by contriving to grow fronds 1.2 m across (nearly 4 feet) and 4 m (13 feet) long. The trunk of this species is not thickened at the base.

On the mainland a few birds are found associated with fern forest, mainly robins (small birds of the genus *Miro*, unrelated to the North American thrushes called robins) and tits *(Petroica)*, both perching birds of the order Passeriformes, insect-eaters similar to the chickadees of North American woods. These birds are probably descendants of ancestors carried by wind from Australia, long subsequent to the separation of New Zealand. There is abundant evidence that small birds are in fact able to cross wide stretches of ocean, but this is a topic for a later chapter. Birds would not, of course, have been found in the earliest fern forests, for they did not evolve until the Jurassic period.

So far as numbers and variety go, the predominant animals associated with fern forests are the insects and, in the soil, the earthworms. There are also carnivorous snails that capture and eat the earthworms, members of the family Paryphantidae, and some fungus-eating slugs of the genus *Athoracophorus*. And on the forest floor some flightless forest birds visit the groves to hunt the earthworms and insects; these are the wekas *(Gallirallus)*, a flightless rail, and kiwis *(Aperteryx)*, small relatives of the moas, emus, and ostriches. The ecosystem probably operates on the lines of Figure 72.

Fern-eating (filicivorous) insects belong mainly to various families of moths. Among the Geometridae species of the genera, *Selidosima* and *Azelina* are known to have caterpillar stages that feed on leaves and hairs of ferns of the genera *Cyathea* (tree ferns), *Todea*, *Polystichum*, and *Thelyp-*

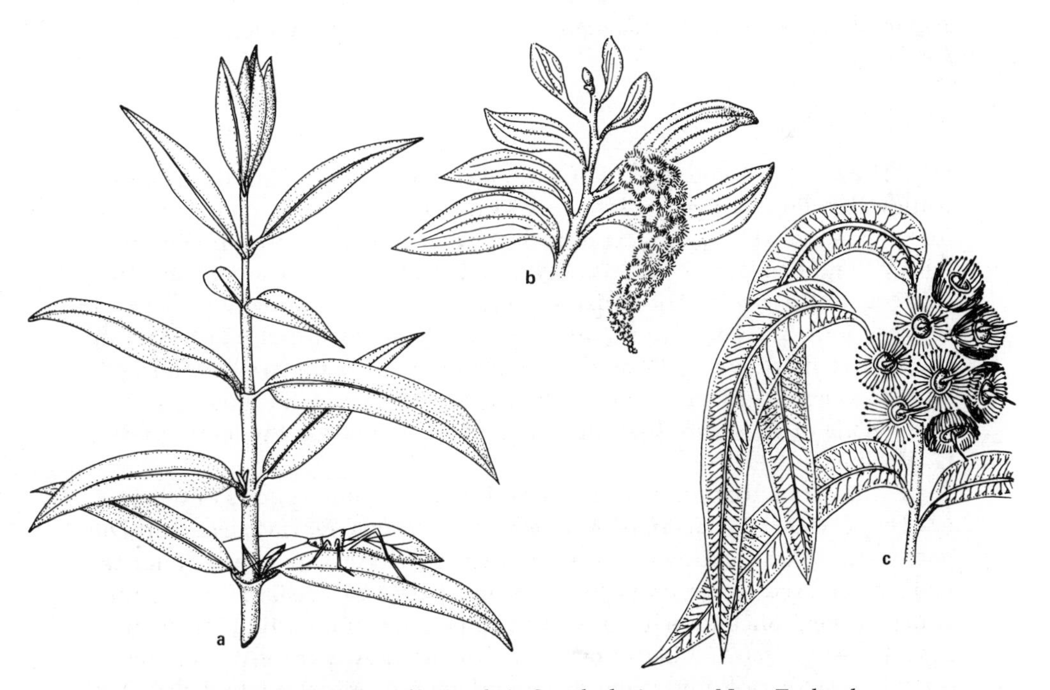

Figure 71. (a) *Hebe salicifolia* (koromiko), Scophulariaceae, New Zealand, drawn by Desmond FitzGerald; (b) *Acacia cunninghamii* (wattle, with phyllodes replacing leaves in adult form), Leguminosae, Australia (23); (c) *Eucalyptus tereticornis* (gum), Myrtaceae, Australia, (23).

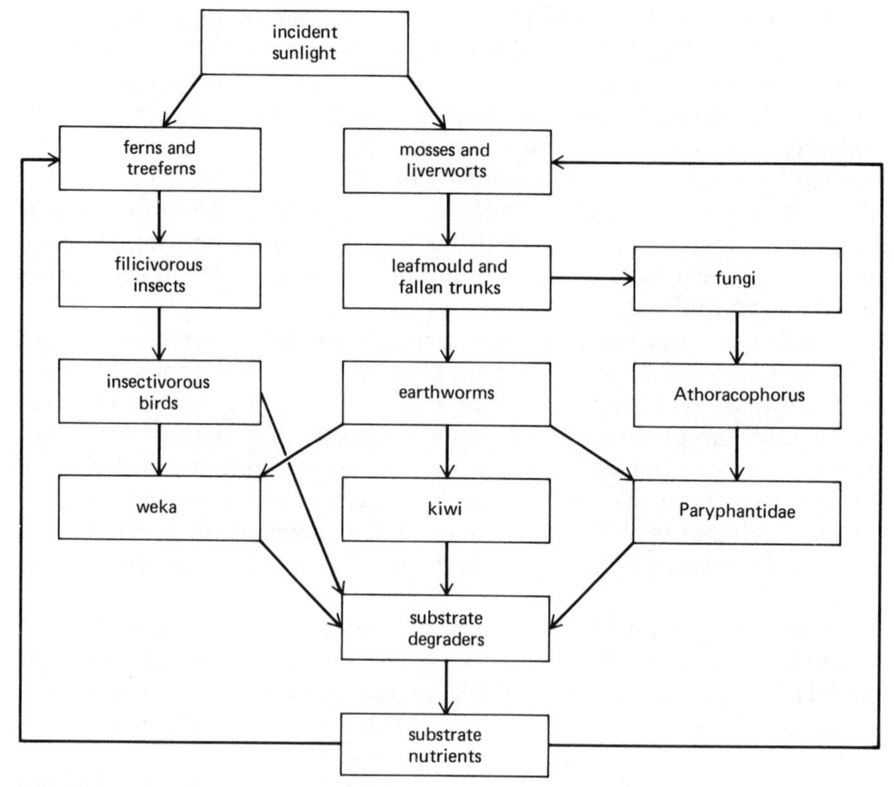

Figure 72. Energy flow and materials transfer in a New Zealand fern forest community (23).

teris. The adult in some species of moth resembles the fern fronds on which it hides by day, flying by night to find nectar in the flowers of a tall rain forest tree called rata *(Metrosideros)*, of the myrtle family. In the moth family Pyralidae, species of *Musotima* feed on the ferns *Adiantum* and *Histiopteris.* In the family Totricidae a species of *Tortrix* is known to frequent the tree fern *Dicksonia*, so its larvae probably feed on that plant. And in the clothes-moth family (Tineidae), a whole series of species belonging to various genera feed on live or dead leaves of ferns in the young stages. The details of these life histories have been elicited by the lepidopterist G. V. Hudson.

Earthworms that are associated with ferns include species of the genera *Dinodriloides, Rhododrilus, Maoridrilus, Neodrilus, Plagiochaeta*, and *Diporochaeta.* These earthworms, some of which occur in South America, have been used as grounds for asserting that New Zealand and South America were once joined, for it was considered impossible for them to disperse across 5,000 miles of ocean. However, they have also been found on subantarctic volcanic islets that must have been completely glaciated during the last ice age, when it is considered all the soil would either have been stripped off by the ice, or else frozen solid for about 70,000 years.

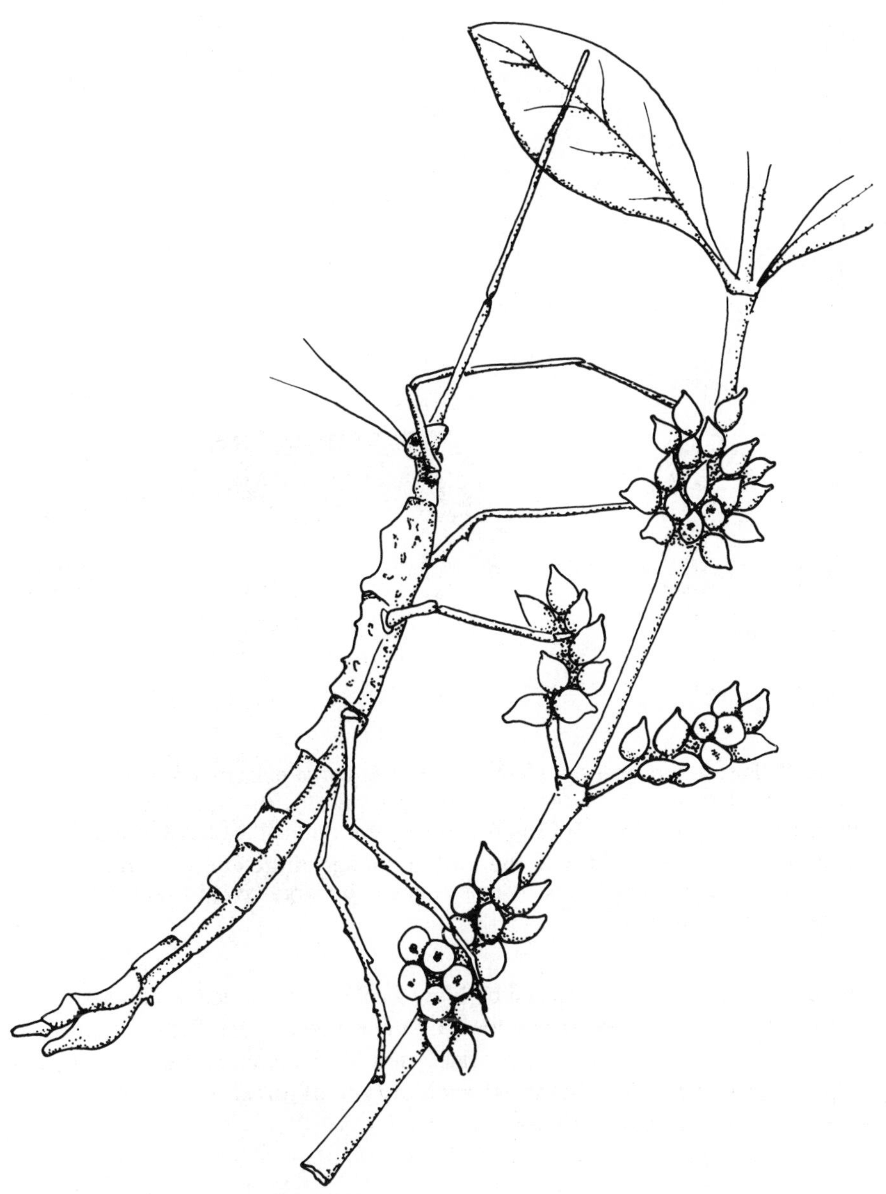

Figure 73. Flightless stick insect (*Argosarchus horridus*) on *Coprosma lucida* (Rubiaceae) in a New Zealand rain forest (23).

Thus it is evident that the earthworms now found on such islets have reached there since the close of the Wisconsin glaciation about 12,000 years ago, a period much too recent for continental drift to have any bearing on the matter of distribution. The earthworms are found often in logs, as also are beetle larvae, and it is most likely that both the earthworms and the beetles that now inhabit the subantarctic islands arrived there as pas-

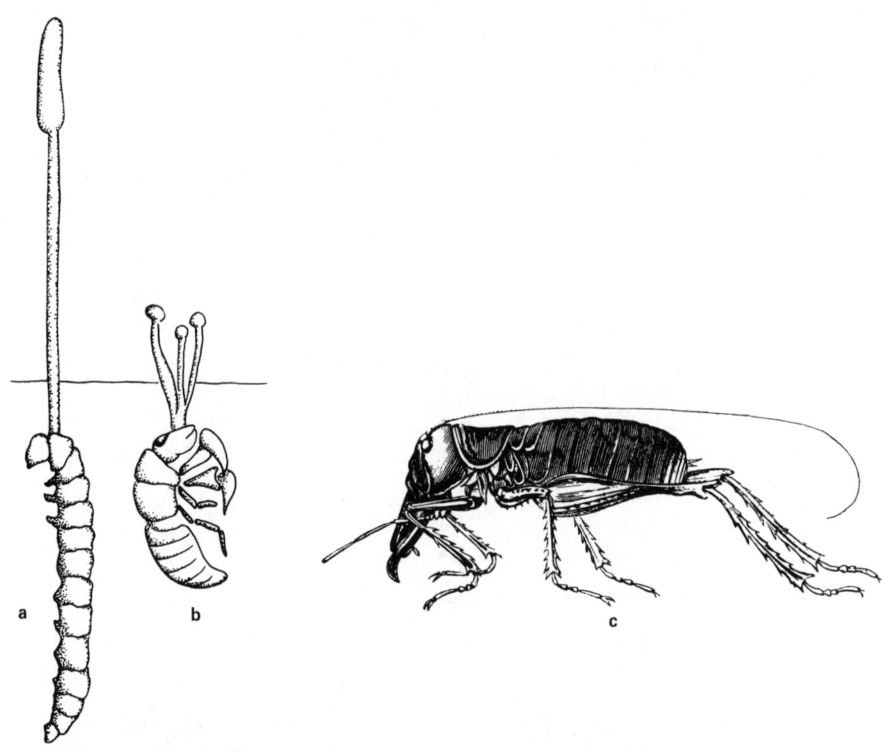

Figure 74. Forest floor insects, New Zealand. (a) Larva of hepialid moth below ground, attacked and converted to woody substance by the fungus *Cordiceps robertsii*, a southern relative of the northern ergot fungus; (b) larva of cicada (*Melampsalta*) similarly attacked; the fruiting bodies of the fungus stand above ground level (23); (c) *Hemideina thoracica*, male, the large-jawed biting flightless weta, Orthoptera (35).

sengers inside logs swept to sea by New Zealand rivers and washed ashore on the islands, as indeed many logs are observed to be at the present time.

The tree-fern forest has been discussed here at some length because of its possible similarity to the most ancient type of forest. The other types of austral forest will now be listed more briefly.

Dacrycarpus SWAMP FOREST. *Dacrycarpus dacrydioides*, or kahikatea, is a forest tree belonging to the gymnosperm family Podocarpaceae. Members of the family have leaves that are like those of the yew, not needle leaves but flattened with a definite midrib, and in some species even flattened so far as to resemble an olive leaf. The seeds are naked, as in all gymnosperms, but are carried at the tip of a juicy, sweet swollen stalk resembling a berry, and taken as food by birds. The podocarps are ancient trees, and *Dacrycarpus* seems to have reached New Zealand in Jurassic times. It occurs as a member of the mixed broadleaf-podocarp forest, but also in pure stands by itself, usually in swampy ground with pools of stagnant water around the roots. Here the trunks stand like a mass of straight ships' masts (and indeed they were felled for that purpose in the old days). Fallen trunks lie like bridges across the pools, and a scrambling mass of screwpine *(Freycinetia)*, a large monocotyledon, usually occurs around the bases of the trees. The insect fauna in these forests is mainly of beetles whose grubs bore into the live or dead podocarp wood. These forests presumably retain the aspect of Mesozoic forests of Podocarpaceae.

THE KAURI FOREST. The kauri *(Agathis australis)*, is a member of the gymnosperm family Araucariaceae, with representative related species in Australia and Chile. The leaves are leathery but decidedly leaflike rather than needlelike; the seeds are borne in cones very much like pine cones. The fossil record shows that kauri was already established in New Zealand in the Cretaceous period. At the present time little remains of the kauri forest on account of ruthless felling, but at the time of the European settlements the forest occupied the northern peninsula of the North Island. Kauri is a tall tree with a stout columnar trunk rising to 60 or 80 feet before the first branch appears. The crown is relatively rather small. It commonly occurs together with two broadleaf dicotyledonous trees, the rata *Metrosiderous robusta* and *Beilschmiedia*. This is often regarded as a subtropical forest, for it does not range south of 38° south latitude. Related trees are the Chilean monkey-puzzle *(Araucaria araucana)* and the Norfolk Island pine *(A. excelsa)*; a third species grows in eastern Australia. Kauri does not form pure stands but is generally mixed with the

Figure 75. (a) *Nothofagus menziesii* (southern red beech), Fagaceae, drawn by Irene Fell; (b) *Elytranthe* (scarlet mistletoe), which parasitizes species of *Nothofagus* in Australia, New Zealand, and South America, drawn by Francis Hutchinson.

trees already mentioned and with tree ferns of the genus *Dicksonia* and ferns of various genera cover the forest floor, a sedge *(Gahnia)* also grows on the ground and *Astelia*, a monocotyledon resembling large clumps of grass.

THE MIXED BROADLEAF-PODOCARP FOREST. The structure of this forest has already been described. Its floristic content varies from one locality to another. In general, the members of the Podocarpaceae associated with rain forest are kahikatea *(Dacrycarpus dacrydioides)*, miro *(Podocarpus ferrugineus)*, matai *(P. spicatus)*, totara *(P. totara)*, and rimu *(Dacrydium cupressinum)*. These are all tall forest trees, forming the ecodominant layer and contributing emergents. According to the locality they are associated with various broadleaf dicotyledenous trees. The rain forest is stratified in approximately six tiers, or layers. Following are representative genera:

1. Tallest emergents, exceptionally 25–30 m (80–100 feet): *Agathis, Laurelia, Podocarpus*
2. Ecodominant layer, 12–25 m (40–80 feet): *Agathis, Laurelia, Metrosideros* (or rata), *Phyllocladus* (or toatoa, a podocarp with flattened leaflike branchlets called phylloclades, each resembling a leathery celery leaf), *Beilschmiedia* (or tawa) ; *Weinmannia, Knightia* (or rewarewa), *Libocedrus* (a cedar with scalelike leaves, also occurring in the Americas), *Dysoxylon* (or kohekohe, a member of the mahogany family also occurring in the tropics), *Alectryon* (or titoki), *Pseudopanax* (or lancewood), *Elaeocarpus* (or hinau, the genus also occurring in the tropics), and others
3. Medium trees, 6–12 m (20–40 feet): *Aristotelia* (or makomako), *Eugenia* (or maire), *Pittosporum* (or karo, now grown in California as an ornamental wind-resistant), *Litsaea* (also in the tropics), *Hoheria* (or houhere), *Paratrophis, Schefflera, Cyathea* (a tree fern)
4. Shrubs, 3–6 m (10–20 feet): *Dicksonia* (a tree fern), *Macropiper* (or kawakawa), *Myrtus, Olearia* (a tree daisy, family Compositae), *Brachyglottis* (or rangiora, another tree daisy), *Coprosma, Carpodetus, Pisonia, Quintinia* and others
5. Small shrubs, 2–4 m: *Melicope, Solanum, Geniostoma,* and also large ferns such as *Marattia, Todea,* and *Blechnum*
6. Field layer of ferns and herbs to 2 m: *Asplenium, Hymenophyllum, Trichomanes, Blechnum, Gleichenia, Adiantum, Pteris,* and a large moss, *Dawsonia superba*

with lianes *(Rubus, Clematis, Rhipogonum)* and epiphytes (*Astelia, Metrosideros*, various orchids and ferns) occurring at all levels.

DIFFERENCES FROM THE NORTHERN TEMPERATE FORESTS. Notable is the complete absence of the needle-leaf pines among gymnosperms, and among the broadleaf dicotyledonous trees notable absentees are the northern families of willows, walnuts and hickories, birches and alders, oaks

Figure 76. *Eucalyptus* savanna, Queensland. Evidence of recent desiccation is the presence in the soil of shallow-rooting systems of former humid forest trees, now replaced by deep-rooting sclerophyll eucalypts. Associated subdominant *Acacia* (23).

and chestnuts, elms, mulberries, magnolias, sycamores or planes, maples, buckeyes, and dogwoods; and absent from the climbers are the grapes and virginia creepers. The occurrence of arborescent types of Compositae (tree daisies) is another difference.

CHILE: AUSTRAL RAIN FOREST IN SOUTH AMERICA. In contrast to the differences from Northern Hemisphere temperate forest, the austral rainforest shows striking convergences with the South American forests of similar latitudes. These similarities extend to various genera not known earlier in geological occurrence than Tertiary epochs. These facts show that theories which attribute the better-known sharing of *Nothofagus* to a supposed former union of the southern continents, and subsequent separation by continental drift, are not only unnecessary, but also seriously lacking by their failure to account for the continuing dispersion of species of forest trees and other plants across the South Pacific into quite late geological epochs. Following are rain forest elements shared by South America. Earliest known geological occurrences are given in parentheses.

1. Tallest emergents and ecodominants	*Laurelia* (Eocene), New Zealand and Chile
	Dacrydium (upper Cretaceous), Malaysia, Australia, New Zealand, and Chile
	Podocarpus (upper Cretaceous), tropics and Australia
	Pseudopanax (-), also known in tropics
2. Medium trees and shrubs	*Aristotelia* (Miocene), Australia, New Zealand, and South America
	Fuchsia (Miocene), New Zealand and South America
	Griselinia (Miocene), New Zealand and South America
	Cordyline (Miocene), Malaysia, Polynesia, New Zealand, and South America
	Mida (-), Juan Fernandez and New Zealand
	Hebe (-), New Guinea, Australia, New Zealand, and South America
	Senecio (-), cosmopolitan genus, here arborescent
3. Epiphytes	*Griselinia* (see above)
	Astelia (Oligocene), New Guinea, Australia, New Zealand, and South America
	Bulbophyllum (-), also Africa, Malaysia, Australia, New Zealand, and South America
4. Lianes	*Muehlenbeckia* (Miocene), New Guinea, Australia, New Zealand, and South America
	Fuchsia (see above)
5. Parasites	*Mida* (-), Juan Fernandez and New Zealand
6. Field layer	*Nertera* (-), Malaysia, Australia, New Zealand, and South America
	Astelia (see above)
	Libertia (Pleistocene), New Guinea, Australia, New Zealand, and South America
	Scutellaria (-), New Zealand and South America

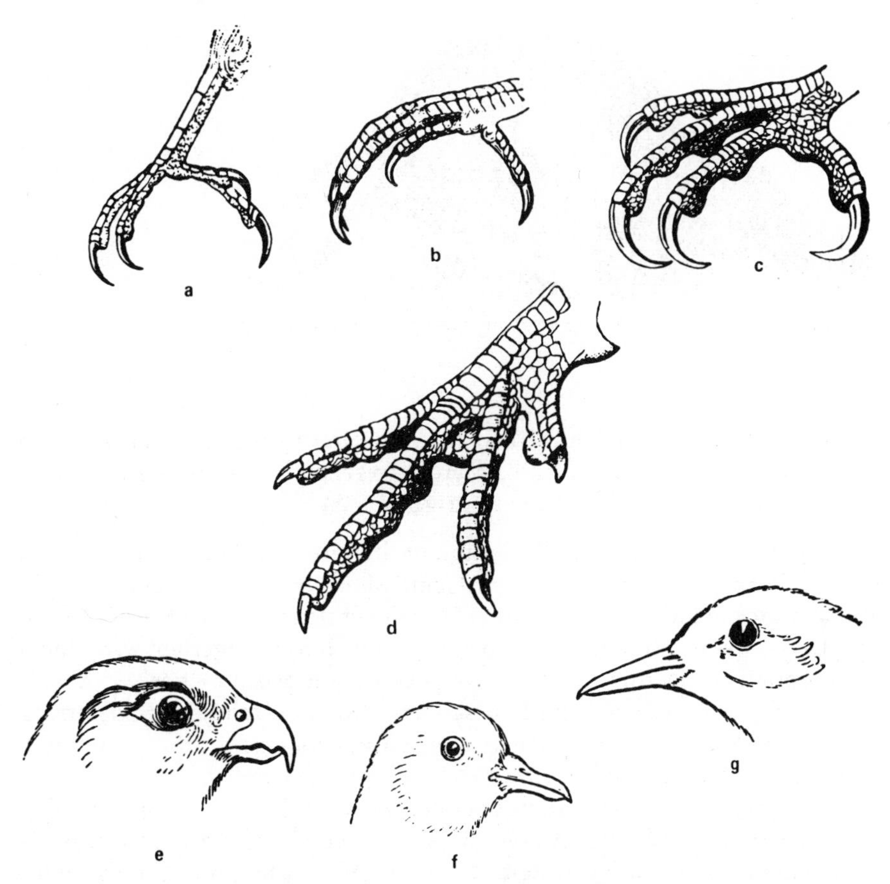

Figure 77. Characters used in Table 35. (a) Foot of Psittaciformes; (b) of Coraciiformes; (c) of Falconiformes; (d) of Galliformes; (e) bill of Falconiformes; (f) of Columbiformes; (g) of Passeriformes (15).

AUSTRAL ELEMENTS REPRESENT TRANSOCEANIC DISPERSIONS. The sharing of elements by the southern continents cannot have been due to former connections by land bridges or continental drift because:

1. The shared elements are of widely different dates of origin in the geological record. As Ernst Mayr has shown for Malaysian birds, heterochroneity contraindicates land-bridge hypotheses (as also continental drift) and points to other types of dispersion.
2. Contemporarily evolving terrestrial animals, such as prototherian, metatherian, and eutherian mammals did not participate in the circumaustral dispersions, even though whole sections of the rain forest ecosystem were dispersed around the Southern Hemisphere. These forms, which would demand a land link, totally failed to use the very link that land-bridge and continental drift theories predicate.

Figure 78. *Ornithorhynchus anatinus* (platypus), Australia (36).

3. Land links would have had to continue through the Tertiary into quite late geological time in order to serve as the dispersion routes for taxa of late evolutionary origin.

Therefore the observed dispersion of the rain forest ecosystem around the southern part of the Southern Hemisphere must have been brought about by some factor that can transfer propagules of trees, shrubs, lianes, epiphytes, plant parasites, and forest ground herbs, together with small invertebrate cryptozoic inhabitants of such plant assemblages.

Whatever the transfer mechanism was, it was incapable of transferring such animals as echidnas, platypuses, bandicoots, and other marsupial mammals of the austral rain forests.

The only mechanism that operates now and that has presumably operated throughout the interval of time since the Cretaceous period is the planetary circulation system, the west-wind drift, and the southern antitrades.

Therefore it is inferred that the observed southern transoceanic distributions are to be attributed to that cause.

The Austral Beech Forest

THE STRUCTURE OF THE FOREST. *Nothofagus*, or southern beech, is represented in the mountains of New Guinea and the highlands of New Caledonia, in the more humid, cooler forests of Australia and Tasmania, in New Zealand, and in southern South America. The genus is composed of evergreen species, except in southern South America where some deciduous forms occur in the coldest part of the range of the genus. *Nothofagus* is classified taxonomically in the family Fagaceae, which includes the Northern Hemisphere beeches, oaks, and chestnuts. The southern genus has small leaves, like those of the European silver birch; the leaves are leathery and sometimes hairy, both adaptations for wind resistance, a desirable feature in the west-wind zone of the southern latitudes. The fossil record shows that *Nothofagus* has been continuously present in the austral forests since the Cretaceous period, about 100 million years ago.

In the Wellington district, in mid-New Zealand, *Nothofagus* occurs as a tree element in mixed rain forest. Where the forest ascends onto higher country the beeches become dominant, and higher still almost the only tree, forming a closed canopy forest in which so little light reaches the forest floor that there are few herbs other than scattered ferns. Lianes are generally lacking, so the forest is easy to walk through. But overhead the canopy in summer is brilliantly colored by the clumps of parasitic mistletoe *(Elytranthe)*, the most conspicuous epiphyte. *Elytranthe*, so intimately associated with the beech forest in New Zealand, is of Malaysian derivation, and it has apparently failed to make the trans-Pacific crossing to South America. Few other plants are especially associated with *Nothofagus* forest, so we are dealing with a rather simple case of transoceanic dispersion, not involving an entire ecosystem as in the case of the mixed rain forests of the austral zone.

It is therefore a curious anomaly that the circumaustral dispersion of *Nothofagus* has occupied reams of published speculation on former land connections and continental drift in order to account for its dispersion.

In the light of the discussion in this work, in which the much more complex case of the mixed rain forest is set out, and its dispersion attributed to the planetary circulation system, there is now little point in examining at length the tedious literature on *Nothofagus*. In my opinion, *Nothofagus* is only one rather simple facet of a more embracing problem, already discussed at sufficient length in these pages. There is, however, an aspect of *Nothofagus* dispersion that has received scant if any attention in the literature, namely, its connection with water. This topic may be briefly noted next.

HOW DOES NOTHOFAGUS DISPERSE? Some authors have inferred that *Nothofagus* seeds must disperse by wind and have attributed this to the supposed presence of a wing on the seed. The "wing" is minute and incapable of aiding flight. John T. Holloway has shown that a falling seed in tumbling from the top of a tall *Nothofagus* in a high wind scarcely traverses more than a horizontal span of about 80 feet before it strikes the ground. The same writer has shown that the aerial survey he carried out

Figure 79. *Sarcophilus harrisii*, Tasmania (23).

a

b

Figure 80. (a) *Trichosurus vulpecula* (Australian opossum) (21); (b) *Thylacinus cynocephalus* (marsupial wolf), Tasmania (21).

on *Nothofagus* forests points very strongly to dispersion by water; the tongues of new *Nothofagus* forest closely match the natural drainage patterns of the streams. Thus it is to be inferred that immersion in water does not harm the seed.

Barber has shown that *Nothofagus* logs of diverse origin are constantly washing ashore on the subantarctic islands, as also on Tasmania. Circumpolar drifts are evidently of regular occurrence. Such logs normally contain crevices or other spaces where the seeds could fall or enter with the aid of the wind during a fall from a ripe capsule.

Tertiary coal beds from New Zealand show that accumulations of *Nothofagus* logs have in former times drifted and become waterlogged and sunk to become coal. The sum of these pieces of evidence suggests strongly that the circumaustral dispersion of *Nothofagus* is brought about by the transfer of occluded seeds in drifting logs, floating downstream from the mountain areas where the forest grows (or falling directly into the sea in the numerous cases where *Nothofagus* occupies fjords and actually overhangs the sea), followed by a west-to-east drift on the west-wind drift, as demonstrated by Barber and his associates in Tasmania. Many invertebrates, especially the larvae of coleoptera, doubtless reached

Figure 81. (a) *Vombatus ursinus* (wombat), Australia (21); (b) *Perameles fasciata* (bandicoot), Australia (21).

the various subantarctic islands by this means, the larvae spending up to five years as tunneling stages in logs that could often be cast ashore before the pupa is due to hatch out and emerge from the wood. In the cases of the subantarctic islands the climate is too cold for the occluded seeds to survive the first winter, so no permanent transfer of *Nothofagus* is possible south of the limiting latitudes determined by the physiology of the beech species.

Sclerophyll Forests: Australian Eucalypts

In marked contrast to the other austral forest biomes, the Australian sclerophyll, or eucalypt, forest has not achieved a general southern dispersion. Since eucalypts flourish when brought to New Zealand and to South America, and reproduce freely in their new habitats, it must be assumed that *Eucalyptus*, and the genera associated with this genus in the gum-tree forests, simply do not have the power of natural dispersion by the means utilized by the other forest species we have considered here. The bark is shed constantly by many eucalypts, so that occluded seeds are likely to be discarded rather than permanently sheltered in some internal crevice. The boles of the trees, in consequence of the bark-shedding habit, tend to be very smooth and unsuited to the occlusion of small objects. The fruits are capsules unattractive to birds, so dispersion by birds is unavailable. Eucalypts are adapted to life in semiarid conditions, and their seed resistance and mode of dispersion is doubtless adapted especially to arid environments. All these factors may have

Figure 82. *Dendrolagus bennettianus* (tree kangaroo), Australia (21).

weighed heavily against dispersion by water. Following are some examples of the temperate aspects of eucalypt forest.

In southern Australia emergents include gigantic *Eucalyptus regnans* (sometimes reaching 90 m, or about 300 feet); sassafras *(Doryphora)*, forming conical trees; lilypilly *(Acmena)*; coachwood *(Ceratopelatum)*. The open nature of the gum forest permits a richer herb cover than in the closed rain forests. Field layer species include members of the genera *Thysanotis, Helichrysum, Dipodium, Correa, Wahlenbergia, Brunonia; Pterostylis* and other orchids, some of them shared with the other forests of Australia and New Zealand; and ferns.

In the montane Tasmanian forest the mixed rain forest meets the sclerophyll forest, bringing together *Eucalyptus gigantea, Nothofagus gunni,* and sassafras; also *Dracophyllum* (a southern heath tree, occurring in New Zealand too), *Hewardia* and *Senecio* in shrub form; the field layer includes *Anemone.*

The southern montane forest includes *Eucalyptus regnans* (here called mountain ash), *Eucalyptus gummifera* (or bloodwood); blue gums such as *Eucalyptus deanei* and *E. saligna*; various other eucalypts (scribbly gum, stringy bark, iron bark); shrubs such as the warratah *(Telopea)*; *Grevillea, Banksia, Acacia.*

The montane and subalpine forest of the eastern mountains includes the emergent *Eucalyptus coriacea,* and shrubs: *Epacris, Pimelea, Senecio* (all occurring in New Zealand in similar situations); the field layer includes many subalpine herbs.

Vertebrate Fauna of the Austral Biomes

The New Zealand and Australian avifaunas are closely related, the former largely derived from the latter, so both may be reviewed together. For reasons given in the next chapter, the South American vertebrates do not form part of the austral biome.

Table 35 relates to forest-inhabiting orders not already mentioned as water birds in the key on page 123. The order Ciconiiformes has been included again as some swamp and forest members were omitted from the review of marine birds.

Table 35. Key characters of the orders of birds in austral forests.

DIAGNOSTIC CHARACTERS	COMMON NAME	ORDER
Wings vestigial; no flight feathers (that is, feathers lack vane)		
Bill long, slender, nostrils at tip	Kiwis	Apterygiiformes (New Zealand)
Ostrichlike birds of New Guinea; horny crest on head	Emus and Cassowaries	Casuariiformes
Wings present; feathers normally developed (with vane)		
Two anterior toes; 1 or 2 posterior toes		
Bill partly covered by a soft crumpled skin (cere)		
Tarsus feathered; nocturnal predators	Owls	Strigiformes
Tarsus naked	Parrots	Psittaciformes
Bill without cere; tail longer than wing	Cuckoos	Cuculiformes
Three anterior toes; 1 or 0 posterior toes		
Outer 2 anterior toes fused together	Kingfishers	Coraciiformes
No so		
Hind toe at the same level as anterior toes		
Bill hooked, toes strongly clawed	Eagles and hawks	Falconiformes
Bill straight; bill, neck, legs long	Herons, egrets, and bitterns	Ciconiiformes
Bill shorter than head, not hooked		
Nostrils covered with fleshy skin	Pigeons	Columbiformes
Not so	Songbirds	Passeriformes
Hind toe at higher level than front toes		
Bill arched; feathers continue on to forehead	Gamebirds	Galliformes
Bill compressed; forehead often covered by a horny flat shield	Rails	Gruiformes

KIWIS [ORDER APTERYGIIFORMES]. Kiwis are the smallest members of the group of paleognathous orders of flightless birds occurring (or recently extinct) in Southern Hemisphere lands. The kiwis are distinguished from the other orders by the elongate bill, with nostrils at the tip (for locating and extracting earthworms). There are four species, all restricted to the New Zealand region; they are nocturnal birds with a shrill whistling call, now restricted to the denser forests. Kiwis have probably undergone reduction in body size; they lay a relatively enormous egg, which the male incubates by more or less leaning on it. The nest is on the ground, often lined with dry fern.

EMUS AND CASSOWARIES [ORDER CASUARIIFORMES]. Emus are ostrichlike birds with three toes (ostriches have only two) on each foot; emus live in the grasslands of Australia and so are not forest birds ordinarily. However, the related genus *Casuarius* ranges the temperate montane forest of Papua. It differs in having a horny crest or casque on the head. Like other paleognathous birds, these assign to the male the task of incubating the eggs.

OWLS [ORDER STRIGIFORMES]. The barn owl *(Tyto alba)* ranges the continents of the world, including Australia, but not New Zealand. The genus *Ninox* includes small owls that take birds, insects, and mice at night; *Ninox boobook* is the boobook of Australia; *N. novaezelandiae* the ruru of New Zealand forests.

PARROTS [ORDER PSITTACIFORMES]. Parrots are numerous and arresting members of the austral forests. In New Zealand two large green and brown species of *Nestor*, called kaka, inhabit the upper canopy, feeding inter alia on nectar, which they collect with a brushlike tongue; smaller parakeets, kakariki (*Cyanoramphus* spp.), also occur, and on the forest floor of the South Island a nearly extinct ground parrot, or kakapo *(Strigops)*, feeds on berries and leaves. In Australia the cockatoos have crests on the head, frequenting open forest and grassland, feeding on the ground in the case of *Kakatoe*, the sulfur-crested white cockatoo, and on forest tree seeds in the case of *Calyptorhynchus*, the red-crested black cockatoo. The superb parrot *(Polytelis)* is another one that feeds on the ground in flocks. The king parrot *(Aprosmictus)* lives in small flocks and feeds on the roots of grasses and on seeds. In open forest clearings the rosella (*Platycercus*) takes seeds and berries. The lorikeet *(Trichoglossus)*, like the New Zealand kaka, takes nectar and also fruit; at seasons when the eucalypts occasionally produce a heady liquor, the lorikeets sometimes fall to the ground intoxicated.

CUCKOOS [ORDER CUCULIFORMES]. Migratory cuckoos, conspicuous with their long barred tail, include the genera *Chalcites* and *Urody-*

Figure 83. *Macropus (Wallabia) agilis* (wallaby), Australia (36).

namis. The species usually winter in the Solomon Islands and in Papua. This means a minimum of a 900-mile nonstop flight for the small New Zealand species to reach the nearest island.

KINGFISHERS [ORDER CORACIIFORMES]. The largest kingfisher is the kookaburra *(Dacelo gigas)*, of the lower tiers of the open Australian eucalypt forest and savanna. It feeds on snakes, lizards, and large insects and centipedes. The extraordinary chorus call is heard on the soundtrack of synthetic jungle movies of the Tarzan genre. *Halcyon sanctus* is the common blue kingfisher, or kotare, of New Zealand and Polynesia.

EAGLES AND HAWKS [ORDER FALCONIFORMES]. *Uroaetus,* the wedge-tail eagle of Australia, is 1 m in length. It feeds on wallabies, and when Australia was first settled mistook clothed whites for kangaroos, which it attacked. Its nest takes three months to build, from branches, and is large enough to fill a truck. *Harpagornis* was a similar extinct eagle of New Zealand that fed upon moas. *Falco* and other genera still survive in the forests.

HERONS, EGRETS, AND BITTERNS [ORDER CICONIIFORMES]. The white heron, or kotuku *(Egretta)* is a cosmopolitan bird, now increasing under protection. Public agitation in 1910 resulted in laws in America that saved the species from extinction under attack from hunters for plumes for women's hats. The bittern *(Botaurus)* is another cosmopolitan, a swamp dweller.

PIGEONS [COLUMBIFORMES]. There are five families of pigeons and all are represented in Australia, the major world center of distribution of the order therefore. Two of the families occur nowhere else, and nearly one half of all known species of pigeons are found in Australia. The crowned pigeons *(Guaridae)* of New Guinea are the largest species known in existing faunas. The fruit pigeons or wonga of the east Australian scrubs *(Leucosarcia)*, and the bronzewings *(Phaps)* are among the well known Australian members of the order. In New Zealand the kereru or wood pigeon *(Hemiphaga)* is represented by forest-dwelling species on the mainland and in the Chatham Islands. All these forms are primarily fruit and foliage eaters.

GAME BIRDS [GALLIFORMES]. Quail *(Coturnix)* are the chief Australian and New Zealand representatives of the order, scratching ground-dwelling insect-eating and seed-eating birds. The New Zealand species became extinct after the European settlement. Pheasants, which represent the order in many lands, are lacking entirely from the region (save as introduced species). Peculiar to Australia are large communal birds known as mallee fowl or megapodes *(Leipoa)*, about the size of a small turkey; they do not hatch their eggs by sitting on them, instead they construct mounds of earth and decaying vegetation, and deposit the eggs in them, several hens sharing a mound. The heat generated by decomposition incubates the eggs.

RAILS [GRUIFORMES]. As noted in Table 35, many of these birds have a flat horny shield on the forehead. Most are long-legged swamp birds, and like other aquatic birds, tend to range widely, with rather similar species on the various continents. The gallinules *(Porphyrio)* are dark blue birds about the size of a duck, with long red legs and a red bill, known in Australia as swamp hens, in New Zealand as pukeko. A larger and more compact relative is *Notornis*, a rare upland bird of the south island of New Zealand, formerly thought to be extinct, and still an endangered species rigorously protected in the mountain valleys, but threatened by introduced mammals. The wekas *(Gallirallus)* are smaller relatives of a brownish color inhabiting wooded areas of New Zealand. Some species of rails became flightless in the absence of predators, but fell victim to man or his client mammals after the human settlements began.

SONGBIRDS [ORDER PASSERIFORMES]. In this very numerous group, notable Australian members are the brush-tongued honey eaters, of which the many Australian birds called honey eater, greenie *(Meliphaga)*, banana bird, and the New Zealand tui *(Prosthemadera)* and bellbird *(Anthornis)* are examples. The silver eye *(Zosterops lateralis)* successfully colonized New Zealand from Australia in 1855 and 1856, flocks flying the Tasman Sea (1300 miles) on the strong west winds of that year. Australian swallows *(Hirundo)* have recently performed the same feat. These events show how New Zealand acquired its avifauna. The bowerbird *(Chlamydera)* has in some species attentive males that build and decorate special grass tunnels adorned with colored materials, silver paper, and so on, for the edification of the females. The lyrebird *(Menura)* is an astounding mimic, the male putting on 60-minute performances in which he will imitate for the female (or any adoptive human substitute) all the sounds of the neighborhood, the distant barking of dogs, the calls of all other birds, human conversation, whip cracks, and so on.

Monotremes (Subclass Prototheria)

EGG-LAYING MAMMALS [PROTOTHERIA]. The platypus *(Ornithorhynchus)* and the spiny echidna *(Tachyglossus* and *Zaglossus)* are mammals that lay eggs in underground burrows, suckling the young with milk after the eggs hatch out. These characters, intermediate between those of reptiles and other mammals, define the subclass Prototheria, of which Australia and New Guinea constitute the only known range. The platypus is semiaquatic, the echidna is a fossorial (burrowing) animal. *Zaglossus*, with a long slender snout, is the New Guinea representative of the group, the others being Australian and Tasmanian.

Marsupials (Subclass Metatheria)

The pouched mammals and their relatives make up this subclass; there are representatives in South America, the family Didelphiidae, characterized by having paired uteri; they are not closely related to the Australian forms, where the uteri are partially fused to form a single Y-shaped structure. The Australian members comprise seven families, which may be distinguished in Table 36.

FAMILY DASYURIDAE. *Antechinus* and *Phascogale* are mouselike or ratlike animals, the latter with a squirrellike tail. They are insectivorous. *Dasyurops*, the spotted marsupial cat; *Sarcophilus*, the Tasmanian devil; and *Thylacinus*, the so-called tiger, like a striped wolf—these are all carnivores, unfortunately becoming rare or extinct.

Table 36. Key characters of the families of Australian marsupials.

DIAGNOSTIC CHARACTERS	COMMON NAME	FAMILY
Toes distinct; incisor teeth numerous, usually 8 in each jaw		
Cursorial, often doglike, catlike, or mouselike	Marsupial cats	Dasyuridae
Fossorial, molelike	Marsupial moles	Notoryctidae
Second and third toes of hind feet joined by web		
Incisor teeth numerous, usually 8 in each jaw	Bandicoots	Peramelidae
Only 2 incisors in lower jaw (Diprotodontia)		
Tail rudimentary		
Arboreal; forelimb with digits	Koala	Phascolarctidae
Terrestrial; fossorial	Wombats	Vombatidae
Tail well-developed		
Arboreal; squirrellike	Phalangers	Phalangeridae
Terrestrial leaping forms	Kangaroos	Macropodidae

FAMILY NOTORYCTIDAE. *Notoryctes* is the only included genus. Adapted for subterranean life as in true moles, the eyes have atrophied, the external ear reduced, and the limbs adapted for digging.

FAMILY PERAMELIDAE. Bandicoots are a widely divergent group, some resembling rats, others are more like rabbits, with large ears. An example is *Isodon*.

FAMILY PHASCOLARCTIDAE. *Phascolarctos*, the koala, is the only included member. These charming animals are familiar to all, at least through television. Now strictly protected, the shocking toll of earlier hunting seems at last to be under repair.

FAMILY VOMBATIDAE. Wombats are rather badgerlike or bearlike terrestrial blunt-nosed animals that dig burrows. *Vombatus ursinus* is a Tasmanian example.

FAMILY PHALANGERIDAE. The phalangers, or Australian opossums, are arboreal long-tailed animals, more or less omnivorous like most marsupials. *Pseudocheirus*, the ring-tail opossum, and *Trichosurus*, the brush-tail opossum, are examples; the latter was introduced into New Zealand where it has now become a serious pest, destroying the eggs and young of defenseless native birds. *Phalanger* is the New Guinea cuscus.

FAMILY MACROPODIDAE. This is a large family, though most of the smaller members are now undergoing extinction at the hands of introduced foxes. The best-known are the large kangaroos, *Macropus*. Also included are *Dendrolagus* (tree-dwelling) and *Petrogale*, rock-wallaby.

12

TROPICAL FORESTS

The tropical forests can be defined as those biomes in which trees are the ecodominant plants and that lie between the Tropics of Cancer and Capricorn. The land area of the tropics is divided into three main continental regions, namely, Central and South America, Africa, and southern Asia with Indonesia; since the three land regions are separated by oceans, it is not surprising that considerable differences exist between the floristic and faunistic content of each region. The African and Asian regions are the most similar and are seen to be much more similar if fossil biota are taken into account. About 25 million years ago, for instance, many more mammalian groups were shared by India and Africa than is the case today, and it seems likely that continuous forests once ranged across the intervening desert areas that now occupy southwest Asia and Arabia. South America, on the other hand has a mammalian fauna so distinctive as to imply very long isolation from the rest of the world. Australia, as already noted, experienced similar isolation, together with New Guinea and those parts of the Australo-Papuan region that lie to the southeast of Wallace's line of

Figure 84. Architecture of a tropical rain forest (Brazil). (a) Dominant emergents to height of 150 feet; (b) ecodominant layer; (c) intermediate tiers; (d) field layer (23).

demarcation of faunas. These latter regions now lie in the tropical belt, and their forest biota must also be taken into account in this chapter.

Wide differences in climatic conditions occur within the tropical zone, so much so as to produce several very distinctive forest biomes. The most important of these are the following: (1) tropical rain forest, (2) monsoon forest, (3) savanna, and (4) montane tropical forest.

TROPICAL RAIN FOREST. This is a tiered rain forest of the same structural type as the austral rain forest, with several tiers, emergent tall trees, an ecodominant layer, intermediate tree layers, a shrub layer, and a forest floor or field layer; in addition epiphytes grow on the trees, and lianes form a scrambling knotted tangle between the trees. The same features—plank buttresses, columnar trunks, small canopy crowns, cauliflory, stilt roots, and so on,—as already noted for austral rain forest also typify tropical rain forest. Additional features include the presence of drip tips on the leaves, the frequency of very large leaves in some families (as, for example, the banana family Musaceae and gingers Zingiberaceae). It is unusual for stands of single species of trees to form dominant communities in tropical forests; the species tend to be very mixed, a fact that fortunately has hindered the ruthless exploitation of tropical forests by milling. Tropical rain forest in general requires an annual rainfall of not less than 200 cm (80 inches) to compensate for the increased evaporation at temperatures ranging from a mean of about 20° C in the coolest month to a mean of 25° C in the warmest month. The rainfall can be as much as 500 cm per annum. Rain forests occur in the Central American region, in the West African region (covering about 10 percent of the African continent), in the wetter parts of tropical Asia, and in Indonesia. These forests are very rich in species; for example, about 11,000 species of flowering plants have been reported from the island of Borneo alone. Some animal groups are largely confined to the tropical forests; about 70 percent of all Amphibia inhabit the damp tropics.

MONSOON FOREST. This forest occurs in regions subject to monsoon seasons, with alternating dry and wet periods. The rainfall is generally less than 200 cm per annum. During the dry season many of the trees lose their leaves, or at least cease to grow. An example of such a biome is seen in parts of the Ganges valley. When the wet season begins in June the soil becomes waterlogged. Grasses germinate, marshes enter their wet phase, and trees bud. By August, when the temperature has a daily range of about 20–35° C, the vegetation is growing at its maximum rate and all types of herb and tree reach the peak of luxuriance. By October the cool dry season begins, with the temperature ranging each day from 6–32° C, and cloudy skies; the grasses now begin to wither, most of the trees flower and set fruit, and winter annuals sprout. By January the hot dry winds begin, with daily temperature range from 18–45° C, and the deciduous trees now shed their leaves. By March new buds are appearing on the trees, and the evergreen trees shed their leaves and immediately replace them by new leaves. The marshes enter the dry phase. By June the thorn bushes have dried up and are likely to catch fire. The rains begin again about June 15, which marks the end of the dry season. Similar annual cycles also occur in monsoon regions of the southern tropics, with the calendar six months out of phase.

In 1848 the British botanist Sir Joseph Hooker began a two-year botanical journey through the northeastern forests of India, Nepal, and Burma. One of the monsoon forests he visited was in Sikkim, Nepal. His diary entries tell us much of what the Asian forests were like before the population explosion occurred. From Sikkim he wrote:

The rains commenced on 10th of May, moderating the heat by drenching thunderstorms. Bhomsong was looking more beautiful than ever in its rich summer clothing of tropical foliage. The little flat on which I had formerly encamped (in December) was now covered with a bright green crop of young rice. The oaks are very grand. I measured one (whose trunk was decayed and split into three however) which I found to be 49 feet in girth at 5 feet from the ground. . . . The valley was fearfully hot, and infested with mosquitos and peepsas (*Simulium*, a gnat). Many fine plants grew in it. I especially noticed *Aristolochia saccata*, which climbs the loftiest trees, bearing its curious pitcher-shaped flowers near the ground only; its leaves are said to be good food for cattle. But the most magnificent plant of these jungles is *Hodgsonia* (a genus I have dedicated to my friend Mr. Hodgson), a gigantic climber allied to the gourd, bearing immense yellowish-white pendulous blossoms whose petals have a fringe of buff-coloured curling threads, several inches long. The fruit is of a rich brown, like a small melon in form, and contains six large nuts, whose kernels are eaten by the Lepchas. The stem, when cut, discharges water profusely from whichever end is held downwards. The Took (*Hydnocarpus*) is a beautiful evergreen tree, with tufts of yellow blossoms on the trunk; its fruit is as large as an orange, and is used to poison fish, while from the seeds an oil is expressed. Tropical oaks and *Terminalias* are the giants of these low forests, the latter especially, having buttressed trunks, appear truly gigantic; one of a kind called "Sung-lok" measured 47 feet in girth at 5 feet from the ground, and 21 at 15 feet from the ground, and it was fully 200 feet high. I could only procure the leaves by firing a ball into the crown. Birds are very rare, as is all animal life but insects, and a small freshwater crab, *Thelphusa*. The commonest of a few snails was a *Cyclostoma*.

SAVANNA. This term is applied to light forest in which the trees are openly spaced with patches of grass between them. It characterizes tropical regions where the rainfall is not adequate to support a luxuriant tiered forest under the present conditions of animal and human occupation. The evidence suggests that, given the presence of grazing animals, especially those associated with human occupation, tropical rain forest cannot regenerate, for the early stages of the subsere are disrupted by excessive grazing, or by human interference. Some botanists think that the equatorial rain forest requires a wetter climate than now exists for its successful regeneration and that savannas are spreading at this epoch in the earth's history. Others believe that all savannas have been man-induced, and that human penetration and burning of natural forests has led to the spread of savannas and the drastic contraction of rain forest areas. There is archeological evidence indicating that rain forest expanded in Africa during glacial phases in temperate regions, and that savanna and desert expanded during interglacial warm periods. However, it does also appear that during pluvial phases in the Congo, savanna phases occurred contemporaneously in East Africa-Kenya. For example, during the Danjerian pluvial period of the Congo, correlated with the third, or Riss, glaciation of Europe, such mammals as giraffe roamed East Africa, indicating a savanna biome. Although there is abundant evidence of the presence of Stone Age man in East Africa at this time, it seems very dubious whether human interference produced or maintained savannas, for during the Riss glacial stade human populations must surely have been sparse, and the giraffes that require savannas range far back into Tertiary time, long before man or his fires could possibly have had any bearing on the distribution of savanna. In other words, though man may create savannas, not all savannas are created by man. Possibly the forest and savanna are in delicate balance, and any change in either climate or grazing pressure or human interference elicits a rapid change in the forest-savanna boundaries. Savanna at present is very extensive to the north and south of the equatorial forest in Africa, with its boundaries set by the Sahara and Kalahari deserts. Savanna areas also occur in Asia and Central America. Savanna also occurs as an ecotone wherever a grassland biome meets a forest, as where the long-grass prairie of North America meets the temperate forest—so savanna is by no means restricted to tropics. However, the tropical savanna may well be included as a tropical forest biome for the reasons already given, which suggest strongly that savanna and tropical rain forest readily mutate one into the other as climatic or other parameters change.

The claim that tropical forest cannot of itself regenerate on the site of a man-made clearing is not supported by the fact that lost cities in Indo-China and Indonesia now lie within tracts of rain forest, matching similar forest-engulfed Mayan ruins in Central America. Here is Sir Stamford Raffles, writing in 1830 of Java:

> The traditions of the country concerning the former seats of government enable us to trace at this day the site of . . . Majapahit . . . and other unequivocal vestiges of former cities. . . . In the centre of an extensive forest is pointed out the site of Setingel, distinguished by heaps of stones and bricks. At Majapahit the marks of former grandeur are more manifest . . . the wall about twelve feet high . . . surrounded by a noble forest of teak. . . . On the right of the enclosure was the tomb of the princess Putri Champa . . . in the Mahomedan style, and having upon it, in ancient Javan characters, the date 1320 . . . the ruins of the palace and several gateways are to be seen, but the whole country for many miles, is thickly covered with a stately teak forest, which appears to have been the growth of ages, so that it is difficult to trace the outline of this former capital. . . .

Evidence of this kind, coupled with early medieval dating of the abandonment of such cities, reminds us that there seems to have been a cooler world climate (to judge by herring distribution) about A.D. 1300 and later, see Fig. 1, Chapter 1. Perhaps the inferred cold period initiated at that time may have caused the tropics to be wetter, and so aided the regeneration of forest and the invasion by trees of areas that nowadays would tend to remain as savanna.

Figure 85. Interior of humid tropical forest (Indonesia). To the left, tree ferns and trunk of ecodominant with liane stems suspended; foreground, field layer; right, a shade palm and trunk of buttressed ecodominant, from which hang aerial roots of epiphytes and against which are other aerial roots and ascending stems of lianes. The swollen stream betokens the rainy season in progress.

Figure 86. Stilt roots developed by a forest tree that began life as an epiphyte similar to that shown in Figure 64; subsequently the host tree was smothered and destroyed, leaving the epiphyte in sole possession, though with other epiphytes now growing upon it. From a sketch made by Alfred Russel Wallace in Borneo (42).

MONTANE TROPICAL FOREST. These forests warrant the epithet tropical only by virtue of their occurrence within the geographic tropics. In other respects they are members of subtropical, or temperate, or cold temperate zones, depending on the altitude at which they occur. On the Himalayas, for example, occur monsoon forests at lower elevations; above these come temperate deciduous forests with oaks and olives and other northern temperate trees; then come bamboo forests, and these are succeeded at higher elevations by spruce and fir. In southern equatorial regions the montane forests yield podocarps and many other trees characteristic of the austral rain forests. These zoned forests clearly match the isotherms of their environments and demonstrate the control that climate exerts on the floristic content of forests. Botanically the montane forests of the tropics are more logically considered with the appropriate temperate biomes that they most closely approximate; zoologically, however, they stand somewhat apart, for much of their included fauna is closely related to the tropical fauna of the lowland forests of the tropics, while a sizable faunal element is of temperate or even subpolar character. In the Himalayas monkeys can be observed in such improbable situations as fir forests, together with such unexpected fauna as migratory European woodcocks and starlings and even gulls. On the southern side of the equator, the high forests of the Owen Stanley Mountains of New Guinea shelter a marsupial fauna like that of Australia, and also the forest cassowary. Animals are evidently much more adaptable to environmental differences than are plants, and this is doubtless the reason why it is easier to define biomes in terms of plants rather than of animals. If animals tend to inhabit certain regions or biomes in preference to others, it is probably to a large extent due to their preference for certain plants or trees or their fruits and seeds.

One way to garb the bleak lists of tropical tree names with more meaning, without entering into tedious descriptions, is to find out the nearest well-known relatives in forests nearer home. Table 37 is a short list of such comparisons, showing the ordinal name of the main groups on the left, the vernacular name of tropical representatives in the middle column, and the names of common Northern Hemisphere trees of the same orders in the right-hand column. The sections that now follow are intended to amplify these comparisons by outlining the roles that particular plants play in the different levels of tropical forests.

Trees of the Tropical Forests

ORDER PALMALES. The general aspect of a palm is well known. Two main types may be distinguished: (a) *feather palms*, in which the frond is divided pinnately, like a great fern frond; these include the tallest species, of genera such as *Archontophoenix*, the tall king palm of tropical Indonesia and Asia; *Roystonia*, the royal palm of the Caribbean, with tall slender smooth trunk surmounted by a crown of fronds whose bases are swollen and sheath one another; and *Phoenix*, the date palm and its relatives; (b) *fan palms*, in which the frond is shaped as the name sug-

Table 37. The principal orders of plants that contribute trees and lower tier elements to the tropical forests, with equivalent northern hemisphere representatives.

NAME OF ORDER	TROPICAL EXAMPLES	REPRESENTATIVE IN NORTHERN FORESTS
Palmales	Feather palm, fan palm, coconut	Palmetto
Casuarinales	She-oak, or ironwood	(Absent)
Juglandales	Engelhardtia	Hickory and walnut
Urticales	Nettle tree, figs	Elm
Ranales	Nutmeg, cinnamon	Magnolia
Rosales	Acacia, rosewood	Apple and robinia
Geraniales	Mahogany	Lemon
Sapindales	Mango	Maple and sumac
Malvales	Hibiscus, cocoa, kapok	Linden
Parietales	Sal, *dipterocarpus*	Camellia and St. John's wort
Myrtales	Clove	Myrtle
Umbellales	Ginseng	Dogwood
Ebenales	Ebony	Persimmon
Gentianales	Curare	Olive and ash
Tubiflorales	Teak	Catalpa
Rubiales	Quinine, coffee	Elder and honeysuckle

NOTE: Notable absentees from the tropical forests are the Salicales (poplar, willow) and Fagales (oak, beech, chestnut, birch).

gests, the whole plant is more squat, often with branching or multiple trunks, as in *Licuala*, a tropical Pacific species.

Palms vary much in their ecology. Some, like *Licuala* and *Chamaedorea* of Central America, grow only in the rainy season and require forest shade and abundant moisture. The tall species will grow in forest, where they contribute to the emergents of the ecodominant layer. Small species frequent the forest margin, clearings, and also the savanna; and some, like the coconut *(Cocos)*, grow along shorelines and on atolls. Some are xerophytic and can withstand severe drought, owing to the dense cuticle covering the leaves, and matted fiber on the trunk; such forms, for example, the date, can thrive in an open desert environment, provided underground water is available.

ORDER CASUARINALES. She-oak, or ironwood, has been mentioned already as part of the Australian flora. *Casuarina* occurs in rain forest, particularly near clearings and on savannas, in Indonesia and in India.

ORDER JUGLANDALES. Of the hickorylike trees that are so conspicous in northern forests, the tropical representatives are few and minor. A notable exception is the magnificent rain forest ecodominant *Engel-*

Figure 87. Savanna ecotone with grasses, shrubs, and exposed tree fern, Madagascar (22).

hardtia, of Java, one species, *E. spicata*, reaching a height of 200 feet. In former times cross sections of the trunk were made into cartwheels, like those shown in rock carvings of the Bronze Age. The wood is highly resinous. Species of smaller size occur in the Himalayan forests and in the Assam valley of Burma.

ORDER URTICALES. Nettles, elms, and figs constitute three families of this order, each with tropical representatives. Nettles (family Urticaceae) reach giant size in *Laportea*, the nettle trees of Asia, Indonesia, and Australia, though considered as trees they are not large. A Himalayan species exudes a noxious vapor, and all species inflict painful stings. The elms (family Ulmaceae) have only a few large representatives in the tropics, for example, *Chaetoptelea*, a hugh elm of Costa Rica, and *Gironniera*, of Polynesian forests. In Samoa the latter genus may yield the larger trees, from which sea-going canoes are cut. Figs (family Moraceae) are repre-

Figure 88. Tropical montane forest, Rungit Valley, at about 3000 feet, Himalayas, lower limit of *Pinus longifolia* (see also Chapter 16). Sketch by Sir Joseph Hooker (31).

sented by numerous species of *Ficus*, some of which reach a great spread on account of their habit of producing ascending and descending aerial stems so that the tree grows outward and becomes a large grove. Best-known as the soursop *(Artabotrys)*; also in this family is *Canangium*, yield-tree *(Ficus elastica)*, both of Asia. Figs are pantropical.

ORDER RANALES, THE SPICES. Tropical shrubs and trees of varying size are members of the families that make up this order. The nutmegs, genus *Myristica*, comprise a family of Asian and Indonesian shrubs whose nuts yield mace. The cinnamon trees, family Lauraceae, are important commercially in Asia and Indonesia, the spice being obtained from the bark of a small shrubby species, *Cinnamomum fragrans*. A larger species, *Cinnamomum camphora*, reaches a height of 40 feet and yields camphorwood. The same family includes the South American lumber tree *(Ocotea)*,

Figure 89. (a) *Cinnamomum iners* (camphorwood), Lauraceae, Asia; (b) *Ficus elasticus* (India rubber tree), Moraceae, Asia; (c) *Engelhardtia spicata*, Juglandaceae, Ceylon, with the orthopterous leaf insect *Pulchriphyllum crurifolium*. Drawn by Desmond FitzGerald.

Figure 90. (a) *Hevea brasiliensis* (rubber), Euphorbiaceae, Brazil; (b) *Acacia caffra*, Leguminosae, Africa; (c) *Mangifera minor* (mango), Anacardiaceae, Asia. Drawn by Desmond FitzGerald.

or greenheart, originally the material from which the Panama Canal lock gates were fabricated; and *Persea*, or avocado, a South American tree with edible fruits. In the family Annonaceae is a South American fruit tree, the custard apple *(Annona reticulata)*; and an Asian species of the family is known as the soursop (*Artabotrys*); also in this family is *Canangium*, yielding in Polynesia trees large enough for canoe hulls.

ORDER ROSALES, THE TREE LEGUMES. The family Leguminosae includes many common tropical genera of trees and shrubs, for the most part members of the middle and lower tiers, or trees of moderate size occurring in stands on the savannas. *Acacia*, *Bauhinia* (butterfly tree), and *Erythrina* (coral tree) are wide-ranging trees of middle size, the latter important as a food species for sunbirds in Africa. *Erythrina*, *Tephrosia* (arrow poison tree), and *Dalbergia* (rosewood), are nearly pantropical. Central American genera include *Haematoxylon* (logwood), *Myroxylon*, *Brya*, and *Eperua*. In West Africa occur *Delonix* and *Colvillea*; in the Far East *Albizzia*, *Pongamia*, *Butea*, and many others.

ORDER GERANIALES, THE MAHOGANIES. These are tall ecodominants of the tropics placed in the family Meliaceae. The largest species rise to 80–100 feet and yield a durable wood. The original, or so-called Spanish mahogany, came from Cuba, and the tree that yielded it, *Swietenia mahogani,* is now nearly extinct, related species on the mainland of Central America being also severely depleted. In West Africa occur several species of the related genus *Khaya,* now largely milled as a substitute for the Caribbean tree. Similar trees include the *Soymida febrifuga* (rohan) of India, and another Indian tree of the same family, *Chloroxylon* (satinwood). The genus *Cedrela* is yet another of the mahoganies, with a wide distribution including Central America, Asia, Indonesia, and Australia. *Oxleya,* the yellowwood of New South Wales, is an extratropical representative of the mahoganies.

Placed in the same order, Geraniales, are several other families that include tropical trees and shrubs. Chief of these is the family Euphorbiaceae, which includes the Brazilian rubber tree *(Hevea brasiliensis),* an

Figure 91. *Ficus elasticus* and other species of fig with descending roots are trained in India to grow into natural bridges across the Himalayan streams. This example was sketched by Sir Joseph Hooker during his explorations of Bhutan (31).

ecodominant reaching a height of 100 feet, the so-called African teak *(Oldfeldia africanus)*, also some species of shrubs of the genus *Acalypha* occurring in New Guinea and in tropical Polynesia. Placed in a family by itself is the coca *(Erythroxylon coca)*, a small shrub of Bolivia and Peru from the leaves of which the drug cocaine is obtained. A further family of the Geraniales is the Rutaceae, including the genus *Citrus*, some members of which are tropical, notably the orange. Hebrew scholars tell us that *Citrus medica* (citron) is the tree whose fruit (mistranslated as apple) induced our first parents to embark on the enterprise of subduing nature; the scriptures disclose no antidote unfortunately.

ORDER SAPINDALES. This order includes some very large tropical trees of the same family as the sumac (family Anacardiaceae).

Figure 92. (a) *Anacardium occidentale* (cashew), Anacardiaceae, South America; (b) *Hibiscus tiliaceus*, Malvaceae, South America; (c) *Pometia pinnata*, Sapindaceae, Asia. Drawn by Desmond FitzGerald.

Best known is *Mangifera* (mango), of Asia, yielding ecodominant forest trees in Ceylon, and with other species in India and Vietnam. The sumacs *(Rhus)* are herbs in some parts of the world but grow into trees in America and Indonesia. *Anacardium* is an American genus that includes cashew and pistachio. *Odina* is a forest tree in the Ganges valley, and *Semecarpus* and *Dracontomelum* are ecodominant in Ceylon and Vietnam, respectively. A second family, Sapindaceae, includes ecodominant *Nephelium* of Ceylon (with second tier representation in Vietnam), and *Pometia*, ecodominant and emergent in Vietnam and the southwest Pacific. Several shrubs and the litchi nut tree *(Litchi)* occur in Asia and Polynesia.

ORDER MALVALES. This order comprises several tropical families. One of them, the Malvaceae, or mallows, is represented in the tropics by *Hibiscus*, with species that may be herbs, shrubs, or small trees; all bear handsome five-petaled flowers in many colors. Lindens of the northern biomes, family Tiliaceae, are represented by the tree *Berrya* in Ceylon. The family is otherwise indicative of cold climates, and the Pleistocene glaciation, for example, was marked by the southward spread of lindens into North Africa. *Elaeocarpus*, placed in its own family, is represented by trees of the highest tier in Vietnam, New Guinea, as also in temperate New Zealand forests. Another tropical family is the Sterculiaceae, with *Sterculia*, a pantropical genus of tall trees yielding a soft wood. To this family belongs the cacao tree *(Theobroma)*, of South America; the large seed pods, borne on the trunk by cauliflory, yield cocoa beans. *Theobroma* grows to about 20 feet in height. An African representative is the kola (genus *Cola*), yielding an extract of that name derived from its nuts. Yet another family of the order is the Bombacaceae, of which two well-known tropical American representatives are the balsa *(Ochroma)*, notable for its low-density wood, and kapok *(Ceiba)*, whose seed pods are crammed with the silky fiber to which the name kapok is attached. In Africa the baobab *(Adansonia)* occurs on savannas, growing to a height of 70 feet and often attaining a diameter of 30 feet, thus making it one of the most massive and peculiar of all trees. It yields edible fruits called "monkey bread."

ORDER PARIETALES, THE DIPTEROCARPS. The largest forest trees of India and adjacent lands are members of the family Dipterocarpaceae. Best-known is the gurjun *(Dipterocarpus turbinatus)*, exceeding 200 feet in height and 15 feet in girth, described by Hooker as "the most superb tree met with in the Indian forests." The dipterocarp forests occur in the humid regions of southeast Asia and Indonesia, but there are also species of the family that occur in dry or monsoon areas. The various species have a variety of local names, none of which are widely applied. The type genus, *Dipterocarpus*, is represented by various species that appear as emergent ecodominants in forests from India eastward to the Philippines, with conspicuous species in Ceylon, Burma, and Vietnam. *Pentacme* occurs in Vietnam and as a tall emergent in the Philippines, where it carries the local name of the luan. The great manggachappui tree of the Philippines, *Hopea*, has species elsewhere, including ecodominant forms in Ceylon. *Shorea*, under various local names, provides ecodominant species in the humid parts of India, Ceylon, and Burma, and emergents in Vietnam and the Philippines, where it is called the red luan; other names applied to *Shorea* are tangileyakal and guijo. The genus *Vatica*, known as the sal in India, includes ecodominant species in India, Ceylon, and Vietnam, and also provides some second tier trees in Vietnam. In Vietnam another genus, *Anisoptera*, contributes emergent ecodominants. In Indonesia *Dryobalanops* is conspicuous, and *Doona* is a conspicuous ecodominant in Ceylon. These great trees, one of the natural wonders of Asia, are totally absent from the African and American tropics. It would seem that these trees are difficult to acclimate, and inquiries directed to botanical gardens and arboreta have failed to disclose any specimens under cultivation in North America.

ORDER MYRTALES, MYRTLES. This order is not conspicuous in the natural forests of the Northern Hemisphere but, nonetheless, some genera are now well known on account of the ease with which they may be grown under domestication. Their characteristic features, the aromatic leaves with a marginal vein, and the flower clusters with numerous stamens, followed by conspicuous hemispherical seed capsules, are features well illustrated by *Eucalyptus*, the most conspicuous savanna tree of tropical Australia. The genus, as already noted, is also dominant in the temperate sclerophyll forests of Australia, but in humid parts of northern Australia the Indonesian elements take over. *Tristania* and *Metrosideros*, also noted in the previous chapter, extend beyond Australasia, with the latter genus reaching to Hawaii. *Eugenia* is pantropical, with austral extension. *Careya* occurs in India and Burma. Among the tropical American representatives are allspice *(Pimenta)*, the brazil nut tree *(Bertholletia)*, *Couroupita*, and guava *(Psidium)*. Cloves of commerce are the immature dried seed capsules of an Indonesian species of *jambosa*. A number of well-known tropical flowering trees belong in this order, including some species of *Eucalyptus* now grown in warmer parts of the temperate zone under cultivation. All the genera noted in this paragraph so far fall in the family of the true myrtles, Myrtacea. Other families of the order include the crepe myrtles (Lythraceae), with a flowering tree of the genus *Lagerstroemia* contributing fine elements to the forests of Vietnam and India. The family Combretaceae includes the genus *Terminalia*, represented by fine flowering species in Africa, Asia, and Indonesia, some of them also important as forest timber trees, as, for example, the limba of Africa *(Terminalia superba)*. In Asia the so-called Indian laurel *(Terminalia tomentosa)* is another important timber species. In the New World the family has developed mangrove trees, growing along the muddy and sandy margins of the continental shelf where rocky platforms are lacking; these so-called white mangroves include *Laguncularia*, *Conocarpus*, and *Bucida*. A comparable genus in Asia is *Anogeisus*.

ORDER UMBELLALES, GINSENGS. This order includes the family Araliaceae, of which some tropical members constitute very conspicuous elements in the third and fourth tiers of the forest. They are usually recognizable by the form of the leaf, which is fanlike, with radiating leaflets carried at the tip of a long petiole or stalk. *Schefflera* forms third-tier elements in Southeast Asia and in the Indonesian archipelago, a few extending southward into the austral region, where *S. digitata* lends a tropical aspect to the rain forests of New Zealand. The genus *Reynoldsia* of Indonesia is notable in providing species in the ecodominant layer. Elsewhere in the tropics the arborescent types are lacking, and the Araliaceae tend to be represented more by lianes, shrubs, or even herbs on the forest floor. In the Northern Hemisphere smaller species of *Schefflera* are now often to be seen as domesticated house plants, favored on account of their attractive fanlike leaves, and the smooth elegant nodes of the stem.

ORDER EBENALES, THE EBONIES. Tropical trees famed from antiquity and mentioned in the Bible fall in this order. A black wood, or wood with black and brown streaks intertwined, is obtained from trees of the genus *Diospyros*, a pantropical genus with especially strong representation in Africa, Asia, and Indonesia. In Ceylon, the traditional source of ebony since the days of Solomon, the tree contributes to the ecodominant layer of the forest. *Diospyros* also has species in Central America, where it is the only representative of the ebony family. The fruits of the tree are known in English as persimmon, but the ancient Greek name was that of the tree, *dios pyros*, or the Pear of Zeus. Forests of *Diospyros* occur on sandy soils in Vietnam. A related tree, *Maba*, is conspicuous in the forests of West Africa. Another tropical family, the Sapotaceae, is placed by botanists in the ebony order. It includes some ecodominant trees, such as *Palaquium* in the Indonesian and Ceylon forests, and *Mimusops*. In the dry monsoon forests of India occur the genera *Madhuca* and *Bassia*. West African members are *Chrysophyllum* and *Aningueria*. In tropical America occurs the sapota *(Achras)*. One type of natural rubber, known in commerce as gutta-percha, is obtained from the Indonesian tree *Palaquium gutta*.

ORDER GENTIANALES, THE CURARES. This order is represented in the tropical forests by two families of trees. One of these, the Loganaceae, includes the genus *Strychnos*, which ranges the tropics of the world and has been notorious since Stone Age times for the deadly poisons its tissues yield. The included species are trees, shrubs, and lianes, with opposite leaves and flowers that produce berries, not all of which, however, are poisonous. Thus *Strychnos spinosa* of Africa yields an edible orange-colored berry about as large as an orange. Strychnine and brucine are two toxins obtained from the seeds of an Indonesian species, *Strychnos ignatii*; the toxins paralyze the respiratory center of the brain and have long been employed by primitive peoples as arrowhead poisons. Curare is obtained from an American species, *S. toxifera*. Some species attain the

Figure 93. (a) *Dombeya rotundifolia*, Sterculiaceae, Africa; (b) *Ochroma lagopus* (balsa), Bombacaceae, South America; (c) *Gordonia papuana*, Theaceae, Asia. Drawn by Desmond FitzGerald.

height of middle-tier trees, as *S. nux-vomica*, of Indonesia, another source of strychnine. Other trees are *Fagraea*, of east Asia and Indonesia, and *Geniostoma*, which ranges into Polynesia. The family Apocynaceae, represented in the Northern Hemisphere by herbs such as dog's bane, yields trees in the tropics. In West Africa conspicuous elements are *Carpodinus*, *Landolphia*, *Mangenotia*, and *Kickxia*. Some ecodominant elements in Ceylon fall in the genus *Alstonia*. In tropical America the family is represented by *Plumeria*, and *Thevetia*.

ORDER TUBIFLORALES, THE TEAKS. This order, in which the flowers take the form of tubular or bell-shaped structures, includes several families of importance in tropical forests. One of these, the Verbenaceae, is best known for the trees of India, Burma, and Indonesia known as teak (genus *Tectona*). Tropical trees usually do not form stands dominated by a single species, but in some of the dry monsoon areas of India *Tectona* is a conspicuous dominant. The lost cities of Java, dating from the Middle Ages, also are described as being overgrown by teak stands. In the wetter regions of India teak is replaced by a related genus *Gmelina*. A representative in Central American forests is the genus *Citharexylon*.

Some of the best-known flowering trees of the tropics belong to the family Bignoniaceae, another member of the order Tubiflorales. In the Americas *Jacaranda* and *Tabebuia* occur, also the genus *Catalpa*, two species of which range into the temperate parts of the United States. *Spathodes*, with bright orange or purple flowers, yields tall trees in Africa, India, and Burma, the African tulip tree *(S. campanulata)* being an example. Another well-known African member is the sausage tree *(Kigelia)*. In the wetter forests of India and Indochina *Stereospermum* yields ecodominants. The family Solanaceae is also classified under the Tubiflorales and although none of its included members is large enough to contribute ecodominants to the forest, nonetheless the family is sufficiently conspicuous in the lower tiers as to warrant mention. The genus *Solanum* includes many herbs of the field layer, such as the nightshades, and a few shrubby trees, such as Tamarillo; besides these, the commercially important herbs include potato, tobacco, and tomato. *Datura* includes temperate as well as tropical shrubs; one of them, jimsonweed, is of philological interest as preserving the seventeenth-century pronunciation of Jamestown, Virginia.

ORDER RUBIOLES, QUININE AND COFFEE. A number of nonarborescent but important shrubs fall in the family Rubiaceae. *Coffea*, or coffee, is native to Africa and Asia and has been cultivated for its berries since ancient times; the Brazilian plantations, thus, derive from imported stock. Coffee requires shade for its successful cultivation, the sensitive plant *(Mimosa pudica)*, being used in New Britain, for example, as the source of the shade. A visitor has the sensation of carrying plague when everywhere one turns and touches the very leaves and twigs draw back and fold up. Native to Central America is the *Cinchona*, the source of natural quinine. Asian members of the family include *Paradina, Stephegyne,* and *Psychotria*, of the forests of Vietnam, the two former also ranging westward into India, where the genus *Gardenia* occurs.

THE FERTILIZATION OF TROPICAL FOREST TREES. The frequent reference to the showy flowers of tropical trees in the foregoing section may serve to stress the importance of insects and birds in cross pollination in the tropical forest. This question is discussed in Chapter 14, where the contrasted character of the northern forests is examined.

Lianes of the Tropical Forest

The tangled creepers that clamber up the trunks and boughs of the forest trees form one of the most characteristic features of the tropics and often render the wetter forests all but impenetrable without an axe or machete. Although some wet tropical forests are reported to have dense lianes only around the margins (as in Costa Rica and near Manaos, Brazil), others not uncommonly are difficult to traverse, for the snakelike stems of the creepers stretch from tree to tree or hang in interlocking catenaries from one trunk to another. These creepers may live for many years; having once achieved the highest tiers, where the flowers and fruits are commonly developed, their own long stems tend to thicken with time, eventually growing so heavy that they break the branchlets on which they were initially supported, thereby releasing a long festoon of jointed or coiled stem, to hang like a writhing python in midair, supported by other, more remote extensions of the same plant.

Not all lianes are flowering plants, nor do they all have the uppermost sunlit tiers as their target. Thus the ferns, which prefer the lesser illumination of the middle or lower tiers, contribute liane genera such as *Lygodium*, capable of climbing to a height of 10 m, mostly in the forests of tropical and subtropical Asia (there is a hardy North American species of the genus, however).

Among the monocotyledons, lianes are contributed by several orders. The Pandanales yield scrambling palmlike species in the genus *Freycynetia*, distributed mainly around the tropical margins of the western Pacific lands. The order Arales yields *Monstera* and *Philodendron*. *Monstera* has various species in South America, lianes of strong habit, with large leathery perforated leaves, and long suspended cordlike aerial roots. The flower is a white conical structure resembling an ice cream cone, called the spathe, from which arises an internal yellow clublike spadix, about a foot long, containing the reproductive organs. The spadix eventually matures into an edible fruit with the flavor of banana. *Philodendron* is also a tropical American climber with large shiny leaves, often perforated irregularly; species of this genus require much water and may grow on the ground, the roots covered by forest pools, while the ascending stem is carried by adjacent trees. The species of the genus are grown as house plants in temperate countries, but they only do well if the atmosphere is humid as well as warm, and the shriveled brown leaf margins of many unhappy captives betray their owner's neglect of this requirement. The Liliales yield a few lianes, among which the beautiful *Gloriosa superba* may be noted.

Most of the orders of dicotyledons contribute lianes of one or other family. Among the Piperales may be noted the cubeb (*Piper cubeba*), a treelike woody vine of southeastern Asia. It has pointed lance-shaped leaves and the flowers yield brownish berries carried in dense cylindrical clusters; from these black pepper is obtained. The mistletoes of the northern temperate forests belong to the order Santalales, which has some tropical lianes, notably *Loranthus*. Some species are trees, others creepers. The berries of the African species form food for the sunbirds. *Loranthus* parasitizes the host tree, after the manner of the true mistletoes, extracting nutrients by means of rootlike structures that penetrate the tissues of the supporting tree. The genus extends as far south as Samoa, but it is replaced by a cool-adapted genus in southern Polynesia, namely, *Elytranthe* of the *Nothofagus* forests of New Zealand and southern South America. The resplendent *Bougainvillea*, with its red or orange bracts, is now well known from garden importations. Its natural home is the tropics; it belongs to the order Centrospermales. Among the Rosales, a notable legume is *Entada*, a pea relative producing gigantic pods, 1 m in length, hanging in midair from the highest branches of the forest giants to which it clings. The pods, if they fall into streams of rivers, are carried to sea and swept thousands of miles to remote coasts. Thus a steady stream of *Entada* beans flows from

Figure 94. (a) *Kigelia pinnata*, Bignoniaceae, Africa; (b) *Diospyros lycoides* (ebony), Ebenaceae, Africa; (c) *Dipterocarpus grandiflorus*, Dipterocarpaceae, Asia. Drawn by Desmond FitzGerald.

the Caribbean islands to England and Scotland, by way of the Gulf Stream current; indeed, the very existence of the Gulf Stream was inferred as long ago as 1696 by Sir Hans Sloane, on the basis of beans collected in Scotland and recognized by him as corresponding to others he had observed growing in Jamaica. The pantropical distribution of *Entada* is undoubtedly due to the ability of the seeds to disperse in this extraordinary manner.

Among the Geraniales, the milkweed genus *Chlorocodon* is a twining liane with long oval leaves and panicles of purple and white flowers, in tropical African forests. Another member of the milkweed family (Euphorbiaceae) is *Allamanda*, which frequents summer-wet winter-dry forest in Brazil, producing both shrubs and climbing species, bearing large yellow or purple flowers and prickly fruits.

The Gentianales yield a number of fine tropical lianes. *Willoughbeia*, in the rain forests of tropical Malaya has several species; they yield a latex capable of manufacture into rubber. *Hoya* in the forests of tropical Polynesia contributes common lianes of a slender habit. The Jasmine and its relatives, genus *Jasminum*, is native to Asia and the species are sometimes shrubs, sometimes lianes. The white flowers yield a perfume.

Among the Tubiflorales the genus *Clerodendron* is one of the best known of all the tropical climbers, on account of its splendid and continuously produced flowers. The genus ranges the Old World tropics,

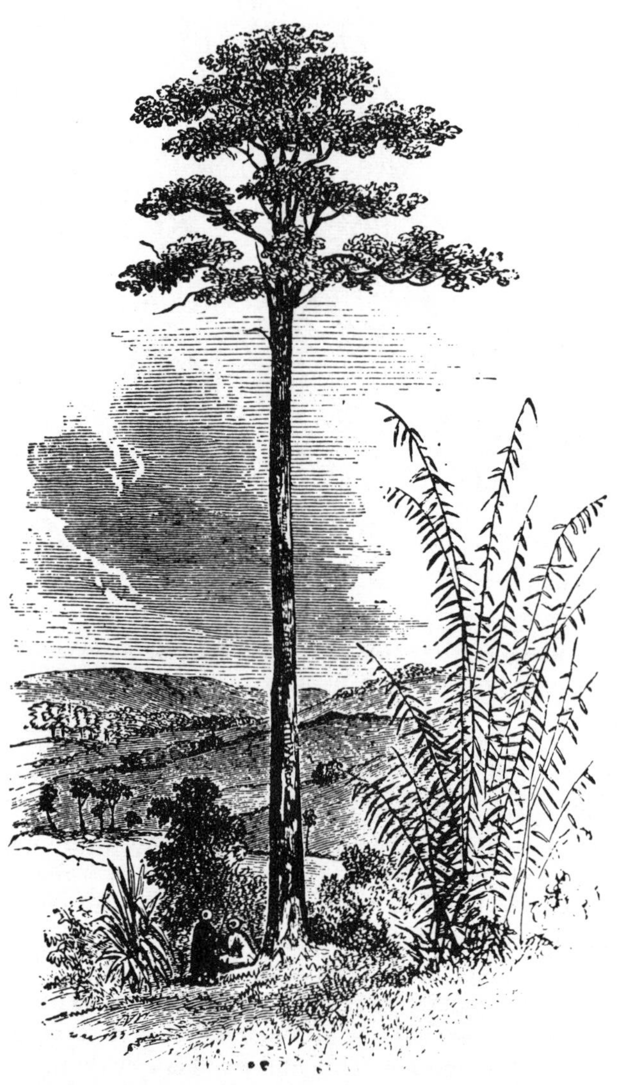

Figure 95. *Dipterocarpus turbinatus* (Gurjun), of India, upward of 200 feet high, Dipterocarpaceae. This specimen was sketched by Sir Joseph Hooker (31).

Japan, and Mexico. Some temperate species of the genus are also known. An interesting circumstance is the clear relationship in the genus between climate and leaf fall. A Javan species of the monsoon forest sheds its leaves in the dry season, as do many monsoon forest trees; species in temperate China may shed the leaves in winter. Other species are evergreen. *Clerodendron* will flower continuously if the temperature does not fall below 20° C. In the same order the genus *Cobaea* of South America may be noted. In its natural tropical habitat it is a perennial creeper, with ovate leaves 6 inches long and pendulous bell-shaped violet and green flowers hanging in pedicels 6 inches long, the individual flowers half that length. This beautiful plant will survive for one season if grown in the temperate region, so gardeners in California treat it as an annual. In the hottest forests grow species of *Clytostoma*, evergreen tendril climbers with funnel-shaped flowers. Brazil also has related genera such as *Adenocalymna*, in wet forest.

The order Campanulales also yields tropical lianes with bell-shaped flowers. The balsam apple *(Momordica)* frequents the forests of Africa and Asia, especially around abandoned clearings and plantations; it bears yellow flowers, later producing warty orange fruits. The gourds (family Cucurbitataceae) also belong in this order, and some of them are tree climbers; an example is *Lagenaria*, whose hard woody-rinded fruits are the decorative gourds sold by florists.

Figure 96. (a) *Tectona grandis* (teak), Verbenaceae, Asia; (b) *Coffea arabica* (coffee shrub), Rubiaceae, Africa and Asia; (c) *Miconia argentea*, Melostomataciae, pantropical. Drawn by Desmond FitzGerald.

Epiphytes of the Tropical Forest

Plants that live suspended above the forest floor, attached to other plants, are collectively called epiphytes. They are especially numerous in tropical forests. Since plants normally obtain their water and mineral nutrients from the soil, epiphytes experience special environmental problems. On the one hand, there is less competition for space, because few plant groups have evolved successful epiphytic members. On the other hand, the very reason for the lack of competition lies in the circumstance that survival in such situations as are available on tree trunks, in forks, and on branches is in itself a difficult matter, requiring special structural or physiological adaptations. We may recognize the following categories of epiphytes:

Xerophytes: with physiological adaptation to drought
Tank epiphytes: with water-storage devices
Parasites: adapted to take nourishment from the host tree
Stranglers: epiphytes that kill the host and usurp its territory
Carnivores: epiphytes that capture small animals
Estivators: epiphytes that rest during the dry season

Some epiphytes exhibit characters of more than one of the above categories. Following are tropical examples of the categories listed.

XEROPHYTES. These are species with adaptations to conserve water and to prevent loss of moisture by evaporation. The leaves are leathery and resist desiccation. The epiphytic orchids include species of this type, in genera such as *Aeridea* of the Asian and Indonesian forests, with curving graceful leaves and long spikes of sweet-scented flowers of pastel colors. The leaves are retained throughout the year. The genus *Cordula* includes drought-resistant orchids of Africa and Asia, producing greenish, yellowish, or white saclike flowers of the lady's-slipper type. Of the numerous species some are very successful in poor conditions, and these flourish as house plants when transported to the temperate zone.

TANK EPIPHYTES. Plants in this category have structural features that collect and store rainwater, thereby relieving the plant body from the necessity of developing thick epidermal coatings. The storage organs most commonly are the bases of the leaves, which grow in such manner that each leaf sheathes the one above, and the leaves themselves are concave above, so as to make a collecting channel leading to the basal sheath, where the water is stored. A pint or more of water may be obtained from such plants if one is thirsty enough to climb up to commit the robbery. The tanks of such epiphytes may serve as breeding chambers for mosquitos, nematode worms, the eggs and young of tropical tree frogs, and may even house venomous snakes such as *Bothrops*. The orchid *Miltonia* has slender sheathing dark green leaves and produces racemes of white or rose or yellow flowers. It ranges the tropics. Most notable of tank epiphytes, however, are the American Bromeliaceae, epiphytic relatives of the better known ground-dwelling pineapple. *Aechmea* has sword-shaped leaves of the pineapple type, with sheathing bases that hold water; it inhabits Brazilian forest, where it produces blue and purple flowers in conditions of humid heat. Another bromeliad, *Neoregelia*, has a manner of life that suggests how the epiphytic habit probably arose in the first place. The plant body has the form of a rosette; it grows in the peat soil found in forks of trees and requires humid shade. Shortly before the flower forms the sheath leaf adjacent to the bud turns red. Another genus, *Nidularium*, of the Brazilian rain forest, has a similar habit. Some ferns have become tank epiphytes, an example being *Asplenium nidusavis* of the Malayan jungle, where the fronds are converted into undivided sheathing leaves, forming a vase-shaped container. These ferns perch very conspicuously far out on overhanging branches and impart a singularly exotic character to the forests where they grow.

PARASITES AND STRANGLERS. The genus *Loranthus*, a tropical mistletoe, has already been noted as contributing to the liane flora. Some species, however, germinate on suitable places on forest trees and thus are deprived of access to the soil. Such species assume the parasitic habit already referred to. The best-known of the strangler parasites are the banyans, or strangler figs. These plants produce seeds that germinate on some other tree, where they grow for a period as perching plants, obtaining their nourishment from the air and from the small accumulations of peat found in tree forks. As they increase in size descending stems are produced that hang in midair until they reach the ground, whereupon a rooting system is formed. In the course of time numerous such descending stems arise, and eventually the host plant is overwhelmed, smothered by the luxuriant foliage of the usurping fig, and after the passage of a century or so, a grove of stems stand like a classical porch, supporting a large fig banyan tree, beneath which is an empty space formerly occupied by the host. These peculiar groves were frequented in ancient times by Buddhist missionaries and contemplatives.

CARNIVORES. Best known of the epiphytes in this category are the pitcher plants, species of the genus *Nepenthes*. About 60 species occur in the tropical forests of the east Asian and Indonesia region as far south as northern Australia. The leaves are so shaped as to form a vase at the base of each. Rain water is trapped in the vase, and nectar is secreted from the adjacent part of the leaf forming the lip of each pitcher. Insects attracted to the pitchers are prevented from emerging by downwardly directed bristles that line the pitcher. Digestive enzymes are secreted into the water at the base of the pitcher, and so the insect is caused to dissolve, the body substances being broken down into the simpler compounds that can be absorbed by plants.

ESTIVATORS. These are epiphytes that have adapted to the uncertain or irregular supply of water in the tree-top environment by adopting a cyclic life style in which the plant grows actively in the wet summer season and then discards further activity during the dry winter. These are monsoon-adapted forms. Examples among the orchids include *Cymbidium*, from the montane monsoon forests of Asia (some occur in high country and hence are to be classified as temperate elements). They require very abundant rain during the growing season, when they produce flowers of rather dull shades of purple, or white or greenish. *Maxillaria* is an orchid genus of which some species grow on the ground; others have become cyclic epiphytic forms; they too require a heavy summer rainfall. They are evergreen in spite of the resting period; the flowers are white, yellow, or deep purple. The genus *Dendrobium* includes orchids of varying habits, some of them being estivators, in Asia and Indonesia; some species are adapted to cool climates, and the range extends southward to include New Zealand. The bromeliads also are represented in this category, the genera *Aechmea* and *Nidularium*, already mentioned, having other species that are estivators. Like the orchids, they require abundant summer rain, followed by a dry cool season. *Billbergia* and *Cryptanthus* are similar bromeliads, confined, like all bromeliads, to the American forests. Begonias are beautiful tropical forest herbs, with the generic name *Begonia*, distributed in the Old World and New World tropics. Some grow on the forest floor, others are epiphytes, and both types tend to be estivators, requiring a wet season alternating with a dry season.

Cacti are plants adapted to desert conditions in the Americas, but one genus, *Zygocactus*, is a forest epiphyte. It has jointed stems, at the tips of which beautiful red clusters of flowers appear. Strangely, this cactus is quite intolerant of desiccation, requiring a constantly humid atmosphere; if taken from its natural habitat and transferred to a dry atmosphere, the joints of the stem break apart and fall off.

The Floristic Diversity of a Tropical Forest

The term *floristic* relates to the systematics, or natural classification, of plants in the context of their evolutionary origin. Thus, members of the same ecological unit—such as the ecodominant layer—may represent very diverse sections of the plant kingdom; but although such members have had different evolutionary histories, in the context of their ecology they may have converged, and because they have developed similar requirements or similar abilities to occupy a given environment, they come to form a recognizable ecological assemblage.

In a book dealing essentially with environmental matters it is obviously impossible to deal with the systematics of organisms. Nonetheless, to ignore the systematics entirely would be to obscure the remarkable diversity of elements that enter into ecological groupings of organisms. So, in the sections that follow, some reference is made to the orders or

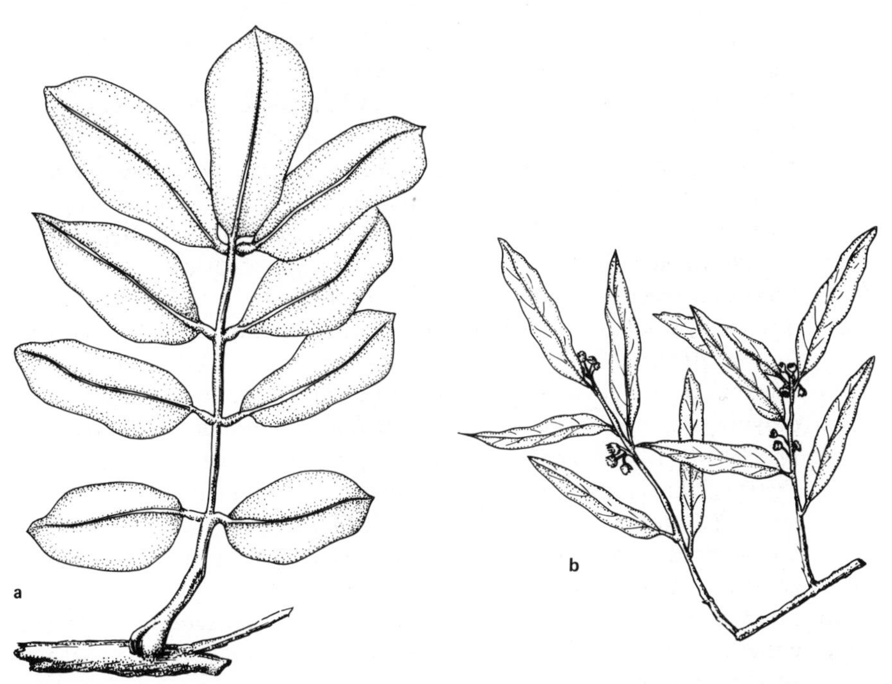

Figure 97. (a) *Trichilia roka*, Meliaceae, pantropical; (b) Mida (Fusanus), Santalaceae, pantropical. Drawn by Desmond FitzGerald.

families of plants in the tropical forests, and examples are cited by generic names, although these taxonomic units are not defined. By using such names, however, it is easier to grasp the inherent complexity of floras and faunas, in much the same way as a list of Vikings given by name for a particular Norse saga does lend additional interest to a historical episode and does throw light on the historical background, in a way much more illuminating than is the case when the saga merely begins with something like "Then Olaf took fifty men and sailed out. . . ." What follows now is a shortened version of a description of a New Guinea rain forest given by E. G. Robbins (1961). It is presented here mainly to illustrate the diversity of a tropical forest, but an attentive reader may find much additional sidelight thrown by it on questions of the origin of Australian and New Zealand forest trees and other forest plants, as discussed in the preceding chapter; and at the same time, relationships to other floras may be seen by comparing it with later chapters.

Robbins found that the forest could be described as a six-tiered entity, having much the same arrangement of the elements as in a New Zealand rain forest, though the floristic content differed in many important respects. The tiers are as follows:

1. Tallest trees (ecodominant layer), rising to 100 feet or more: *Podocarpus, Elaeocarpus, Sloanea, Opocunonia, Schizomeria, Alstopetalum, Cryptocarya, Syzygium, Albizia, Astronia, Alphitonia, Elmerilla,*

Fagraea, Ficus, Guioa, Garcinia, Himantandra, Ilex, Planchonella, Pygeum, Perrotetia, Sterculia, Xanthoxylum

2. Tall trees, to 40–80 feet: *Elaeocarpus, Weinmannia, Quintinia, Litsaea, Rapanea, Pittosporum, Phyllocladus* (mainly upper montane, however), *Ackama, Pullea, Gillbeea, Ficus, Syzygium, Sphenostemon, Sloanea, Sericolea, Myristica, Daphniphyllum, Dillenia, Diospyros, Casearia, Couthovia, Zanthoxylum, Discocalyx, Ardisia, Helicia, Timonius, Evodia, Eurya, Gordonia, Ternstroemia, Laportea*
3. Medium trees, to 20–40 feet: *Pittosporum, Litsaea, Paratrophis, Schefflera, Cyathea* (a tree fern, mentioned in Chapter 11), *Casearia, Bubbia, Aglaia, Dichria, Mearnsia, Decaspermum, Xanthomyrtus*
4. Shrubs, to 10–20 feet: *Dicksonia* (tree fern), *Piper, Olearia* (a tree daisy), *Rapanea, Ascania, Carpodetus, Quintinia, Amaracarpus, Psychotria, Mussaenda, Gardenia, Medinilla, Everettua, Poikilogyne, Pygeum, Eurya, Symplocos, Acronychia, Rhamnus, Evodia, Perrotetia, Chloranthus, Ascarina, Pandanus, Marattia* (fern), *Rhododendron*
5. Small shrubs, to 5–10 feet: *Marattia, Todea* (both ferns), *Geniostoma, Solanum, Melicope, Chloranthus, Ascarina*
6. Field layer, to 5 feet: there follows a list of seven widely dispersed genera of ferns, *Begonia*, an arum, three sedge genera, a ginger, and several other herbs

In addition, Robbins reported lianes (including a bamboo) and epiphytes at all levels.

Comparable lists, with different included genera, are given by other botanists for various examples of tropical forests in different parts of the world. Just as the New Guinea list includes various genera shared with adjacent southern lands, so also a list from Northern India, say, will be found to include genera shared with the temperate lands to the north.

Among points that may be made are these:

1. Tree genera of the ecodominant layer occur also in the lower tiers. This is because the several species of a genus are commonly of different size modules; also, the immature stages of an emergent necessarily lead it successively through the various lower tiers, and if such is not found to be the case, then it must mean that the forest is not reproducing itself, or is still undergoing some changes in the course of its sere.
2. Similarly, elements of the lower tiers occur in several tiers.
3. The New Guinea tropical forest includes elements that occur widely throughout the tropics. Most tropical forests have such pantropical elements. Examples of such are *Ficus, Sterculia,* and *Diospyros.*
4. Some montane elements from both hemispheres are here coming into contact on the equator; thus the southern *Phyllocladus* occurs together with the northern *Rhododendron.* This is a peculiar feature, presumably reflecting a former colder climate that permitted such extensions of range, in a manner analogous to the inferred passage of the Southern Hemisphere kelp *Macrocystis* into the north Pacific during a Pleistocene glaciation.

13

TROPICAL FAUNAS

All three subclasses of Mammalia are represented in tropical forests. The egg-laying Prototheria and the pouched marsupials, Metatheria, of the New Guinea forests have already been noted (page 230). There remain the one family of marsupials commonly distributed in Central America, and some obscure and very rare forms from Colombia and Ecuador. Otherwise the tropical mammals are all members of the subclass Eutheria, or placentals. Table 38 sets out the classification of tropical mammals.

Geological investigations have shown that a seaway existed through the Panama region until about a million or so years ago. The seaway seems to have been continuously in existence from at least the Cretaceous period, and this means that South America was an isolated island continent through all of Tertiary time, these past 100 million years. Whereas many plant species can become distributed through wind-blown seeds, by seeds adhering to the feathers of migrating birds, and by other passive means, these methods of dispersal are not open to land animals. Thus, for mammals especially the existence of the Panama seaway had a profound

Table 38. The classification and distribution of the tropical orders of mammals.

DIAGNOSTIC CHARACTERS	ORDINAL NAME	COMMON NAME	DISTRIBUTION
Pouched mammals (Metatheria)	Marsupialia	Opossums	South and Central America
Placental mammals with no pouch (Eutheria)			
No teeth; body covered by horny scales, reptilelike	Pholidota	Pangolins	Africa and Asia
No teeth; body furred above and below; head elongate	Xenarthra	Anteaters	South and Central America
Teeth present but all small and similar			
Body covered by bony carapace; burrowers	Xenarthra	Armadillos	South and Central America
Body furred; snout blunt; arboreal	Xenarthra	Sloths	South and Central America
Body piglike; ears rabbitlike	Tubulidentata	Aardvarks	Africa
Teeth differentiated into incisors, canines, and molars			
Forelimbs modified as wings	Chiroptera	Bats	Africa, Asia, and America
Proboscis (trunk) present, tusks	Proboscidea	Elephants	Africa and Asia
No hoof developed			
Thumb opposable to fingers	Primates	Monkeys and apes	Africa, Asia, and America
Thumb not opposable to other digits			
Canine teeth present			
Canines larger than incisor	Carnivora	Lions, wolves, and so on	Africa, Asia, and America
No so; snout pointed	Insectivora	Moles, shrews, and so on	Africa
Canines replaced by a toothless region (diastema) of jaw			
Long-eared leapers	Lagomorpha	Rabbits and hares	Africa and Asia
Short-eared runners	Rodentia	Cavies, rats, and so on	Africa, Asia, and America

Table 38. The classification and distribution of the tropical orders of mammals.

DIAGNOSTIC CHARACTERS	ORDINAL NAME	COMMON NAME	DISTRIBUTION
Hoof present (ungulates, that is, walking on nails)			
Odd-toed ungulates (monaxonic, that is, the third toe of each foot fully developed, with or without a toe formed on either side)	Perissodactyla	Horses and tapirs	Africa, Asia, and Central and South America
Even-toed ungulates (paraxonic, that is, the second and third toes of each foot best and equally developed, with or without a toe formed on either side)	Artiodactyla	Giraffes, cattle, and hippopotamuses, deer, camels, and pigs	Asia, Africa, and Central and South America
Three toes in front, 4 toes on hind limbs, small mammals in rocky habitats	Hyracoidea	Hyraxes	Africa and West Asia

influence upon the evolution of Tertiary faunas, for South America was cut off from the rest of the world by an almost impassable barrier of open sea.

South American Mammals

The survival of marsupials in South America was undoubtedly due to the long Tertiary isolation of the continent, the Panamanian isthmus having risen to connect North and South America relatively recently in geological time. Thus, like Australia, the South American fauna shows the effect of independent evolution in isolation, with the preservation of ancient mammalian types that elsewhere were extinguished and replaced by eutherian mammals. One species of opossum occurs in North America, and it is closely related to some South American species. Other genera of opossums also occur in South America; these facts show that the North American opossum is a recent entrant from the south, evidently from stock that crossed the isthmus at the end of the Tertiary. Similarly the North American armadillo, and the recently extinct ground sloths and glyptodonts of North America, came from the south by the same route at the same time. So, before the elevation of the isthmus, these groups were exclusively South American.

Figure 98. *Ateles cucullatus* (hooded spider monkey), Cebidae, South America (21).

ORDER MARSUPIALIA [OPOSSUMS]. The Didelphiidae comprise several genera of opossums with a more reptilian type of reproductive system, the two oviducts each forming a separate uterus, unlike the case in the Australian marsupials. *Didelphis* is the one genus that also ranges into North America. *Metacheirus*, with ratlike species of arboreal insectivorous opossums, *Philander*, with woolly-furred species, *Marmosa*, with mouselike species (some of which occasionally are transhipped with fruit to the United States), *Monodelphis*, with shrewlike species, *Chironectes*, a water opossum, or yapok, with webbed toes and a naked tail, are among the special adaptations of marsupials in the South American fauna. All the forms noted have numerous incisors. Paralleling the diprotodont forms of Australia are some rare and little-known animals believed to be marsupials and referred to a separate family the Caenolestidae, three genera all from the forests of Colombia and Ecuador. Some of these small marsupials may lie close to the Mesozoic ancestry of mammals.

Figure 99. *Lagothrix infumatus* (woolly monkey), upper Amazon forest, Cebidae (23).

ORDER INSECTIVORA [SOLENODONTS]. The best-known members of this order are the moles, shrews, and hedgehogs, yet none of these three families occurs in South America. The only representatives of the order are the Solenodontidae of the Caribbean region, and found nowhere else, animals like very large rats with a long tubular snout, large claws, and a scaly tail that they can use after the manner of a kangaroo in tripod support with the two hind legs. They are secondary heterotrophs feeding on insects, lizards, and carrion. They are found in Cuba, Haiti, and Hispaniola. These peculiar animals, and the absence of the northern forms represented in North America, again are witness to the isolation of Tertiary South America.

ORDER CHIROPTERA [BATS]. South America lacks the large fruit-eating bats (Pteropodidae) of the rest of the tropical region. Instead it

Figure 100. *Cyclopes didactylus* (little anteater), arboreal, Mexico to Peru, Myrmecophagidae (21).

has two distinctive families, the blood-eating vampires (Desmodontidae) and the leaf-nosed bats (Phyllostomatidae); the latter, although structurally allied to the insectivorous bats, have adopted the fruit-eating habit of the absent Pteropodidae.

ORDER PRIMATES [MONKEYS]. The New World primates are restricted to tropical America, save for man himself. The distinctive feature is the absence of all groups found in the Old World and their replacement by two families of New World monkeys, the marmosets and Cebidae. The latter family includes the spider monkeys (*Ateles* et al.), howler monkeys *(Alouatta),* capuchins *(Cebus),* and others. These forms have in general occupied the upper canopy of the forest, where they are fruit- and insect-eating; the equivalents of the ground apes of the Old World are not found here, save for the hominid *Homo sapiens,* for whom carbon dates imply occupation of territory as far south as Patagonia since at least about 11,000 B.C., and possibly much earlier. Thus the South American biomes, unlike those of the Old World tropics, were not subject to hominid hunting pressures during the past half million years.

ORDER XENARTHRA [EDENTATES]. Here are another group of families concentrated in South America, the entire order having evolved here in isolation during the Tertiary. Large species developed, one of them larger than an elephant, but unfortunately these remarkable animals seem to have been hunted to extinction by early man. In addition to skeletons, remarkably well-preserved dry dung occurs in certain caves, together with portions of the hide and hair of giant sloths. Carbon dates for the extinction in South America are not available, but see, however, dates for extinction of the species that entered North America given in Chapter 15. Species of the three living families are smaller than the fossil Edentata. The armadillos, or Dasypodidae, include both forest and plains species, one of the latter still inhabiting the southern part of North America where it entered from the south. The tree sloths, family Bradypodidae, are now

Figure 101. *Choloepus hoffmanni* (unau, or two-toed sloth), arboreal, Costa Rica and Panama, Bradypodidae (21).

Figure 102. *Dasypus novemcinctus* (nine-banded armadillo), Central and South America, Dasypodidae. Drawn by Desmond FitzGerald.

restricted to Central and South America. The anteaters, family Myrmecophagidae, include a large savanna species of *Mymecophaga*, which is terrestrial, and some smaller species of other genera that are arboreal forest-dwellers.

Orders Not Restricted to South America.

The South American fauna includes four other orders of Eutheria that are not restricted to the continent, namely, the Rodentia, Carnivora, Perissodactyla, and Artiodactyla. These may be summarized briefly.

ORDER RODENTIA. Squirrels (family Sciuridae) evolved in the Northern Hemisphere forests, so the flying squirrel *(Glaucomys)* of South America is evidently an immigrant from the north. Similarly the tree porcupine *(Coendou)* must be a derivative of the North American porcupines (Erethizontidae). A number of other rodent families, including the agoutis Dasyproctidae, the tuco-tucos, the hutias, and others associated more with aquatic or Pampa environments, are peculiar local evolutionary stocks.

ORDER CARNIVORA. These in general match North American groups and belong to the same families. Bears (Ursidae) are lacking from the forest but do occur on the subalpine Andean tundra. Cats (Felidae) are represented by several species of *Felis*, including jaguar *(F. onca)*, puma *(F. concolor;* same species as is called mountain lion in North America), ocelot *(F. pardalis)*, and jaguarondi *(F. yaguarondi)*, plus some smaller species. The jaguar formerly ranged widely in North America; the puma still does, despite long-standing bounty programs only recently abandoned. The cats, including the great cats, have been present in North America since at least mid-Tertiary times; the uniformity of the great cats through the entire length of the Americas, south to the Magellan Straits, is

Figure 103. *Coendou* sp. (tree porcupine), arboreal, Central and South America, Erethizontidae (21).

evidence of their tolerance for habitat diversity and their dispersive powers. Some zoologists seem to consider them as South American entrants into the North American fauna, a view that raises some problems as to the Tertiary history of American cats. The wolves (Canidae) apparently lack the immigrant powers of the cats, for the South American species constitute an assemblage of jackallike animals placed in separate genera such as *Speothos*, a forest species, and *Dusicyon*. Here isolation seems evident. The otters (Mustelidae) have evolved a very large form, *Tayra barbara*, peculiar to South America; this animal, which can be very fierce, is apparently not necessarily aquatic. And the raccoons (Procyonidae) range both continents, with some peculiarly South American types such as kinkajou and coati; both occur in southern North America, evidently as immigrants from the south.

ORDER ARTIODACTYLA. The most distinctive South American members are the camels of the New World, namely, the llama and its relatives; however, these are not forest-dwellers. Of the forest-inhabiting species, the brockets, or wood deer (*Mazama* spp.) and the small pigs called peccaries (*Peccari* and *Tayassu*) are apparently southern in origin. The peccaries have entered southern North America, and the collared peccary is common in Texas and Arizona, the numbers here on the increase because of wise game management.

ORDER PERISSODACTYLA. The South American horses included mid-Tertiary mountain forms, long extinct. Rhinoceros do not seem to have occurred in South America, though North American forms lingered on until the Pliocene. Only the archaic tapirs now represent the order, three of the species of *Tapirus* being forest-dwellers. Tapirs also occur in Malaysia, a remarkable case of discontinuous distribution, indicating that surviving tapirs are remnants of formerly widespread stock.

Biogeographic Implications

The foregoing paragraphs show clearly that the geological history of the dispersion of terrestrial mammals has led to the observed present content of South American and Australian faunas. Each arose independently, and neither fauna is related closely to any other. Each is more or less the unique outcome of unique circumstances. The dispersive powers of plants, by way of their seeds or other propagules, have led to the phenomenon of the austral plant biomes, in which the Australian, New Zealand, and South American temperate forests are clearly related for they share common genera of trees, expecially *Nothofagus* (Chapter 11). On the other hand, the terrestrial mammal fauna of South America is not part of a common temperate Austral biome, but largely unique, such elements as are not unique being related to those of the other tropical biomes. For example, comparable great cats constitute the major carnivores. Tapirs are shared with the forests of tropical Malaysia. Primates are conspicuous. Among the Artiodactyla, pigs and deer are shared, also camels in the broad sense. For these reasons, Alfred Russel Wallace, when he recognized the major zoogeographic regions of the world, named the South American region the *neotropical realm* and set it apart from the Australian and New Zealand region, which he recognized as distinctive in itself.

Botanists seem at first to have accepted these divisions as applicable also to plants. Later, when the *Nothofagus* and other anomalous trans-Pacific distributions became apparent, extravagant theories were proposed to account for the plant distributions by means of land bridges now vanished, or radial bridges linking each of the southern continents to Antarctica, or more recently by continental drift hypotheses. But what applies to plant dispersion obviously does not apply to terrestrial mammalian dispersion. The various explanations offered in the preceding pages

Figure 104. *Tupaia tana* (malayan tree shrew), Tupaidae (17).

have depended on the dispersive powers and adaptability of the group concerned. None requires any upheaval of the earth's surface or collapse of supposed land bridges. All that seems necessary is a continuing flow of the presently observed air and water masses, under the control of the forces produced by the earth's rotation and the energy fed into the atmosphere by the sun.

Continued exploration of the geological record of the southern continents may throw new light on the question. As to the reliability of conclusions based on the distribution of organisms, it should not be forgotten that the now universally accepted view that the Panama seaway

Figure 105. *Galago crassicaudatus* (galago, or bush baby), tropical Africa, a lemuroid or primitive primate (21).

separated North and South America during Tertiary times actually took its origin in observations by Alfred Verrill at Yale in 1870, when he drew attention to the occurrence of identical species of sea urchins on the Pacific and Atlantic coasts of the Panama isthmus and deduced that a sea had once extended through the isthmus. Forty years elapsed before Charles Schuchert finally provided the geological evidence that this interpretation was correct.

Mammals of the African and Oriental Forests

THE FORMER UNITY OF AFRICAN AND ASIAN FOREST FAUNAS. In contrast with the isolation exhibited by the South American tropical mam-

Figure 106. *Homo sapiens*, Dyaks in Borneo forest, visited by F. Boyle in 1864 (12). These hunters probably retained a forest culture inherited from Neolithic times.

mal fauna, the faunas of Africa and Asia can be seen to be nearly related. So similar are they that it is easier to make a list of the major differences, for these are much fewer than the many similarities. A short list of differences would look like this:

AFRICAN REGION	ORIENTAL REGION
Hippopotamus present	Hippopotamus absent
Giraffes present	Giraffes absent
Bears absent south of Sahara	Bears widespread
Deer absent south of Sahara	Deer present throughout
Tapirs absent	Tapirs present
Aardvark present	Aardvark absent

As against these differences, the similarities extend to the species level—lion, leopard, and cheetah occurs in both regions, for example. Again, groups that are known from no other regions in the world are shared—for example, the pangolins (order Pholidota).

Now if we take into account the evidence of fossil faunas from the middle and late Tertiary, it becomes apparent that groups not shared at the present epoch were in fact shared by Africa and tropical Asia in the recent geological past. Thus, giraffes occurred in the Indian region formerly, as did ostriches, so the birds apparently also follow the same rule. Hippopotamus was present in India during the Tertiary and did not become extinct there until the Pleistocene. It looks as though most of the supposed differences between the African and Oriental tropical faunas are due to quite recent extinctions in one of the regions of groups that were of common heritage for millions of years. A likely explanation for the changes in the recent past might include such possibilities as (1) the savanna lands probably extended across northeast Africa and Arabia and southwest Asia, to provide a continuous open forest or habitable grazing area from India to Africa, and (2) human activity or climatic deterioration or both have led to the contraction of the forests and savannas, and also to extinctions of animals such as hippopotamus in India.

ORDER INSECTIVORA [INSECTIVORES]. The order, which includes some mammals not unlike early fossil mammals from the Jurassic period, is well represented in Africa. One family, the tenrecs (Tenrecidae), is peculiar to Madagascar. Large water shrews up to 2 feet long and resembling otters occur in equatorial African rivers and constitute a separate family, Potamogalidae. Hedgehogs (Erinaceidae) occur in both regions, as do shrews (Soricidae). Moles (Talpidae), on the other hand, do not occur in either Africa or the Oriental tropics.

ORDER CHIROPTERA [BATS]. The large fruit bats, or flying foxes, make up the suborder Megachiroptera; in the Oriental region (and also northern Australia) the family Pteropodidae is conspicuous. The

Philippines fruit bat *(Pteropus jubatus)* reaches 1.5 m in wingspread. Pteropodidae are represented in Africa by *Epomophorus*. Numerous insectivorous Microchiroptera occur in both regions.

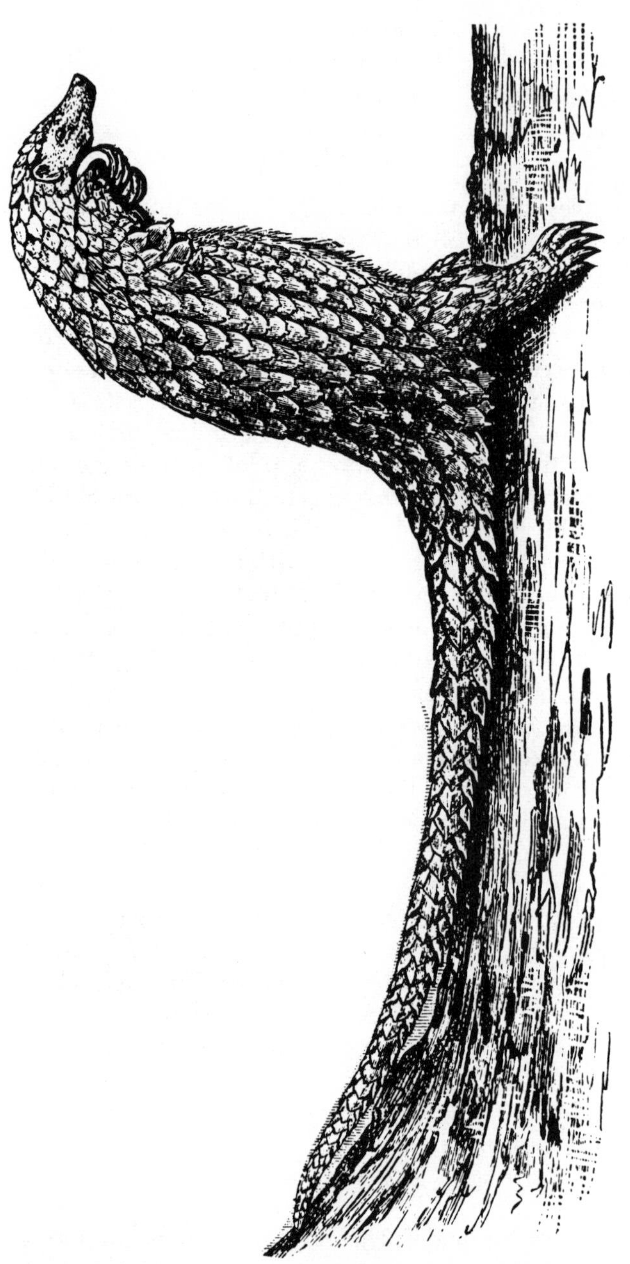

Figure 107. *Manis longicaudata* (long-tailed pangolin), West Africa, Pholidota (36).

Figure 108. *Tapirus indicus* (Malayan tapir), southeast Asia, Perissodactyla (18).

ORDER PRIMATES [APES AND MONKEYS]. The great apes (family Pongidae), characterized by lacking the tail, are represented in Indonesia by the orang *(Pongo pygmaeus)* and by the siamang *(Hylobates)*, and in Africa by the gorilla *(Gorilla gorilla)* and the chimpanzee *(Pan troglodytes)*. These apes are omnivorous and take young monkeys as food as well as much vegetable material; it was formerly thought that they eat only fruit, a mistake that led to vitamin deficiency and sterility in zoo animals.

One of the areas in Africa where gorillas occur is Rio Muni, on the equatorial west coast. Jungle occupies the lowlands on the coast. The lowland gorilla lives here and still raids the plantations and is killed by the villagers of the Fang tribe. The plantations that attract the gorilla comprise plots of banana and coffee. These are imbedded in and completely surrounded by the primal forest, the upper canopy about 100 feet above; the forest clearings are totally covered by herbs, 2-m-high shrubs, and elephant grass. The gorillas make nests of ferns, leaves, and sticks. They eat wild fruits, stems, buds of *Aframomum* (ginger family) and Musanga (an umbrella-shaped tree), and banana plants. Upward of 5,000 gorillas inhabit this dense rain forest in an area of about 25,000 square kilometers.

There are numerous monkeys in Asia and Africa, classified in several families. The more archaic primates are the lemurs of Madagascar and the Oriental region, and the lorises and tarsiers. Man is classified in the family Hominidae, which formerly included the extinct apemen *Australopithecus* and related forms from Africa, and various forms of pithecanthropus now classified as *Homo erectus*, and best known from sites in Java and China. The hominids all have a broad ilium in the pelvis for the origin of the gluteus muscles that operate the tendon that holds the hind limb in a

Figure 109. *Elephas indicus* (Indian elephant), southeast Asia (here seen in the Ceylon jungle), Proboscidea (30).

Figure 110. *Felis marmorata* (marbled cat), overall length 1 m, southeast Asia, Felidae, Carnivora (40).

semicatatonic state for maintenance of erect posture. Carbon dates show that *Homo sapiens* and *Homo neanderthalensis* were coexisting contemporary species or subspecies for tens of thousands of years; as intermediate forms exist, it now seems likely that modern man is of mixed derivation.

ORDER RODENTIA [RODENTS]. Squirrels (family Sciuridae) represented here are *Sciurus* and *Ratufa*, forest squirrels of Asia; *Petaurista*, the flying squirrel of Asia; *Heliosciurus*, tree squirrels of Africa.

ORDER PHOLIDOTA [PANGOLINS]. Family Manidae, with genus *Manis*, pangolins, the only known genus, has species in both Africa and the Oriental region. These animals, whose characters are shown in Table

Figure 111. *Herpestes urva* (short-tailed mongoose), southeast Asia, frequenting marshes, feeding on fish, crabs, and frogs, Viverridae, Carnivora (23).

38, feed on termites and ants by night, spending the day in a burrow. However, they can learn to reverse the activity cycle, for the late Jawaharlal Nehru records in his prison diaries that a pet pangolin shared his cell and fed from his hand. Pangolins have fur on the lower surface of the body.

ORDER PERISSODACTYLA [ODD-TOED UNGULATES]. The order comprises three families: (1) Tapiridae, with the Malayan tapir *(Tapirus indicus)* the only representative outside of the South American species; the Malayan species inhabits deep rain forest. (2) Rhinocerotidae, with several species of *Rhinoceros* now surviving; the Sumatran rhino *(R. sumatrensis)* inhabits forest and savanna, the Javan *R. sondaicus* prefers deep forest; the others, *R. unicornis* of India and two African species, the black rhino *(Diceros bicornis)* and white rhino *(D. simos)*, are found mainly in savanna country; all except the black rhino are becoming rare, and the Indonesian species are in danger of extinction. (3) Equidae, horses and zebras; these are essentially grassland forms, with limbs adapted for fast galloping.

Figure 112. *Giraffa camelopardalis* (giraffe), African savannas, Artiodactyla (36).

Savanna Mammals

The remaining orders of African and Asian mammals are mainly, though not exclusively, found in the more open type of forest called savanna. Where monsoon seasons occur, the animals may have an annual cycle, retreating to the forest during the dry season when insufficient forage is available in the open country. The carnivores will follow the herbivores under such conditions. Animals such as elephants, which prefer savanna country, will inhabit the forest if no open country is available, as in the West African rain forest, where a smaller race of elephant occurs. Animals that frequent water, such as water buffalo and hippopotamus, are found in both forest and savanna. Animals that tolerate a wide range of environments are termed *eurytopic*. The tiger is a good example here, for it ranges from the hot equatorial forest of Indonesia, across the savanna, monsoon forest, and rain forest of southern Asia into the montane forests of the Asian mountains, and northward into the Siberian steppe and summergreen forest near Lake Baikal.

ORDER PROBOSCIDEA [ELEPHANTS]. The surviving elephants, *Elephas indicus* of Asia and *Loxodonta africana* of Africa, are only a remnant of the formerly varied and widespread order. The extinct genera of the northern lands were largely destroyed by early human overkills, the last to disappear being the American mastodon, possibly as late as 4000 B.C., though an earlier date of about 8000 B.C. has been proposed. The oldest proboscidians date from the early Tertiary of North Africa, at which time they were pig-sized animals. In some parts of Africa elephants are becoming extinct; in other parts they are too numerous and are destroying the forest. Planned ecology programs aim to redistribute the herds more efficiently.

ORDER HYRACOIDEA [CONEYS]. The coneys, or *Hyrax*, are rat-sized animals with the characters indicated in Table 38. One species frequents rocky country; the other enters forests. Both are restricted to Africa and the adjacent part of western Asia Minor. They are not related to the North American pikas, sometimes called coneys in Bible-oriented communities. David, the shepherd-king of early Israel, and a good observer of nature, makes reference to the habits of the rock coney in one of his nature poems, apparently the oldest record of the order.

ORDER TUBULIDENTATA [AARDVARKS]. The single species, *Orycteropus afer*, the aardvark of Africa, is of uncertain relationship. Specimens can reach a length of 2 m, weighing up to 45 kg. They are strictly nocturnal burrowing insectivores but flourish on a diet of ground meat, eggs, and milk when fed on such fare in zoos. They were formerly classified with the American anteaters, which have the same dietary prejudices,

Figure 113. *Okapia johnstoni* (okapi, or forest giraffe), heavy West African forest, Artiodactyla (36).

the Copenhagen anteaters, however, preferring their eggs to be scrambled. By such feeding expedients zoos may succeed in saving many threatened animals from extinction when natural habitats and food resources are destroyed.

ORDER CARNIVORA [CARNIVORES]. The smaller forest and riverside predators are the civets and mongooses (family Viverridae). Bears, which are lacking from Africa south of the Sahara, form the family Ursidae, more or less omnivorous animals: *Helarctos malayanus*, the Malayan sun bear, is a tropical example of this otherwise cool-climate family. The dominant carnivores of the region are the great cats (Felidae), the wolves (Canidae), and the hyenas (Hyaenidae), animals that are somewhat intermediate between cats and wolves.

Spotted cats now face extinction, and in some areas the striped cats are endangered, chiefly through more or less illegal fur traffic. In 1969 three specimens of the Balinese tiger were known to survive. Efforts by interested conservation societies to save the tiger failed and it became extinct in 1970. Chita is a Sanskrit word for any spotted great cat but in the form *cheetah* is now applied to *Acionyx*. Except for possible breeding stock in the hands of Indian princes, the wild Asian form has just become extinct. The African form is apparently the same species, so replacement is still possible, though the Asian genetic line is probably lost. In Gujarat in northwest India the last surviving Asian lions exist. The Barbary lion was

Figure 114. *Hyemoschus aquaticus* (water Chevrotain), West Africa, Artiodactyla (26).

extinguished during World War II, seemingly by off-duty military personnel of the allied nations. This was the subspecies with the very large mane in the male, admired by the medieval crusaders. Tigers are still common, and even a menace, in parts of India, but in general the great cats are threatened animals, man having assumed the role of dominant predator in most regions where they occur.

Figure 115. *Redunca redunca* (waterbuck, or reedbuck), marshy regions in Africa, Antilopidae, Artiodactyla (26).

ORDER ARTIODACTYLA [UNGULATES]. The order is very varied and its members numerous. Among the families are the Hippopotamidae, now restricted to Africa; the Suidae, or pigs; the Giraffidae, now restricted to Africa with *Giraffa* on the savanna and *Okapia* in the densest equatorial rain forest; the Cervidae, or deer, with no African representatives south of the Sahara, the Asian genera including *Cervus, Muntiacus, Axis*. The cattle family Bovidae is represented with the following: in Africa, eland *(Taurotragus)*, gemsbok *(Oryx)*, gnu *(Connochaetes)*, buffalo *(Syncerus)*, and many others; and in Asia, gaur and banteng *(Bos)*, buffalo *(Bubalus)*, and others. The African buffalo is considered to be the most dangerous African beast if molested. The family Ovidae (sheep) does not presently range south of the Sahara, and in Asia is mainly found in montane country. Tragulidae (chevrotains) are small deerlike animals found in the forest or near swamps in Africa and Asia, believed to be related to the early Artiodactyla. Antelopes, with many groups, are common African grazing animals, with fewer representatives in Asia.

The Biogeography of the African and Tropical Asian Regions

The Oriental zoogeographic region as defined by Wallace comprises tropical Asia as far south as Java and Borneo. During the past quarter million years, when pithecanthropine hominids occupied parts of this region, the sea levels were reduced at times of glaciation, to produce continuous land from Asia as far south as Bali (the Indonesian historians state that Bali was still joined to Java at the time when the oceanic peoples began their occupation of the region during the millennium before Christ). The land apparently ceased along a line passing between Borneo and Celebes, and between Bali and Lombok. This line, described by Wallace as separating the eutherian mammal faunas of Asia from the marsupial mammal faunas of Australia and New Guinea, is taken as the boundary between the Oriental region and the Australian zoogeographic region.

The plant dispersions depend on different circumstances, and since the sea gaps though deep and permanent were never very wide, and indeed still are relatively narrow, wind dispersion of seeds, and bird dispersion of seeds in swallowed fruits, occurs easily. So there is no abrupt change in the nature of the forest biomes on either side of Wallace's line; the line is more in the nature of an ecotone, for some animal groups have no difficulty in crossing it, particularly birds and bats. And it is primarily a zoogeographic feature rather than a biogeographic generalization.

On account of the barrier to migration evidently presented by the Sahara Desert, zoogeographers usually associate North Africa with Europe as a unified zoogeographic region. The rest of Africa, called the Ethiopian region, then has the characters imparted to it by its past history of intercommunication with Asia, moderated by its present isolation by desert from Asia, coupled with the added differences resulting from late extinctions.

Birds in the Tropical Forest

Birds occupy much the same ecological roles in tropical forests as in austral forests: that is to say, the orders represented are largely the same, and they serve as pollinators and seed-dispersion agents, as primary and secondary heterotrophs, and frequent all the tiers of the forest. The greater uniformity of bird faunas is evidently due to their powers of dispersion by flight and also to the fact that many migratory birds of northern and southern forests pass through the tropics in the course of their semiannual movements. Following is a summary of the orders represented.

COLIES AND NIGHTJARS. Birds of this assemblage have anterior toes. The Colies (order Coliiformes) have a pointed tail, longer than the body. They are diurnal, like most birds, and are usually seen in flocks, creeping up and down the trunks of trees, and much resembling mice in this context. Colies occur only in Africa; an example is the mousebird *(Colius)*. In contrast, the nightjars (order Caprimulgiformes) are nocturnal predators, mainly insectivorous, and the tail is blunt and shorter than the body. An example is the Abyssinian nightjar *(Caprimulgus)*. Nightjars are widely distributed throughout the world, though lacking from New Zealand.

TROGONS. Birds of the order Trogoniformes have anterior toes and posterior toes, the outer hind toe being shorter than the inner hind toe. They are forest birds of medium size, green above and red below, usually sitting for long periods without moving, thus hard to see in the forest greenery. An example is the genus *Apaloderma*.

OWLS AND PARROTS. These are two orders in which the foot structure resembles that of trogons, but the outer hind toe is longer than the inner one, and there is a patch of soft skin (cere) partly covering the base of the bill. Owls (order Strigiformes) are nocturnal predators, birds of silent flight (on account of the soft plumage), and with the ankle joint (tarsus) covered by feathers. The barn owl *(Tyto alba)*, of North America, Eurasia, South America, and Australia, is also pantropical. Parrots (order Psittaciformes) are diurnal and have a naked tarsus; their diet is very varied, mainly seeds and other vegetable matter. They are often brilliantly colored, most notable being the macaws of South America; the type of the order is *Psittacus*, a genus that includes the West African gray parrot, often mentioned in the literature of mariners.

CUCKOOS. The foot here is like that of the preceding group, but the bill lacks a cere and is curved downward; the tail is barred trans-

Figure 116. Mimicry in African butterflies. The upper and middle specimens are the male and female of *Papilio merope*. The lower specimen is an unrelated species, (*Amauris niavius*), also a female. The females that are so similar have adopted a common coloration and pattern and are distasteful to eat. The common pattern is one that birds learn to recognize and to avoid (34).

versely and is longer than the wing. In the tropics, as elsewhere, the order Cuculiformes exhibit parasitic habits and are vigorous predators on small animals, including larvae of insects considered unpalatable by other birds. *Cuculus canorus*, the cuckoo of classical literature, is widely distributed in the tropics during the northern winter, when it leaves its summer breeding range in the summergreen woodlands.

WOODPECKERS, TOUCANS, AND HONEY GUIDES. The order Piciformes has similar foot structure to the preceding orders, but the bill is straight, has no cere, and the wing is usually longer than the tail. These birds seek prey in the wood of decaying tree trunks, where the larvae of beetles live, and the nest may be built in tunnels in the wood. The honey guides feed on the nests of bees.

The remaining tropical woodland orders are characterized by having anterior toes.

HORNBILLS, KINGFISHERS, AND MOTMOTS. All members of the order Caraciiformes have the outer two anterior toes fused together. Mainly predators, the kingfishers are worldwide, the hornbills range Africa and tropical east Asia, and the motmots occur in South America.

Figure 117. *Kallima inachis* (Indian dead-leaf butterfly); (a) flying; (b) becoming instantly inconspicuous on settling, posing as a leaf, which the underside of the wing resembles (34, 35).

The following four orders have a hind toe placed at the same level as the anterior toes.

EAGLES AND OTHER BIRDS OF PREY. The order Falconiformes, with a hooked bill and strongly clawed toes or talons, is widely distributed in all tropical forests. The tropical forest eagles may prey on animals as large as monkeys, such forest species having a characteristic head crest or crown of feathers. Other members of the order are important as scavengers of dead animals.

HERONS. Of the order Ciconiiformes, they are distinguished by the long straight bill and the long neck and legs. *Pigeons* (order Columbiformes) have a cere covering the nostrils, the bill straight and shorter than the head: extinct large ground pigeons of Mauritius and other islands in the Indo-Pacific include the dodo *(Didus ineptus)*; these flightless pigeons fell victim to dogs and rats and other animals introduced by Europeans and prone to destroy their nests and young. The immense group of songbirds (order Passeriformes) includes more than half of all known species of birds and cannot be discussed here.

RAILS, CRANES, AND SUNGREBES. This order, Gruiformes, comprises birds with three anterior toes, a hind toe at a higher level, a laterally compressed bill, the forehead often also covered by a colored horn plate; members are pantropical, many also occurring on other biomes.

JUNGLE FOWL. The game birds (order Galliformes) have a foot similar to that of the preceding order, but the bill is arched, and the feathers of the head continue across the forehead. The peafowl occurs in Asia, and a recently discovered smaller species of the same genus *(Pavo)* occurs in the central African forest. The domestic fowl *(Gallus bankiva)* occurs as a wild species in Indonesia and Southeast Asia.

TINAMOUS. Superficially similar to jungle fowl are the tinamous of South America, but their internal anatomy betrays a skull structure linking them with the large flightless ostriches and their kin. The order Tinamiformes is therefore placed by zoologists with the orders comprising the flightless southern birds, of which mention has already been made. Such large birds are mainly grassland animals, though in New Guinea the genus *Casuarius* ranges forests of that island.

14

SUMMERGREEN FORESTS

The *summergreen forest biome* embraces all deciduous forests in which the trees form new leaves in spring, the green phase spans the summer, the leaves are shed in the fall, and the trees stand devoid of foliage through the cold, hard months of winter. The biome occupies the nonarid portions of the Northern Hemisphere temperate zone where continentality produces warm summers followed by hard winters when the soil freezes and prohibits growth. Most summergreen forests lie between the latitudes of about 30° north and 45° north, though they are displaced northward to about 55° north in western Europe, where the climate is moderated by the north Atlantic drift current, and there is a compensating southward displacement on the eastern seaboard of North America, where the Labrador current exerts its chilling influence.

FORESTS IN HUMAN HISTORY. The tropical savannas were, seemingly, the home of man's earliest hominid ancestors during most of the last 5 million years. Only in the latter part of the old Stone Age did human communities extend to other biomes, with forest occupation coming last of all, perhaps about 10,000 years ago. During the Bronze Age settlement extended through most of the western European woodlands.

The oldest extant written references to a deciduous forest that I have been able to find occur in a poem attributed to Hesiod, a Greek who lived in Boeotia, east of Athens, about 800 B.C., the closing era of the Mediterranean Bronze Age. Like many poets after him, Hesiod sang of the annual cycle of growth, maturity, and reproduction that characterizes a deciduous forest. His wonder was aroused by the mysterious correlation of events in the forest and the successive changes in the constellations visible in the sky by night as the seasons revolve. In one of his poems, called "The Days," he catalogs the events in the life of a country farmer—the shooting of the buds and the arrival of the swallows in spring; the song of the cicada in summer; the ripening of the corn and the maturation of the grape in autumn; then the fall of the leaves, the winter rains and snowdrifts, and the sad plight of the forest beasts during the cruel months when Boreas prowls the frozen land. All these events he dates by a special calendar that we can understand, for he correlates each event with the rising or setting of a given star or constellation simultaneously with sunrise or sunset. A poet for scientists and intellectuals, it is not surprising that Hesiod was honored more in Athens than in Sparta, where Homer's martial illiterates could inspire a Leonidas but gave little to learning.

Hesiod lived in a land where deciduous forests obviously were commonplace, yet Boeotia today has none, and rice grows instead. This illustrates the endless swing of the biomes northward and southward as the earth's climates change. By 800 B.C. Europe had entered the cold rainy epoch that climatologists call the sub-Atlantic stage, an age of cold and famine and tribal displacement that brought the Homeric period to a close and plunged Greece into a second barbarism. Other climatic swings were to follow, as may shortly be traced in these notes. But before exploring this interesting field, we are bound first to make some effort to place the deciduous forests in their proper setting with respect to the long evolutionary sequence of forests whose course we have been following. Where, and when, and under what circumstances did the first deciduous forests take their origin?

ISLAND SANCTUARIES. Chapter 11 shows how archaic forms of life have survived and evolved independently in the lands of the remote southwest Pacific, protected by isolation and undisturbed by the more competitive later products of evolution in northern lands. Suppose now that there were some place, equally protected, that had been open to settlement by migrant northern plants since far back in early Tertiary or Cretaceous time. Such a refuge, to be effective, would have to provide various kinds of climate (for different plant communities); destructive grazing animals would have to be excluded; and to meet our present needs, the refuge would have to be available to Northern Hemisphere plants.

Figure 118. Deciduous summergreen forest in which the predominant species is ash (*Fraxinus excelsior*), with subdominant beech (*Fagus sylvatica*); this is the Eurasian analog of the American white oak-red oak-maple forest. The canopy is sufficiently open to permit abundant growth in the field layer; lianes and epiphtes are lacking, and plank buttresses do not form (32).

A VALLEY OF RELICT FLORAS? One of the few large areas still awaiting the attention of the biologist explorer is New Guinea. The island covers an area of about 300,000 square miles, roughly the size of Texas and Louisiana together; it touches the equator at its northwestern extremity and sprawls southeastward through some 10° of latitude. One of the world's great mountain systems traverses the land from end to end. The slopes are steep, the mountain passes very few, and the precipices of unbelievable proportions; indeed, there is one cliff that drops over 10,000 feet on Mount Leonard Darwin. In 1961 R. G. Robbins gave a preliminary report on the ecology of a secluded mountain valley in the central highlands of eastern New Guinea. The flora he described seemed to comprise delegations of plants from widely separated parts of the world. The valley in question is some 75 miles long, and its floor lies at a height of 5000 feet above sea level.

To the north it is walled in by a mountain range that rises to over 15,000 feet; to the south it is hemmed in by a similar range whose peaks top 10,000 feet. The western end of the valley is blocked by a volcano called Mount Hagen, and the eastern end is drained by the Wahgi River. Robbins had served as ecologist with an Australian expedition that had explored the region. The floor of the valley when Robbins saw it comprised grassland and gardens built up over a long period of time by the forest-clearing activities of the Stone Age people who live there. On the dry parts of the valley kangaroo grass *(Themeda)* had become established and gave a peaceful pastoral aspect to the scene. On the lower slopes of the investing mountains were forests of evergreen oaks, which have small, simple (that is, not lobed) leaves and large acorns. Above the oak forest came sometimes a mixed podocarp-broadleaf forest, sometimes a *Nothofagus* forest, both types of forest rich in plant genera shared by New Zealand and South America, though to each had been added a host of other genera that belong rather to the humid tropics. The greater part of Robbins' report dealt with details of these forests.

But two other matters arise from the facts he reported, both relevant to our present topic. First, the matter of oaks. Oaks, as already stated, form no part of the austral forests. Robbins' oaks are the outermost wanderers of that great northern genus, if indeed they came down from the north. The other concerns the presence in New Guinea of plants hitherto thought to be strictly austral. Robbins, for example, points out that *Carpodetus* was previously thought to have only one species (monotypic genus) and to be restricted to New Zealand, whereas it now turns out that there are at least six species of *Carpodetus* native to the New Guinea highlands. Robbins' own comment was: "This does not necessarily signify that the center of origin is in New Guinea, but only that the high mountains of New Guinea may well represent a more important area for relicts of the ancient southern flora than New Zealand itself!" And where do the oaks come in? If one looks over the content of *Quercus* in the North American flora, as an example, it turns out that the American oaks that come closest to the ones Robbins mentioned are species like *Q. hypoleuca, Q. agrifolia, Q. wislizenii,* and *Q. myrtifolia.* Now these species live in southern and western parts of the United States, where no severe winter occurs and where summer drought is not unusual. Very similar species occur in the Mediterranean region. These forms belong to a type of evergreen broadleaf forest known variously as Mediterranean or evergreen-hardwood or chaparral and are associated with a climate that has cold-moist winters and long dry summers. Growth mainly occurs in the spring and may resume again in the autumn, the ground herbs often appearing in spring and autumn flushes, too. From Joseph Hooker's descriptions of his Himalayan journeys, and from Hasskarl's explorations in the mountains of Java, it would seem that these chaparral oaks extend wherever the altitude and climate permit, and always at the appropriate altitude, from the warm-temperate northern latitudes (at sea level) to the equatorial region (at 6000 to 8000 feet). And if one traces the genus *Quercus* further northward beyond the confines of the Mediterranean, then the species become deciduous, with larger leaves that are active for only the warm part of the year, then cast off during a dormant cold-dry winter. In North America it is a similar story, save that for local reasons the extreme type of temperate climate lies in the east coast and over the eastern third of the United States.

It is therefore not impossible that the oaks of New Guinea may also represent relict forms, examples of what broadleaf trees and forests were like in Cretaceous times when, perhaps, the deciduous habit had not yet evolved. The secluded hanging valleys of the New Guinea interior may contain the survivors of ancient floras whose members met on and above the equator, spreading from the Northern and Southern Hemispheres, subsequently surviving there in isolation. This idea would merely be an extension to the northern floral elements of Robbins' suggestions for the southern elements he found in the Wahgi valley.

THE ORIGINS OF THE DECIDUOUS HABIT. The habit may have arisen independently many times over in the geological record, rather than as a coordinated response by natural selection during some period of intensified temperature variation or humidity variation. The ferns of the Northern Hemisphere include many species that die down in winter, whereas species of the same genera in the Southern Hemisphere are evergreen. The gymnosperms include various unrelated deciduous forms, such as *Ginkgo*, the ancestry of which goes back to Permian times, about 250 million years ago. *Metasequoia* and *Taxodium* and *Larix* are deciduous gymnosperms, the first two at least of Mesozoic origin. Among the broadleaf angiosperms, oaks are not the only trees that yield both evergreen and deciduous species—*Magnolia* has both; some evergreen *Magnolia* species from the southern United States become deciduous if planted in the more northern states. One species of *Nothofagus* in New Zealand drops its old leaves just as the new ones are forming, so that it may be regarded as almost deciduous. European deciduous trees when transferred to Australia and New Zealand are found to have a much shorter leafless period; some, like weeping willow, begin to form the new leaves almost as soon as the old ones have been shed. One of the swamp-dwelling hollies of the

United States and another species of the Appalachians are deciduous, whereas most members of the genus are evergreen.

These seemingly random examples suggest that it depends on the adaptability of a genus or family as to whether or not it is able to occupy habitats subject to severe climatic variation in the course of the natural year. Forms that are variable can respond to natural selection and evolve deciduous species. Families or genera that are not so adaptable are excluded from participation in the floras of regions with such severe annual variation in climate. Some genera, such as *Ficus*, which include deciduous species adapted to monsoon variation in humidity, have failed to produce corresponding deciduous species in areas where severe temperature variation occurs. It is not possible to tell from a fossil whether or not it had deciduous leaves. Fall color may be as ancient as the Ginkgoales or it may have come much later, during the widespread Tertiary coolings.

The plants that inhabited Greenland during the Eocene, about 70 million years ago, must surely have been deciduous, for the North Pole at that period was not too greatly displaced from its present position, which means that Greenland must have had a long polar night, irrespective of whether or not an ice cap existed (it almost certainly did not). Leaves would be a useless encumbrance to a plant in long-continued darkness, so they would tend to be shed, for natural selection would favor those plants that tended to shed the leaves at the onset of darkness. For these reasons we may think of the deciduous forests as being probably about as old as the angiosperms, if not older; their distribution, however, may have varied greatly during the past, for in mild epochs, and during phases of continental seas, such as the Cretaceous Tethys and Sundance seas, there must have been ameliorated climates and much less advantage in the deciduous habit.

THE DISTRIBUTION OF DECIDUOUS FORESTS. The deciduous broadleaf forests form a discontinuous belt around the Northern Hemisphere, for the most part between about 35° north and 45° north latitudes. The discontinuities are caused by inequalities in rainfall, which favor the development of vast grasslands in the dry interiors of the continents, and by inequalities in the altitude of the land, higher mountainous areas tending to carry boreal (needle-leaf) forests. Thus, in North America the deciduous biome is mainly an eastern feature, bounded by savanna ecotones along the eastern limit of the prairies. In Eurasia the forests occupy (or formerly occupied) Europe north of the Mediterranean and south of Scotland, Stockholm, and Moscow, parts of the Russian plain, and on the eastern margin of temperate Asia, Tien-Shan, northern China, Manchuria, and Japan. The steppes separate the east Asian forests from the European occurrence. During interglacial warm periods the east Asian and western North American regions would probably have carried a virtually continuous deciduous forest; at the present epoch there are many similarities between the east Asian forests and the North American equivalents, while the European forests stand somewhat alone.

Figure 119. A thousand years of moderation in lumber harvesting created naturally balanced forests in western Europe, in which man removed mature or dying trees and seedlings replenished the forest. The beech in the foreground illustrates another characteristic of the summergreen trees, the development of large boughs near the ground and the resultant absence of towering emergents. Undergrowth is thickest under the thin-canopied ashes in the background (32).

Ecotone Forests

BOREAL ECOTONES. They appear in northern New England, where broadleaf deciduous forest interdigitates with the needle-leaf forests, especially in hilly country, where the needle-leaf elements occupy the slopes and the broadleaf elements penetrate the valleys. Over the northern New England region, and also New Brunswick and Nova Scotia, maples, birches, and beeches occur together with evergreen needle-leaf trees such as hemlocks, larches (tamaracks) and spruces, pines, and balsam firs; this is sometimes termed the *Acadian forest region.* In western New York, Pennsylvania, and Ontario the ecotone involves such trees as maples, beeches, tulip trees and walnuts, yellow birches, white pines, and hemlocks. In Europe there are extensive mixed needle-leaf broadleaf ecotone forests on the west Russian Plain (northern Ukraine, Latvia, Lithuania, and Estonia) where the mixed forests comprise pines, firs, birches, alders, aspens, and oaks. On the Russian plain proper, coniferous forests lie to the north of Moscow, and to the south of that city are broadleaf forests of oaks, lindens, maples, mountain ashes, and hazels; mixed forest ecotones lie between these two forests. On the eastern margins of Asia similar ecotones occur. Around Lake Baikal a remarkable mixture of subtropical broadleaf forest and boreal conifers is found. On the east coast of the Caspian Sea and in Kazakhstan mixed forest may involve apples, apricots, and walnuts intergrading with fir forest.

SAVANNA ECOTONES. The eastern prairie states from Illinois to Texas present marginal savannas, and from Illinois northwest to Manitoba the maples, beeches, and birches form islands on the prairie, and patches of grassland occur in the forest. In Saskatchewan and Alberta aspens predominate in the marginal ecotone savannas. A remarkable savanna occurs on the east coast of the Kamchatka peninsula where the Siberian birches form isolated gnarled trees in a savanna of which the constituent grasses include species growing to 3 m high, according to Avetisian et al. (1955). The numerous miniature savannas of Europe are essentially parklands and grazing or crop lands created by man since Roman times, and the constituent trees depend on the whim of the landowner, often involving species brought from other parts of the world.

THE STRUCTURE OF A DECIDUOUS FOREST

1. The forest is rich in tree, shrub, and herbaceous species.
2. Particular species may be predominant, and the nature of the soil and its water content often have a bearing on the matter.
3. The forest is tiered, with tall and short trees, shrubs, and herbs.
4. The distinctive features of rain forest (plank buttresses, cauliflory, drip tips, stilt roots, pneumatophores, woody epiphytes, tree ferns, and so on) are absent.
5. Lianes are infrequent.
6. Epiphytes are mainly mosses, liverworts, and lichens.
7. There is a pronounced annual rhythm of growth.
8. The floristic content is mainly restricted to a limited number of arborescent genera that recur from one forest to another, in particular genera such as: oak, beech, chestnut, birch, ash, elm, hornbeam, walnut, hickory; with aspen, willow, alder, and elderberry

on damp ground and bordering the banks of streams flowing through the forest.
9. The flowers of the forest trees are pollinated by the wind.

THE RHYTHM OF A DECIDUOUS FOREST. A diurnal rhythm occurs during the months when the ground is not frozen. But a much more conspicuous *annual rhythm* is exhibited. The annual rhythm is controlled by the sinusoidal variation in the light intensity and temperature of the environment; that is, it is a solar cycle. Four episodes characterize the cycle. These are:

1. In early spring, before the leaves have appeared on many of the trees, the herbs of the forest floor begin to grow, taking advantage of the available light while the leaf canopy is lacking. These are plants such as *Viola, Anemone, Dicentra,* and others that flower, set seed, and complete their life cycle before the shade becomes pronounced.
2. Later in the spring the leaves of the tree canopy are fully formed, and thereafter the forest trees enter their main period of growth and reproduction. The arboreal animals and birds in particular select and defend territory and begin their reproductive cycle.
3. In the summer the forest floor herbs produce a second wave of taller herbs adapted to grow under low illumination intensity.
4. In the fall, after a summer resting period, a third crop of forest floor herbs appears, utilizing the increasing light intensity as the leaves begin to fall from the dominant trees. Late flowering species, such as autumnal crocus, complete the life cycle before the winter begins.

The Identification of Northern Deciduous Trees

Trees, like other flowering plants, are classified by botanists into families and orders mainly on the basis of the structure of the flowers. But most of the time trees are not carrying their flowers, so the ordinary naturalist has to rely on the leaf and stem characters to decide what species of trees he is encountering. Most trees can be identified with not too much difficulty or uncertainty by means of these vegetative characters, as they are called. Following are the characters of stem and leaf most commonly used.

LEAVES. Leaves may arise in opposite pairs along a branchlet, or may arise one by one in an *alternate* sequence to left and to right. Most trees have alternate leaves, so if you find one with opposite leaves, it will be easy to determine; the common trees with opposite leaves are the maples, the ashes, the dogwoods, and the buckeyes (or horse chestnuts); in America *Catalpa* is another common genus with opposite leaves; among the shrubs, the elderberry, usually by fresh water; and in warm

Figure 120. Mixed savanna and open forest characterizes much of the northern summergreen belt. Grazing stock, either domesticated (*Bos, Ovis*) or wild (*Cervus*), keep down the undergrowth and permit penetration of grasses into the field layer, as under these oaks. These relatively stable conditions were established in western Europe by Roman times and continued with little alteration until the population explosion in the twentieth century (20).

coastal districts at the southern limit of the broadleaf deciduous forest come the evergreen broadleaf trees, with the mangroves standing in or beside the sea. The other common northern trees have alternate leaves. This is the first character to establish on your specimen.

LEAF SHAPE. The keys that follow require that next you determine the shape of the leaf. A leaf may be *simple* (if it is not subdivided in any way); or lobed, which is self-explanatory, a maple leaf being an example; or the leaf may be *compound,* that is to say, subdivided into smaller leaflets. Be careful here, because some trees, like ashes, have very large leaves in which the leaflets look like whole leaves. To see whether you have a simple or compound leaf, examine the stalk of the leaf. If there is an obvious joint between the stalk and the branchlet from which it arises, so that when you pull on the leaf (let it dry out) separation occurs easily at that point, the leaf stalk is indeed the end of the leaf (and is termed then the *petiole*). But if these conditions are not met, follow the supposed branchlet down to the next junction; here you are likely to find the joint and separation point—in such eventuality you are dealing with a compound leaf.

A compound leaf has its petiole developed sometimes like a central axis, with leaflets coming off on either side; in such case the axis is called the *rachis* and the leaflets are often called *pinnae,* the whole leaf resembling a fern frond and being termed *pinnate.* Occasionally the compound leaf may have all the leaflets arising from one point, like a fan—a buckeye is an example; such leaves are termed *palmate.* If the leaflets are like fingers, the leaf is sometimes termed *digitate* or *digitate-palmate*; again the buckeye is an example. A maple leaf commonly (in North America) is *palmately lobed,* but in Asian species the maples may have the lobes extending to the center of the leaf, when it becomes *compound palmate*; conversely, trees

Figure 121. The presence of so-called standing stones or druidic circles in numerous locations, surrounded by open grassland and adjacent to inhabited modern villages or towns, betokens the influence of man in maintaining savanna land continuously since Bronze Age times in Europe. The resilience of the broadleaf deciduous forest when the hand of man is removed is shown in Figure 122 (13).

such as the mountain ash, which have *pinnate compound* leaves, occasionally include species in which the pinnae fuse together, thus producing a *pinnately lobed leaf*. The angle between the petiole and the branchlet that carries it is called the *axil*. Buds for new branchlets or flowers normally arise in the axils. Also, in many deciduous trees and shrubs, special *winter buds* are produced (carrying the embryonic tissue for next season's leaves), and winter buds may occur in the axils, or also at the tips of the branchlets, *terminal winter buds*. The presence of winter buds in the axils or at the tips of branchlets is a helpful diagnostic character.

The margin of a leaf can be *entire* (plain-edged or not subdivided in

Figure 122. Structures abandoned after the medieval period, as also abandoned Roman cities in northern Europe, were rapidly overwhelmed by regenerating summergreen forest, in marked contrast to the frequent incapacity of tropical forest to reinvade man-made savannas (Chapter 12). This feature is also discussed in its historical context in Chapter 18 (13).

any way) or bluntly or sharply *dentate* (toothed). The veins of the leaf may enter the teeth at the edge of the leaf; or the veins may approach the edge and then arch over without entering the teeth, uniting instead to make a marginal vein. These are useful characters to note, and may require a magnifying glass.

Note whether the base of the blade of a leaf is symmetric with respect to the petiole or lopsided (asymmetric).

STIPULES. This word is used for scalelike wrappings that sometimes envelop young leaves. Look at partly opened leaves to check whether they have stipules and, if so, whether there is one stipule per leaf, or two stipules, or more. Trees that do not have stipules on their young leaves are sometimes said to be *exstipulate*, especially in the jargon-riddled writings of the nineteenth-century botanists, who contrived to put the simplest statements in the most incomprehensible forms in much the same way as physicians have done in all ages.

FLOWERS AND FRUITS. Some trees, such as willows, are well known for the peculiar pendulous flowers they produce in spring, called *catkins*; the presence of these or of remains of them often solves an uncertainty left by leaf characters. The presence of fruiting bodies like little pine cones is a useful character, for example, in identifying a birch. Oaks *(Quercus)* have exclusive rights to acorns, which they carry most of the year, so these are first-rate characters for quick, reliable determination. Before eliminating oak as the possible genus of an unknown tree, examine the ground beneath for acorns, for some species do not carry the acorns in spring and summer. Maples similarly are obliging in carrying their characteristic winged keys for most of the summer, or you may find them on the ground beneath. The prickly nuts of beeches are distinctive; so are the big nuts of walnuts and hickories, the latter with four seams. Sumacs, which have large pinnate leaves, can be distinguished from other trees with leaves of this shape by the great panicles of florets or seeds that they carry, and by the milky sap.

Field Identification of Northern Forest Trees

The following sections are addressed to those readers who may wish at this stage to gain a working acquaintance with the common broadleaf trees of the northern summergreen forests. Table 39 sets out the 11 groups into which the trees may conveniently be divided on the basis of leaf characters, defined in the preceding section.

MAPLES, DOGWOODS, AND CATALPAS. This group includes trees with opposite leaves, and in which the leaf is simple, that is, not com-

Table 39. Key characters of 11 groups of northern forest trees.

Leaves opposite	Leaves simple		Maple, dogwood, catalpa
	Leaves compound		Horse chestnut (buckeye), elder, ash
Leaves alternate	Leaves compound	Simply pinnate	Locust, walnut, hickory, rowan, sumac
		Doubly pinnate	Acacia, mesquite, honey locust
	Leaves simple	Leaves palmate	Plane (sycamore), liquidambar
		Leaves smooth-edged	Magnolia, sassafras, tulip tree, willow, fig
		Leaves with dentate margins	Nut-trees: Oak, beech, chestnut
			Spines on leaves or twigs: Hawthorn, cherry, holly
			Seeds in cones: Hornbeam, birch
			Leaves asymmetric at base: Elm, linden, witch hazel
			Trees of damp or cold areas: Willow, aspen, alder

NOTE: The characters are chosen for ease in field identification (further explained in the text) and do not relate to formal systematic classification, which depends on flower structure.

pounded from smaller leaflets. Maples comprise species of the genus *Acer*, in which the seeds are winged keys, usually formed in pairs. The leaf is nearly always palmate, that is, of typical maple-leaf shape, though in a few Japanese species the lobes are separated so that the leaf is technically compound. Maples range North America, Eurasia, but the only British species is a shrub, and tree maples are important only east of Moscow and in eastern North America. Dogwoods (genus *Cornus*) have oval pointed (ovate) leaves, flowers with conspicuous white or colored sepals (four outer petallike structures), and the seeds form inconspicuous clusters, without seed wings and without pods. The pink-flowered species now becoming common in North America are introductions from Japan. Dogwoods range the Northern Hemisphere and also, strangely, occur in Peru. *Catalpa* is mainly tropical, ranging eastern Eurasia, eastern North America, and the Caribbean. The seeds are produced in pods, and the leaves are large and heart-shaped.

CHESTNUTS, ELDERS, AND ASHES. These are the trees with opposite compounded leaves. The horse chestnuts, or buckeyes (genus *Aesculus*), are easily distinguished by the fanlike arrangement of the five or seven leaflets, which all arise together at the tip of a long petiole. The fruit is a large nut encased in a spiny capsule. The genus ranges the Northern Hemisphere, with about six native American species. The elders have pinnate leaves with dentate margins. They are shrubs, found near water, with hollow pithy stems and bearing clusters of edible berries. The species fall in the genus *Sambucus*, distributed through the Northern Hemisphere temperate regions, and on mountain ranges of the tropics. The ashes are tall trees ranging Eurasia and North America, producing clusters of winged seeds; the leaves are large and have the same pinnate form as elders. The species of ash are placed in the genus *Fraxinus*.

All other common forest trees of the Northern Hemisphere have alternate leaves.

HICKORIES, WALNUTS, SUMACS, ROWANS, AND LOCUSTS. This assemblage comprises the trees with alternate compound simply pinnate leaves. The locusts *(Robinia)* stand alone in having spines at the bases of the leaf stalks and in having a beanlike fruit pod. The flowers (May–June) resemble pea, to which it is related. Locusts are restricted to the United States and Mexico. The other genera lack spines and do not form pods. Rowans, or mountain ashes *(Sorbus)*, have serrated leaflets and produce bright red clusters of cherrylike fruit, rich in vitamin C and much eaten by birds; it ranges the Northern Hemisphere and, as the name suggests, favors colder or higher regions. The sumacs are the tree and shrub species

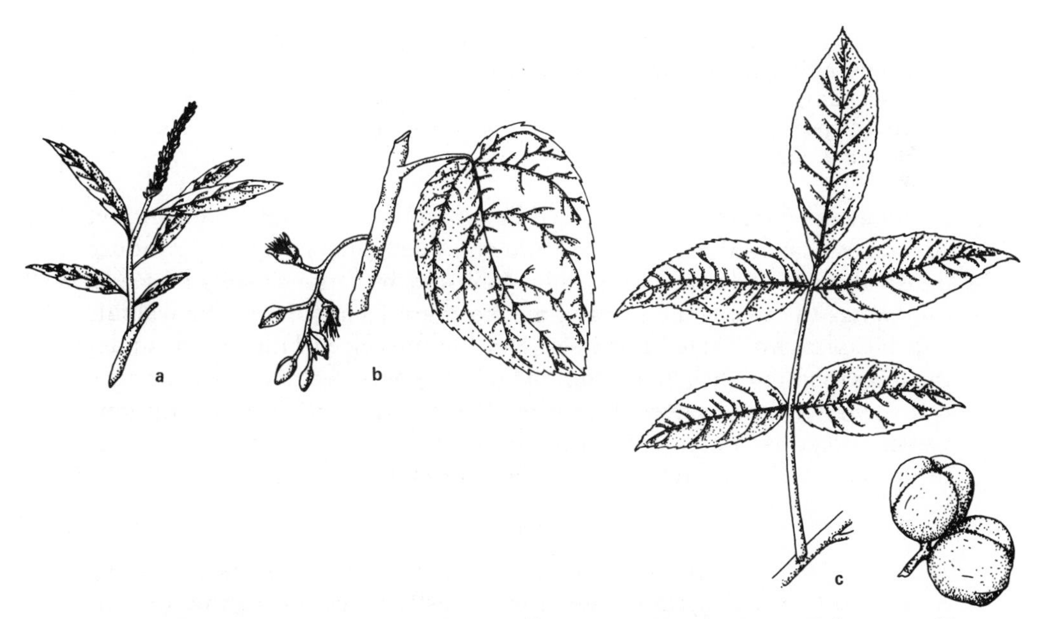

Figure 123. (a) *Salix inferior* (sandbar Willow), margins of streams, Salicaceae; (b) *Populus heterophylla* (swamp cottonwood), river swamps, Salicaceae; (c) *Carya ovata* (shagbark hickory), Juglandaceae; all North America. Drawn by Desmond FitzGerald.

Figure 124. (a) *Fagus grandifolia* (American beech), Fagaceae; (b) *Quercus rubra* (northern red oak), Fagaceae; (c) *Castanea dentata* (chestnut), Fagaceae (hybrids between native American and Eurasian species are now replacing the native American species, as the latter succumb to disease); all North America. Drawn by Desmond FitzGerald.

of the genus *Rhus*, which also includes some herbs, notably poison ivy. All sumacs have soft stems that exude milky sap (latex), and they produce dense clusters of flowers in spring, followed by large clusters of seeds; they range the temperate parts of the Northern Hemisphere. The walnuts and hickories are related nut trees, and the presence of nuts on or under the trees distinguishes them from the other genera; walnuts *(Juglans)* have nut shells with two valves; hickories *(Carya)* have similar nuts with four valves. Hickories range eastern North America, Mexico, and China; walnuts occur in Eurasia and North and South America.

TREE LEGUMES. Trees with doubly pinnate leaves belong to the pea family (Leguminosae). These mostly range the warmer regions of the world, but some species grow in the temperate zone. The leaves resemble fern fronds, and spines are commonly present in the leaf axils. In *Gleditsia*, the honey locust, the leaflets are crenulated along the edge, and simple and double pinnate leaves occur on the same tree. *Acacia* has leaflets with entire edges, and the leaves have six or more pinnae; these are mainly Australian and African, but some grow as natives in North America. *Prosopis* has leaves with two or four pinnae, and numerous large wound-inflicting spines occur; these are the wait-a-bit thorns of Africa, the thorn scrub of Asia, and the mesquite of North America.

The remaining trees have simple, uncompounded leaves. alternately placed on the twigs.

PLANES AND LIQUIDAMBAR. The leaves are palmate, that is, resemble maple leaves. Planes, called sycamores in the United States; not to be confused with sycomore, page 391 *(Platanus)*, have no terminal bud; the leaves are thick and hairy below, adapted to resist wind; they range the drier parts of Eurasia and North America. Liquidambar, with terminal buds on the twigs, have a similar range but are presently lacking from Europe, though occurring there before the late Tertiary coolings began.

TREES WITH SMOOTH-EDGED LEAVES. A very mixed assemblage having entire-edged leaves include, among others; *Sassafras*, with mitten-shaped leaves, the number of "fingers" varying from one leaf to another. Some willows fall here, genus *Salix*, with long elliptic pointed leaves, catkins in spring, usually growing near water. Willow ranges the Northern Hemisphere; sassafras occurs only in eastern North America and eastern Asia. *Magnolia* has oval pointed leaves, and *Liriodendron*, or tulip tree, has peculiarly bifid leaves, looking as if two leaves had grown to make one; both genera favor eastern Asia and eastern North America. Figs *(Ficus)* have a few temperate representatives that fall in this category; the temperate species have leaves that are hairy below and wind resistant.

The following genera comprise species that nearly always have dentate leaves.

OAKS, CHESTNUTS, AND BEECHES. These are readily recognized by their distinctive nuts. All oaks *(Quercus)* bear acorns, and no other tree has such nuts. Chestnuts (genus *Castanea*) bear coarsely toothed long straplike leaves and prickly-cased nuts. Beeches *(Fagus)* also have prickly-cased nuts, but the leaves are ovate and finely toothed. All these trees range the Northern Hemisphere. Under chaparral conditions the oaks may have small unlobed leaves, which may be evergreen, but the acorns always disclose the genus. If nuts are lacking from a tree, they may be found on the ground underneath during most of the year, so they are valuable identification characters.

FRUIT TREES. This conspicuous group of genera with alternate, simple dentate leaves makes up the fruit trees, mostly of the family Rosaceae. Spines are commonly present, though not always. Apples *(Malus)* have the leaves rolled up in the young stage before the leaf opens; spines, if present, are at the tips of the branchlets. Hawthorns *(Crataegus)* have spines in the axils, and the leaves are folded lengthwise in the bud. Plums and cherries *(Prunus)* are distinguished by the fruit, with a stone-like seed at the center. Hollies *(Ilex)* have spines on the edges of the leaves and are often evergreen. All these genera range widely in the Northern Hemisphere.

BIRCHES AND HORNBEAMS. These are genera with alternate, simple dentate leaves and bear their seed in cones. In the hop hornbeams *(Ostrya)* the cones are pendulous, papery, and persistent through

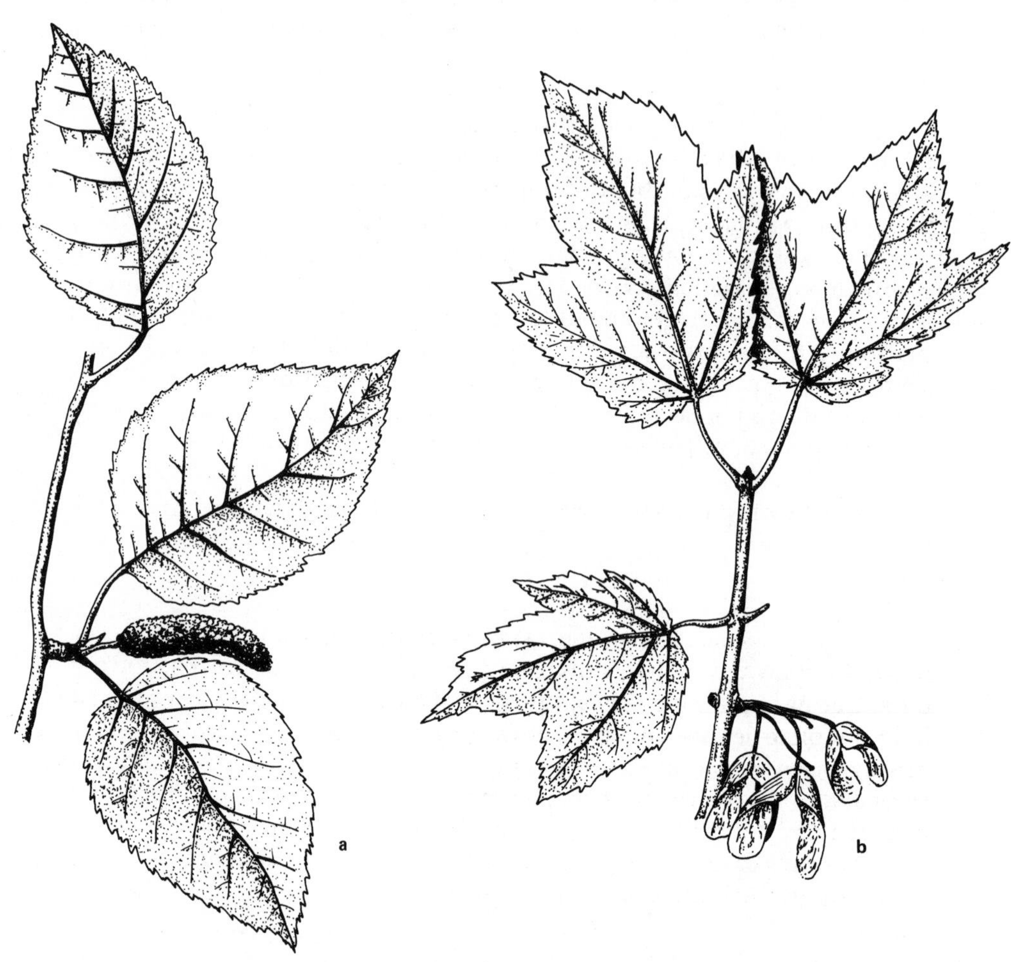

Figure 125. (a) *Betula papyrifera* (canoe birch), open woods and banks of streams, Betulaceae; (b) *Acer rubrum* (red maple), swamps and wetter hill slopes; both North America. Drawn by Desmond FitzGerald.

the year. The two following genera have more rigid, erect cones: hornbeam *(Carpinus)*, with hairy leaves and smooth bark; and birches *(Betula)*, with smooth leaves and flaky bark, often found near water. All these genera range the Northern Hemisphere.

ELMS, WITCH HAZELS, AND LINDENS. In this group of genera the leaves are conspicuously asymmetric at the base, broader on one side of the petiole than on the other. The linden *(Tilia)* has heart-shaped or broadly oval leaves, with a single row of teeth around the edge. Witch hazel *(Hamamelis)* has coarsely dentate leaves and is distinctive in bearing yellow flowers during the fall (instead of the spring, as usual with northern forest trees). Elm *(Ulmus)* has doubly serrate leaves, that is, leaves in which minute dentation occurs around coarser dentation. Witch hazel occurs in east Asia and in North America; the other two genera range the northern continents.

WATER-LOVING TREES. This group of trees has alternate, simple dentate leaves and may be distinguished by their common association with water, either running or standing. The alders *(Alnus)* have coarsely serrate dentation, the veins of the leaf entering the teeth; they produce catkins in spring. They are found in the swamps and high mountains of North America, Eurasia, and North Africa. Such willows (see above) as have dentate leaves are distinguished by the fact that the leaf vein does not enter the tooth opposite it; catkins occur in spring, as in all willows, genus *Salix*. Poplars (genus *Populus*), called cottonwoods in North America, usually have heart-shaped leaves with conspicuously elongated leaf stalks; catkins form in spring. Aspens constitute great forests in subarctic regions. Willows and poplars range the Northern Hemisphere.

SYSTEMATIC CLASSIFICATION. The foregoing pragmatic arrangement of the northern broadleaf trees groups them according to so-called vegetative characters commonly found during most of the year. The formal classification of botanists depends on flower characters not normally observable save in the short spring flowering season and hence are of little use to the field naturalist or ecologist. But to permit comparison with formal groupings of exotic trees mentioned elsewhere in this book, Table 40 sets out the ordinal and familial names and indicates the inferred evolutionary relationships of the trees.

DOMINANCE AND FOREST STANDS. As already noted, particular species of trees may be predominant in an area covered by summergreen forest. This is a notable difference from tropical forests, and it is probably due to the fact that tropical forests have a nearly uniform environment, whereas in the northern continental latitudes in which deciduous forests occur, quite wide differences in environmental parameters commonly occur. In latitudes where the temperature fluctuates widely during the course of a year, relatively minor differences in exposure to sunlight or to winds or to frost may become quite important—so much so as to lead to situations where one can see all the northward slopes of hills carrying one type of forest, and all the south-facing slopes carrying another type, with perhaps a third type on the tops of the hills, and a fourth in the valleys. Another difference is sometimes introduced by the nature of the bedrock from which the forest soil is derived; the latter effect is often called *edaphic*, that is, soil-related as opposed to the former, which is *climatic*.

EDAPHIC FACTORS. Chief of these factors are the response of certain trees (and other plants) to the pH value of the soil and the concentration of calcium carbonate; or, in other words, whether the forest comprises certain tree species or not is sometimes determined by whether or not the soil is acid or alkaline. Examples follow:

TREES FAVORING ACID SOILS	TREES FAVORING ALKALINE SOILS
Hickory *(Carya)*	European poplar *(Populus canescens)*
Oaks *(Quercus)*	European ash *(Fraxinus excelsior)*
Sourwood *(Oxydendron)*	Hazel *(Corylus)*
Trees and shrubs of the heath family Ericaceae (for example, *Rhododendron*)	

In England, where considerable areas of chalk-derived soils occur in the south as also in France and Germany and Denmark, the strong influence of the soil on the character of the forest is often very evident. Chalk soils are usually poor in humus because the excessive calcium carbonate destroys the organic content of the soil. These conditions, unfavorable to many trees, are tolerated by the European species of ash that, accordingly, becomes the dominant forest, so much so as to change the whole character of the vegetation. Ash trees do not cast much shade, so their dominance in a forest means that the nature of the undergrowth and the herbs of the field layer also changes; and these plants also of course have physiological requirements related to the soil as well as to the amount of light reaching them through the forest canopy. Characteristic subdominant species accompanying the ash include hazel *(Corylus)*, blackberry or bramble *(Rubus)*, aspen *(Populus)*, ivy *(Hodera)*, hawthorn *(Crataegus)*, and in the warmer regions a liane *(Clematis)*, plus some needle-leaf trees such as yew *(Tauxus)* and juniper *(Juniperus)*. The edaphic factors may be more potent than climatic in such a case, so that a European ash wood may extend to the limit of forest on a mountainside and be replaced at the upper limit by scrambling or humpy hawthorn scrub.

The amount of soil moisture or water can be considered an edaphic factor, but this is often climatically related. Thus the same trees that favor the moisture-rich mountain valleys, where cloudbursts and frequent rain-

Table 40. Formal botanical classification of the genera of northern broadleaf trees mentioned in this chapter.

ORDER	FAMILY	NOTABLE NORTHERN GENERA
Salicales	Salicaceae	*Salix, Populus*
Juglandales	Juglandaceae	*Juglans, Carya*
Fagales	Fagaceae	*Fagus, Quercus, Castanea*
	Betulaceae	*Betula, Alnus, Carpinus, Ostrya*
Urticales	Ulmaceae	*Ulmus*
Ranales	Magnoliaceae	*Magnolia, Liriodendron*
	Lauraceae	*Sassafras*
	Platanaceae	*Platanus*
	Hamamelidaceae	*Hamamelis, Liquidambar*
Rosales	Rosaceae	*Malus, Sorbus, Prunus, Crataegus, Ilex*
	Leguminosae	*Robinia, Acacia, Gleditsia, Prosopis*
Sapindales	Aceraceae	*Acer*
	Hippocastanaceae	*Aesculus*
	Anacardiaceae	*Rhus*
Malvales	Tiliaceae	*Tilia*
Umbellales	Cornaceae	*Cornus*
Gentianales	Oleaceae	*Fraxinus*
Tubiflorales	Bignoniaceae	*Catalpa*

storms keep the watercourses flowing, may also be found thousands of feet below in a much more sheltered and warmer situation if there is an abundance of water, as in a swamp, for instance.

CLIMATIC FACTORS. The table that follows illustrates by examples the varying character of the dominant trees in forests of the summergreen zone, according to climatic parameters or, in some cases, the combination of climatic and edaphic factors.

PALYNOLOGY: THE ARCHIVES OF THE BOG PEATS. During the evolution of a lake into a swamp there is a period, usually of several thousand years, when an annual layer of silt is deposited on the lake floor by the spring floods. Between each annual layer of silt is the layer of dust blown into the lake by the winds. The dust includes the minute spores of lower plants and the pollen grains of trees and herbs. The outer coat of a pollen grain is very hard and does not decay easily; it also happens to be elaborately sculptured in a pattern that is usually characteristic for each species of plant. Thus, by driving a tube into a bog deposit we can recover the successive time layers of pollen and silt laid down year by year over the whole period of the existence of the lentic deposit, from the time it was a lakebed until the time the bog eventually dried up, or became a swamp forest. Some bogs and swamps are still in a functional growing phase. So putting together all the available evidence for a given district, it is theoretically possible to reconstruct a time chart, layer by layer, showing the different kinds of pollen that fell into the lake over a very long span of time until the present. In this way it has been possible, for example, to tabulate the different species of trees and other plants that lived in a particular district over a span of time reaching back 10,000 to 15,000 years into the past. This study is a branch of paleobotany called *palynology*,

Table 41. Correlation of broadleaf tree distribution with climate.

ENVIRONMENTAL CONDITIONS	REPRESENTATIVE BROADLEAF TREE SPECIES	
	NORTH AMERICAN FORESTS	EURASIAN FORESTS
Subtropical to warm temperate	Live oak, *Quercus virginiana* Cedar elm, *Ulmus crassifolia* Florida maple, *Acer barbatum*	Holm oak, *Quercus ilex* Chinese elm, *Ulmus parvifolia*
Warm temperate with dry soils	Blackjack oak, *Quercus marilandica* Bluejack oak, *Quercus cinerea*	Cork oak, *Quercus suber* Durmast oak, *Quercus robur sessiliflora*
Temperate with damp clay soil	River birch, *Betula nigra* Willow oak, *Quercus phellos* Red maple, *Acer rubrum* Alder, *Alnus* spp. Black ash, *Fraxinus nigra*	Poplar, *Populus alba* Pedunculate oak, *Quercus robur pedunculata* White willow, *Salix alba* Goat willow, *Salix caprea*
Cool temperate: warm summers and cold winters	White oak, *Quercus alba* Northern red oak, *Quercus rubra* Canoe birch, *Betula papyrifera* Gray birch, *Betula populifera* Sugar maple, *Acer saccharum* Silver maple, *Acer saccharinum*	Beech, *Fagus sylvatica* Ash, *Fraxinus excelsior* Silver birch, *Betula pendula* Siberian birch, *Betula albosinensis* Norway maple, *Acer platanoides* Japan maple, *Acer palmatum*
Mountain exposures: windy, cold	Mountain maple, *Acer spicatum* Mountain ash, *Sorbus americana*	Rowan, *Sorbus aucuparia* Siberian elm, *Ulmus parvula*
Subarctic: cool summers and cold winters	Arctic willows, *Salix exigua*, and so on Dwarf maple, *Acer glabrum* Aspen, *Populus tremuloides* White birch, *Betula alaskana*	Alpine willow, *Salix reticulata*, and so on Aspen, *Populus tremula* Lapland birch, *Betula tortuosa* Amur maple, *Acer ginnala*

and its methods were first developed about 70 years ago. Table 42 is simplified from results given by a Danish investigator (K. Jessen) and shows the forest succession as revealed by Danish bog cores covering the last 12,000 years of elapsed time.

If Table 42 is compared with Table 41 it becomes clear that we observe the effects of an ameliorating climate, beginning with a polar environment some 12,000 years ago and reaching a maximum of warmth during the oak forest phase, followed by some deterioration after 1,000 B.C., when the oak forest was displaced by beech. This illustrates one of the methods of studying the former environments of the earth. A much fuller picture can be reconstructed if we introduce the equivalent evidence for the contemporary faunas.

POSTGLACIAL FOREST SUCCESSION. Evidence from many different northern regions is now available to show that the kind of forest succession that occurred in Denmark also occurred all over the northern temperate zone, though the details vary from place to place, and sometimes there seem to be local or regional reversals in the general warming trend. In North America the needle-leaf forests are more conspicuous than in corresponding latitudes of Eurasia, and they also tend to be more varied

Table 42. Broadleaf succession in Denmark.

DATE	FOREST AND OTHER VEGETATION PRESENT IN DENMARK
A.D. 1,000	Beech and heather
Birth of Christ	Beech and heather
B.C. 1,000	
	Mixed oak forest
2,000	
	Mixed oak forest
3,000	
	Mixed oak forest
4,000	
	Mixed oak forest
5,000	
	Mixed oak forest
6,000	
	Linden, alder, oak, and elm
7,000	
	Fir and hazel
8,000	
9,000	Birch and fir, ash
10,000	
	Polar willow, dwarf birch, and *Dryas*
11,000	

NOTE: Simplified from K. Jessen, in E., Albrectsen, *Danmark i Oldtiden* (Copenhagen, 1949).

and more intergraded with the broadleaf forests. Thus in virtually all the postglacial successions reported from the broadleaf forest regions of North America, evidence of former occupation by needle-leaf forests is overwhelming.

An example may be cited from Ohio, where presently the summergreen forest is predominant. The postglacial succession discloses about seven discernible forest phases that followed the tundra period of about 10,000 B.C. Reading forward in time the Ohio sequence is (1) arctic tundra, (2) spruce forest, (3) pine forest, (4) beech and hemlock forest, (5) oak and hickory forest, (6) beech and maple forest; there is a reduplication of the spruce-pine phases early in the sequence, so that if that is interpreted as a climatic reversal, then seven phases are represented.

From the New England region a similar sequence is disclosed, but the supposed earlier reversal of the Ohio series is not apparent, probably because the climate was so severe in New England that no forests developed at all during the initial withdrawal of the glacial tongues which covered the land. The New England sequence simplifies, then, to: (1) tundra, (2) spruce forest, (3) pine forest, (4) oak and hemlock forest, (5) oak and hickory forest, (6) oak and chestnut forest. Further discussion at this point is inappropriate, for we have not yet studied the characteristics of the needle-leaf forests to which the spruce, pine, and hemlock belong; neither have we yet encountered the tundra.

POSTGLACIAL SUCCESSION IS NOT A SERE. It may be stressed that the foregoing sequences of forest types cannot be the successional phases of a sere. As already noted, the sere for a modern maple-oak-beech forest will begin with grasses, then pass through blackberry-hawthorn-aster stages, then a hickory-oak-elm subclimax, and then the climax. In the case of the postglacial succession, we are dealing with a higher order of magnitude, the time scale is at least 10 times longer, and obviously all the stages we can detect must themselves be climax stages. So the conclusion is inevitable that we are observing a climatic series of climax forests.

THE ROLE OF THE WIND IN THE NORTHERN FOREST ECOSYSTEM. In following the recent geological history of northern summergreen forests we have been brought face to face with some other aspects of their ecology. These may be briefly stated:

1. The history of the northern summergreen forest is apparently inextricably tangled with the concurrent history of the needle-leaf forest. The one readily is replaced by the other and, as earlier noted, they have extensive ecotones in common.
2. Both biomes give clear indications of having moved through latitudinal parameters during the last 12,000 years, the displacement of the one being always related to that of the other forest type.
3. Both types of forest have left abundant records of their former dispersion by way of pollen-grain fossils.

The first two of these implies that no satisfactory biogeographic study can be made on either one of the biomes without reference to the other. Therefore we may postpone a biogeographic discussion until after we have examined the needle-leaf biome in a later chapter. The third aspect, concerning pollen, also applies equally to both biomes but demands attention here since it involves an essential characteristic of the northern forests, and one that separates them very markedly from the forests of the tropical belt. The character in question is the almost universal dependence on the wind for the pollination of the flowers of northern deciduous trees and gymnosperms, including all the conspicuous conifers. This means that without the wind the northern forest trees could not reproduce, for their flowers are not adapted to permitting pollination by insects or birds. Plants that are so dependent are termed *anemophilous*.

CHARACTERISTICS OF ANEMOPHILOUS TREES. For successful utilization of the wind in the reproductive cycle the following conditions must be met.

(1) Very large quantities of pollen must be produced since only a very small fraction of what is produced will reach its ultimate destination, the stigma of a female flower of the same species.

The amounts produced are so great that at the season of flowering great clouds of pollen are swept up like dust into the atmosphere and later deposited over whatever surfaces happen to lie beneath. The pollen is commonly colored yellow, so a massive pollen shower may resemble sulfur. In fact, when massive plantings of northern coniferous trees were first undertaken on the barren hills of New Zealand in the 1930s, the local people were subsequently alarmed when the first pollen showers occurred, for none had seen this phenomenon before. It was immediately inferred that the substance was sulfur from one of the local volcanos, and eruption alarm spread through the community.

(2) The pollen grains must be light, easily carried by the wind, not sticky or clumping (as in insect-dispersed pollen). Some types produced by certain conifers—*Pinus sylvestris*, for example—have a pair of winglike sacs, permitting the grain to remain aloft longer in the air currents. In the Urticaceae the structure of the pollen sac may be such that it is under tension until the moment it ruptures, whereupon the pollen grains are thrown out in a cloud into the air, already wind-borne therefore at the time of their production.

(3) The flowers must be of such a nature as to favor the transfer of the pollen by wind. The male flowers commonly take the form of pendulous *catkins*, a kind of open-structured cone or branchlet that hangs exposed to the breeze, with the pollen carried on a large number of microsporophyls. Grasses also utilize wind pollination, and in their case the anthers are long and slender and hang out of the flowers at the extremities of filamentous stamens.

(4) The female flowers must be constructed on such a plan as to favor the successful acquisition of randomly drifting pollen carried past the female flowers by the wind. The structures employed to this end are some-

what analogous to the devices evolved by marine filter-feeding animals. In fact, a marine biologist would be inclined to classify pollen of anemophilous plants as aerial planktonic larval stages! The female huntress is equipped with long feathery stigmata that are exposed to the wind, and pollen grains of all types are entangled in the hairs on the stigma. Among the random catch will normally be a sample of the same species, leading to a successful fertilization. A modification of specifically terrestrial character is found also—some conifers equip the stigma of the female flower with a droplet of sticky fluid on which pollen grains are captured on the flypaper principle. Since insects and birds are not involved in these sexual transactions, natural selection has seen to it that no evolutionary energy has been wasted on producing flowers with colored petals or other attractive devices to signal nectar to potential fertilizer agencies. Nectar there is none, and none is advertised. So the colors of a deciduous forest are those of the leaves in the fall rather than those of the few insect-pollinated trees that produce bright flowers in the summer.

(5) The flowering season must occur when little else will offer physical obstruction to the wind-blown pollen. Thus almost all the common trees of the deciduous forest, and the needle-leaf trees of the coniferous forest, produce their flowers early in the spring or even at the end of the winter, just as temperatures are beginning to rise. Thus the poplars, alders, hazels, and elms all produce their flowers in February and March, well in advance of the opening of the leaves. This procedure avoids the massive deposition of pollen on leaves, and the consequent overproduction of pollen that otherwise would have been needed. Conifers, apparently less sophisticated in the matter of conserving energy, are evergreen and consequently are obliged to produce much greater quantities of pollen. Beech, birch, walnut, and oak trees produce their flowers at the same time as the foliage is appearing, but the reproductive formalities are brief, and fertilization is achieved before the leaves are fully expanded. Maples belong to several schools in this matter; some species, such as the silver maple, start producing their flower buds in the previous fall and get off to an early flowering even before the snows have melted in the spring. Sugar maples, less precipitous, are content to follow the fashionable beech-birch-oak timetable. Conifers have learned to utilize ascending air currents, and the male cones are produced relatively low down on the trees, while the female cones await the arrival of the pollen from perches high up in the tree.

THE ROLE OF ANIMALS IN THE SUMMERGREEN ECOSYSTEM. The preceding sections have illustrated the relative independence of the northern forest trees from reliance on animals. That this proposition has much to be said for it is further emphasized by the fact that when European exiles in New Zealand pined for English trees in the evergreen wilderness they set about destroying, they imported seeds of oaks, ashes, and all the rest and planted them in such numbers that veritable forests arose, especially later when the Californian gymnosperm *Pinus radiata* was added to the list of imports. These new forests arose where none of the European or American forest mammals existed, and where only a few of the northern bird species had been introduced. The new forests are shunned by most of the native birds of New Zealand, for they can find there no nectar-bearing plants. The fly catchers frequent them in search of the insects, though not in large numbers, for the relatively sterile forest floor, littered with dead leaves, does not favor the rich fern flora of the natural New Zealand forest. Yet these new forests, devoid of their natural northern fauna, are among the most productive in existence, and some trees mature much faster than in their native northern haunts. So whatever role the northern fauna plays in the maintenance of the northern forest ecosystems is not yet entirely plain. The question, however, is not one to be dismissed lightly and may well now occupy our attention in the next chapter.

15

FAUNA OF THE NORTHERN FORESTS

The Holarctic Zoogeographic Realm

Wallace defined this faunal region as comprising the Arctic and temperate parts of the Northern Hemisphere, the southern limits being set by the Sahara Desert, the Himalayas, the Tibetan plateau, and the Yucatán peninsula. The existing fauna includes some characteristic northern elements that are not found elsewhere, such as beavers and moles; but in general it has rather the aspect of an impoverished version of the tropical fauna. For example, some groups that still have relicts in the tropics, such as elephants, rhinoceroses, and hippopotamuses, were present in the Holarctic region during Tertiary times and became extinct there only in recent geological time. Carbon-dating has shown that in many instances the extinction of a given group can be correlated with the first appearance of physical evidence of human predation. These human hunting pressures became particularly severe during the last 15,000 years. Some of the extinctions occurred before man appeared on the scene—for example, rhinoceros in North America—and such faunal losses are thought to have been caused by the climatic deterioration that began at the close of the Pliocene period, about 10 million years ago.

PALEARCTIC AND NEARCTIC FAUNAS. These terms were applied by Wallace to the Eurasian and American portions of the Holarctic region. The Palearctic fauna is distinguished by including some elements that are absent from North America, such as camels, pandas, and horses, together with some immigrants from the adjacent paleotropical realm, such as tigers, hedgehogs, and Old World porcupines. The Nearctic fauna has correspondingly endemic elements, such as musk oxen and pronghorn antelopes, together with immigrants from the neotropical region such as opossums, armadillos, peccaries, and New World porcupines.

However, we now know that camels and horses were present, and indeed evolved, in North America until their extinction apparently at the hands of Amerindian hunters 11,500 years ago; and both the Palearctic and Nearctic faunas formerly shared such animals as mammoths, mastodons, saiga antelopes, and musk oxen. So the distinction between the two

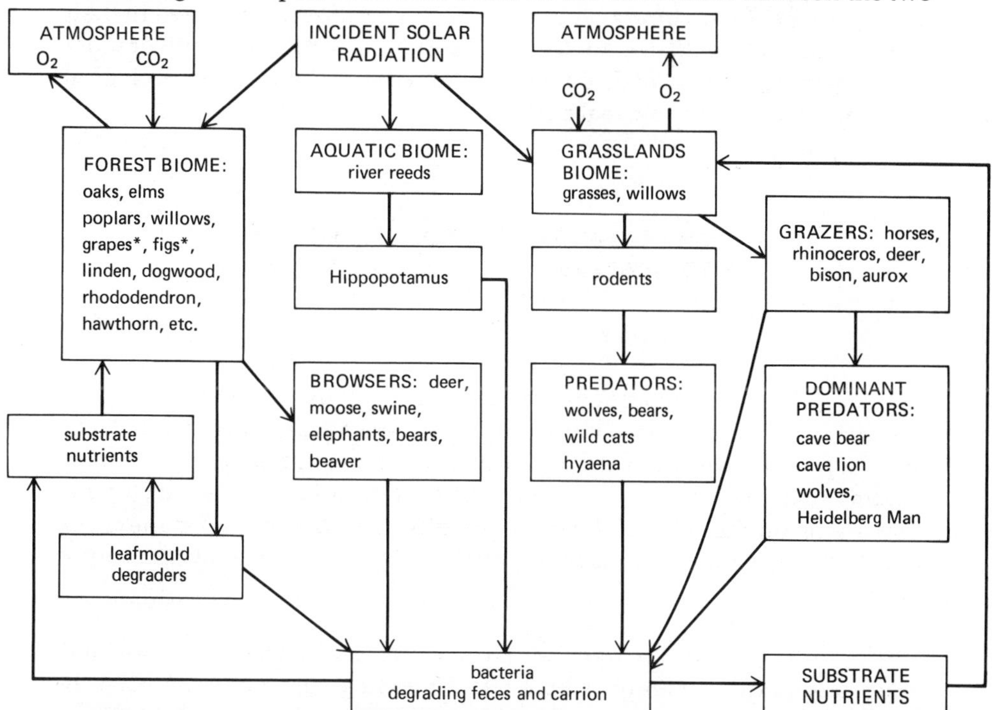

Figure 126. Flow of energy and materials in a fully developed north temperate summergreen forest community with grassland ecotones; based on the Thames valley, England, in the second interglacial warm stade (Mindel-Riss, equivalent to North American Yarmouth), about 1 million years ago. Climatic indicators marked with asterisk, fig implying mild winters, grape implying warm summers. Extant northern faunas may be viewed as the shattered remnant of this assemblage, devastated by human hunters (23).

faunas is much less evident now than it seemed to be when Wallace first defined them. In this text the Holarctic fauna is treated as a single unit therefore.

THE RELATIONSHIP OF THE FAUNA TO THE BIOMES. The Holarctic faunal region occupies six biomes, namely: the summergreen forest, the taiga or boreal forest, the Arctic and alpine tundras, the prairie and steppe grasslands, the chaparral, or Mediterranean, biome, and the arid deserts. Those animals whose manner of life or food requirements tie them to particular types of vegetation have distributions that more or less correspond to the biomes. On the other hand, carnivores roam freely from one biome to another, and migratory birds may spend half the year in a biome far removed from their breeding range. To avoid repetition, the fauna of the summergreen and boreal forests are treated in this chapter, and animals that are more particularly associated with the other biomes are noted in the appropriate chapters. The primary divisions are based on the position occupied by the animals in the trophic cycle, the subsidiary divisions are those of formal systematics. A few animals, such as the pandas, where the trophic niche differs from that of their closest relatives, are treated in their systematic context, that is to say, pandas appear with the Carnivora, despite their vegetarian diet. Table 43 summarizes the inferred ecological roles, and Table 44 the ordinal characters.

PRIMARY HETEROTROPHS: THE BROWSERS AND GRAZERS

Order Lagomorpha

The morphological characters of the order Lagomorpha are given in Table 44. The strong and continuously growing incisor teeth stand up to the wear and tear of grazing on coarse vegetable material, including grass with more or less siliceous content; and the diastema (toothless region on the sides of the jaws) permits food to be stored in the cheek pouches before further mastication by the grinding teeth. Only one family occurs.

HARES AND RABBITS [FAMILY LEPORIDAE]. The terms *hare* and *rabbit* originated in Europe, where two different genera occur; *hare* rightly relates to members of the genus *Lepus*, surface-dwelling and building a rough nest or form in which the furred young are born. The term *rabbit* correctly applies to burrowing lagomorphs that give birth to naked young underground and are placed in the genus *Oryctolagus*. Only the former occur naturally in North America, but the terms rabbit and hare have been indiscriminately applied to them. The snowshoe rabbit, or varying hare *(Lepus americanus)*, ranges from the deciduous forests northward to the

Table 43. Ecological roles of northern forest mammals.

ECOLOGICAL ROLE	ORDER	FAMILY	VERNACULAR NAME
Primary heterotrophs (browsers and grazers)	Lagomorpha	Leporidae	Hares, rabbits
	Rodentia	Erethizontidae	American porcupines
		Hystricidae	Eurasian porcupines
		Aplodontidae	Mountain beavers
		Castoridae	Beavers
		Sciuridae	Woodchucks, squirrels, and chipmunks
		Muridae	Rats, mice
		Cricetidae	Voles
	Artiodactyla	Suidae	Swine
		Cervidae	Deer
		Bovidae	Bison, cattle, sheep, and goats
Secondary heterotrophs (predatory carnivores and insectivores)	Marsupialia	Didelphidae	American opossums
	Insectivora	Erinaceidae	Eurasian hedgehogs
		Talpidae	Moles
		Soricidae	Shrews
	Chiroptera	Vespertilionidae	Bats
	Carnivora	Canidae	Wolves, foxes, and jackals
		Procyonidae	Raccoons
		Mustelidae	Otters, skunks, and weasels
		Felidae	Cats, leopards, and cougars
Dominant omnivores	Carnivora	Ursidae	Bears
	Primates	Hominidae	Apemen and men

tundra lands. A similar species occurs in Europe *(Lepus timidus)*. These tend to be blackish in summer, replacing the coat by white fur in winter. The food is vegetable materials, grass, berries, and leaves of conifers and aspen. Many similar species occur and are usually called jack rabbits in America, the more northern species turning white in winter. Similar species in Eurasia, which remain colored through the year, include *Lepus europaeus*, the common hare of Europe. *Sylvilagus*, or cottontail, includes species with a white tail ranging much of North America; they have the habits of hares. The marsh rabbit, a species of *Sylvilagus*, has the habit of running in the same manner as a dog or other running mammal, using the legs alternately instead of hopping on two legs simultaneously as in leporids generally. The earliest leporids date from the late Oligocene, about 30 million years ago. Leporids are the natural prey of most of the middle-sized carnivores, such as foxes and lynxes. Man also hunts them, without harm to the population, as the reproduction rate is very high. Uncontrolled rabbit populations, as in Australia and New Zealand, have severely damaged the land through overgrazing. Chemical and biological control methods have been used, with considerable success, but most success seems to attend simple legislation, prohibiting the sale of any rabbit by-products; such laws have the immediate effect of inducing farmers to rid their land of rabbits and to replace them by salable stock. This illustrates the overpowering role of man as a predator whenever he chooses to assume that role.

Order Rodentia

The ordinal characters of Rodentia are given in Table 44. The various northern families may be distinguished by the following key:

Spines (quills) developed on the dorsal surface of body	
Quills mingled with fur and bristles; American porcupines	Erethizontidae
No fur mingled with quills; Eurasian porcupines	Hystricidae
No quills on the body	
Tail vestigial; burrowing animal found near streams; North American mountain beaver	Aplodontiidae
Tail flattened, scaly; aquatic; beaver	Castoridae
Tail furry or brushlike; squirrels, chipmunks, and woodchucks	Sciruidae
Tail cordlike; animal resembles a mouse or rat	
Tail longer than body; ears conspicuous; rats, mice	Muridae
Tail not longer than body; ears inconspicuous; voles	Cricetidae

MOUNTAIN BEAVERS [FAMILY APLODONTIIDAE]. The family comprises beaverlike animals, differing in having a furry tail and making burrows in dry earth beside streams. They feed on vegetation. Only one species now survives, *Aplodontia rufa*, restricted to the west coast of North America, where it inhabits mountain forests and dense thickets near water. Formerly the group was more widespread. Aplodontiidae are the most ancient type of rodent known and range back in time to the late Eocene, about 40 million years ago.

BEAVERS [FAMILY CASTORIDAE]. True beavers are restricted to the northern lands; they include the largest northern rodents. There is a single genus, *Castor*, with *C. canadensis* in North America and *C. fiber* surviving in a few places in Eurasia. Beavers, as is well known, feed on the bark and sapwood of trees that they fell by gnawing the base—generally birch or aspen. The dens are constructed from the wood they fell, are entered from below the water line, but are occupied at a level above the water line. Dams are also constructed to create pools in which food fish may live, thus favoring otters and fish-eating birds and creating flood-control zones where food plants and trees may flourish. Beavers live in colonies, apparently derived from a founding pair. To prevent overcrowding of a colony, the two-year-old members are periodically expelled to found new colonies. Beavers became extinct in England during the twelfth century when monks, with papal authority, hunted them on Thursdays to eat the following day, as beaver were declared fish by the church; the last known surviving British beavers were in Wales, where Giraldus saw them in the year A.D. 1188. Beavers in the eastern United States were reported extinct by John Goodman in 1830, save for colonies west of the Missouri. A partial reduction in hunting pressures, coupled with an enlightened policy of reintroduction of stock from Canada, carried out by Audubon societies and the like have led to a gradual increase in beaver populations and a restoration of the eastern range.

PORCUPINES [FAMILIES HYSTRICIDAE AND ERETHIZONTIDAE]. Differences between the Eurasian and American porcupines are given in the key. The American porcupines are represented in the north temperate biomes by the one species, *Erethizon dorsatum* now restricted to the northeastern United States, Canada, and the western half of the United States. Bark is the chief food, obtained by climbing the trunk of a tree where the animal may be observed asleep, coiled in a ball on a fork, or lumbering through the woods in the daytime, or along the sides of salted roads at night. The Old World porcupines occur mainly in warm regions, but a biomes by the one species, *Erethizon dorsatum*, now restricted to the north-Africa, a nocturnal, solitary herbivore. Porcupines are known from the Oligocene period, about 30 million years ago.

SQUIRRELS, WOODCHUCKS, AND CHIPMUNKS [FAMILY SCIURIDAE]. Except for the nocturnal flying squirrel, members of the family are active during the day and are therefore relatively conspicuous denizens of the northern forests. They store food supplies, such as nuts and dried vegetable materials; some have cheek pouches, and most species sit up on their haunches to spy for the approach of intruders on the ground. Woodchucks (*Marmota*) are short-eared, plump burrowers with a fat, furry tail, hibernating, feeding off forest ground vegetation, occasionally damaging crops in agricultural districts. Other small mammals utilize the burrows as refuges. Most other species of the genus are called marmots and inhabit steppe and prairie biomes, the bobac of Siberia being an example.

Chipmunks are squirrellike forms from Asia and North America. Most are placed in the genus *Eutamias*, though the well-known chipmunk of the eastern states of North America is put in a separate genus, *Tamias*. Chipmunks have longitudinal stripes on the body and raised hairs on either side of the tail. They usually frequent rocks, in the forest or on open plains. A favorite haunt in eastern North America is provided by the drystone walls left from earlier farmland now reverting to second-growth forest.

The genus *Sciurus* includes the tree squirrels of the Holarctic forests. Among the best-known forest animals, they have flourished in the company of man, their attractive appearance and boldness making them favorite park animals. Their habits of collecting and burying nuts in open ground must tend to favor the outward spread of broadleaf forests. The

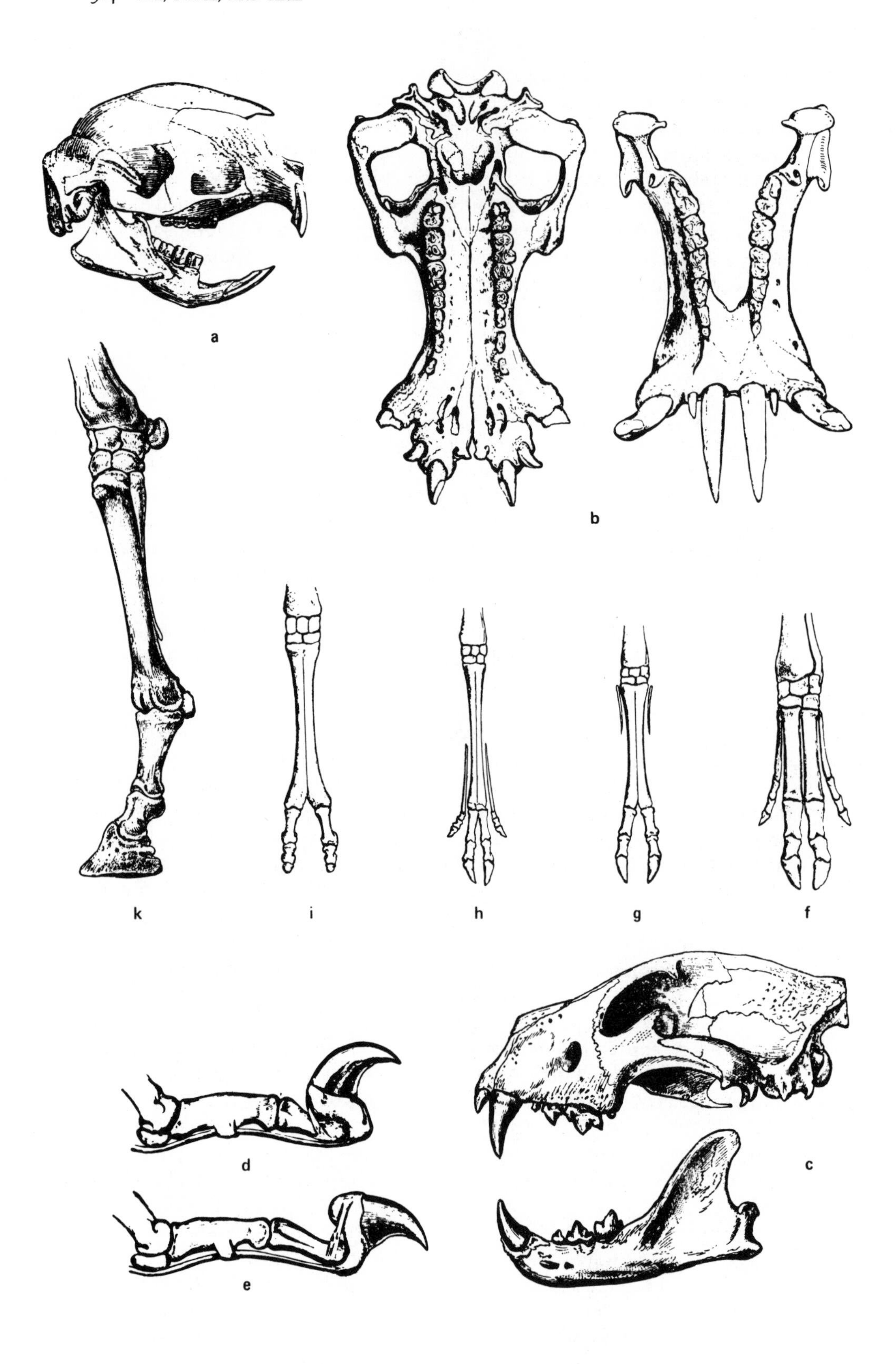
a
b
k
i
h
g
f
d
c
e

Figure 127. Diagnostic structures of mammalian orders mentioned in Table 44 and in the text. (a) Skull of rodent (porcupine), showing diastema between incisors and cheek teeth; (b) skull of hippopotamus (Artiodactyla); (c) skull of lion (Carnivora), showing enlarged canines and carnassial tooth and absence of diastema; (d, e) claws retracted and extended, lion (Carnivora, character of family Felidae); (f) paraxonic foot of pig; (g) of deer; (h) of cattle; (i) of camel (all Artiodactyla); (k) monaxonic foot of horse (Perissodactyla) (18, 26, 40).

flying squirrels, *Glaucomys* of North America and *Pteromys* of Eurasia, have a skin flap between the anterior and posterior limbs, enabling them to parachute in the same manner as the flying phalangers of the Australian forests (their Southern Hemisphere ecological equivalent). Like the latter, they are nocturnal. The other members of the Sciuridae are denizens of the steppes and tundras. Sciurids date from the Miocene about 25 million years ago.

OLD WORLD RATS AND MICE [FAMILY MURIDAE]. The family is now usually restricted to the Eurasian rats *(Rattus)* and mice like the house mouse *(Mus)*, including species that are clients of man, following his settlements throughout the world, traveling as stowaways on canoes and ships, and feeding on stored food products accumulated by man. They also carry diseases shared by man, pig, and rat and are infected by fleas that spread plague. Population densities of 200,000 rats per million human beings occur in major cities.

DEER MICE, PACK RATS, VOLES, AND MUSKRATS [FAMILY CRICETIDAE]. This family, formerly included with the preceding one, can generally be recognized by the characters given in the key, though the systematic distinctions are based on details of the ridges on the molar teeth. Generally speaking, one can say that any mouselike animal in a forest biome having a haired stringy tail and inconspicuous ears, plus, of course, the rodent dentition of very conspicuous incisors with diastema, belongs here. The small members are called voles, and most of the common species of Eurasia and North America fall in the genus *Microtus*. Some voles live in forests; more are found in prairie-steppe environments. Similar animals frequenting streams in Florida are called water rats and placed in the genus *Neofiber*.

The corresponding water voles of Eurasia are placed in the genus *Arvicola*. The musquash is the same animal as the muskrat (the former name being Algonkian) and comprises a single species *(Ondatra)*, readily distinguished by its tail, which is *compressed*, that is flattened in the vertical plane (a beaver's tail is *depressed*, that is flattened in the horizontal plane). The woodrats, commonly called packrats or traderats, are also classified here; they much resemble the house rats (Muridae) but can be distinguished by the hairs on the tail (house rats have scaly tails). They collect miscellaneous items, especially things unusual in a natural habi-

Figure 128. *Hystrix cristata* (Eurasian porcupine), Rodentia (17).

tat; if they find another item, and already have one in mouth, the latter may be discarded and the former taken away. Related species occur on the plains and caused occasional contretemps among the westward-bound pioneers of a century ago by removing coins or bills by night, substituting an acorn or cactus fruit as a trade item.

Order Artiodactyla

Ordinal characters of Artiodactyla are given in Table 44. The three families occurring in the northern forests may be distinguished as follows:

No horns or antlers; 4 toes; snub-nosed; pigs of Eurasia	Suidae
Horns or antlers present, at least in male; 2 functional toes sometimes with vestigial lateral toes; stomach subdivided into a storage chamber (rumen) and digestive chambers; food regurgitated for chewing before digestion (chewing the cud)	Ruminantia
With branched antlers, shed annually; deer; all biomes	Cervidae
With unbranched horns, never shed; cattle, sheep, and goats	Bovidae

SWINE [FAMILY SUIDAE]. New World swine or peccaries do not now enter the northern forests, but in the Pleistocene period a long-nosed peccary, *Mylohyus*, occurred in the North American woods. The Old World swine *Sus* range the Holarctic now, having been introduced to North America, but the natural range was Palearctic and Oriental; they occur in north Africa north of the Sahara. The characters are as given in the key, the stomach having the simple undivided form of other orders

Table 44. Morphological characters of the orders of northern forest mammals.

ORDER	MORPHOLOGICAL CHARACTERISTICS	INCLUDED FORMS
Marsupialia	Pouched mammals	Opposums
Chiroptera	Forelimbs modified to form wings	Bats
Primates	Limbs with fingers, thumb opposable to other digits; large brain	Monkeys (Eurasia) and men
Artiodactyla	Paraxonic limbs (cloven hoof); ungulates (walking on tips of nails)	Pigs, cattle, and deer
Carnivora	No hoofs, digits usually clawed; large canine teeth, no toothless region (diastema) in jaw	Pandas (Asia), bears, wolves, cats, raccoons, otters, skunks, and so on
Insectivora	Canines similar to adjacent teeth, posterior teeth with sharp cusps; snout pointed	Hedgehogs (Eurasia), moles, and shrews
Lagomorpha	Canines replaced by toothless region (diastema), permitting food to be stored in cheek pouches; two pairs of upper incisor teeth	Hares and rabbits
Rodentia	Like Lagomorpha, but only one pair of upper incisor teeth	Rats, mice, and voles

of mammals, so that pigs are not ruminants. The diet is omnivorous; they root up the soil in search of animal and vegetable food and may do serious damage to the environment or cause erosion on steep country. Large boars are dangerous and kill unsuspecting intruders. Domestic pigs may also be dangerous to children, whom they will eat if allowed the opportunity. In the native environment pigs serve as useful scavengers, since the useless surplus beechnuts and acorns are converted into fecal material that enriches the forest soil. In medieval London, as doubtless in other cities, herds of pigs ran wild in the streets and served as the city scavengers. They still do in some tropical villages.

In nature a food chain exists in which man eats pig and pig may eat man but more usually rats, and rats eat dead pigs and dead men. Through this involute cycle the parasitic nematode worm *(Trichinella)* passes from host to host. In later, more civilized, times the parasite passed from pig to man by way of sausage casings, which were made from pigs' intestines. The disease, trichinosis, was formerly widespread in North America but has now been controlled by rat control, by meat inspection, and the substitution of plastic sausage casings. In Roman times pigs ranged the forests of Europe as far north as Sweden, where Tacitus mentions the animal as a cult totem animal for one Swedish tribe. Britannia was noted on the Roman market for the export of ham cured by oak smoke. It is commonly repeated in parasitology texts that observance of the Mosaic law protects the Islamic countries from trichinosis; it is sad to have to say that modern health statistics in fact disclose high intensity of trichinosis infection among the Islamic peoples; the polyxenous (multiple-host) character of *Trichinella* is doubtless the reason, together with inadequate sanitation.

DEER [FAMILY CERVIDAE]. With very few exceptions (none in the northern forests) all male deer carry antlers. Cervidae range the entire Holarctic region, including North Africa. The largest surviving deer *(Alces alces)* is the animal called moose in North America and elk in Europe, frequenting northern forests in the vicinity of water and feeding in the warm months on water plants; in winter when the lakes and rivers are frozen over they eat the young buds of deciduous trees, the tips of needle-leaf branches, and break down saplings for the edible parts. Despite statements to the contrary, moose also clear snow away from areas where dried grass lies beneath, kneeling on the forelegs to graze. The species is a strong swimmer, one being observed to land in Jutland after swimming the Skagerrak from Sweden in 1953. European males stand 2 m high at the withers (point midway between the shoulder blades) and weigh over half a ton. East of the Scandinavian herds lie others in Finland, Russia, eastern Siberia near Lake Baikal. Further east still, in Tartar Siberia occurs a larger race that appears to be identical with the American form, ranging thus from either side of the Bering Straits across North America to the Atlantic coast of Labrador and Newfoundland, also northern Maine. In 1699 Edward Ward wrote familiarly of the

> Moose-deer of New England . . . his flesh lushious, extreamly palatable . . . their Fawns also delicious food, highly commended by all such who are more than ordinary Nice in obliging their Voluptuous Appetites; much said by the Phisicians of that Country concerning the Excellent Virtues of the Horn of this Creature: Being looked upon as an incomparable Restorative against all inward weaknesses.

The wapiti *(Cervus canadensis)* is commonly called elk in America; it is the same deer as is called Altai deer and Turkestan deer in Eurasia. Its range in North America has become very restricted. The woodland caribou *(Rangifer caribou)* is the taiga equivalent of the tundra reindeer; it encounters deciduous forests only in the aspen belt of Alaska. Red deer *(Cervus elaphus)* range the woodlands and temperate savanna ecotones of the Palearctic region, now also established in New Zealand. Sika deer are a small species of *Cervus* occurring in Japan, Taiwan, Manchuria, and China. Whitetail (or Virginian) deer and the mule deer are both species of *Odocoileus,* a genus restricted to North America and South America, ranging all the temperate forests (mule deer lacking from the east) and constituting the major big game animal of North America. Comments on deer have been made under wolf.

CATTLE, SHEEP, AND GOATS [FAMILY BOVIDAE]. The native cattle of the original forests of Eurasia have been much bred by man so the systematics are difficult. The wild cattle of Europe were the urox *(Box primi-*

Figure 129. Large forest and grassland herbivores. (a) *Dicerorhinus antiquitatis* (woolly rhinoceros), Rhinocerotidae, Font-de-Gaume; (b) *Bison bonasus* (wisent, or forest bison), Font-de-Gaume; (c) *Bos primigenius* (aurox, or wild forest cattle), Grotte de la Mairie; all Artiodactyla. All engraved or painted by contemporary European artists about 10,000 B.C. (1).

genius), a very large beast hunted by Stone Age man. The wild species was extinguished in the Middle Ages. The domesticated docile descendants are usually called *Bos taurus. Bison* includes two species, one in America *(Bison bison),* and one in Europe called the wisent *(Bison bonasus),* now extinct as a wild animal, a few small herds thought to remain in a fairly pure state in captivity. The European species was a woodland animal; the American included several races, some that formerly inhabited the eastern forests but are now extinct, some that inhabited prairies and survive in a few controlled herds, and some that inhabit woodlands in the northwest of North America. Sheep and goats that also belong in this family are more closely related to grasslands and semiarid environments and will be considered in a later section.

SECONDARY HETEROTROPHS: THE PREDATORS

Order Marsupialia

OPOSSUMS [FAMILY DIDELPHIDAE]. The North American opossum (*Didelphis marsupialis*, order Marsupialia) is the only marsupial known from the Holarctic region. It ranges the summergreen and evergreen broadleaf forests but has developed a preference for farming savannas. It is nocturnal, feeding on insects, carrion, birds' eggs, also fruits and nuts. Apparently it is not now seriously harmed by human contacts. Young are born about 2 weeks after the mating, about 15 young in the litter, the entire litter weighing about 1 oz. Young spend about 2 months in the pouch. Afterward they accompany the mother, traveling on her back, their tails coiled around hers.

Order Insectivora

The ordinal characters of Insectivora are given in Table 44. Following is a key to the three families occurring in the biome:

Dorsal surface covered by spines; hedgehogs; Eurasia — Erinaceidae
Dorsal surface covered by fur; no spines
  Eyes vestigial; forefeet modified for burrowing; moles — Talpidae
  Eyes functional; mouselike, with slender feet; shrews — Soricidae

HEDGEHOGS [FAMILY ERINACEIDAE]. Hedgehogs also have adapted well to the presence of man, favoring the outer suburbs of cities, where home gardens provide plenty of insects and snails. The best-known species is *Erinaceus europaeus*; related forms occur in eastern Asia and North Africa, but no representatives are known in North America. They are nocturnal, emerging at twilight, in the breeding season accompanied by about half a dozen young. Shy of illumination, they coil up into a tight ball if a flashlight is played on them, or if a dog noses them. When coiled up the spines are an effective protection. They are easily tamed and will visit regularly for milk or bread soaked in milk. Except for taking the eggs of ground-nesting birds, they are beneficial in most habitats. The nuptial ceremonies involve protracted rotary ambulation around a fixed point for an hour or so with accompanying snuffling sounds.

MOLES [FAMILY TALPIDAE]. Moles are subterranean insectivores, short-tailed, with a plump soft-furred cylindrical body and short, broad forelimbs with shovellike claws. The eyes are covered over by skin in some species. Examples are European mole (*Talpa*) and eastern American mole (*Scalopus*).

SHREWS [FAMILY SORICIDAE]. Mouselike animals with a pointed snout, soft fur, and hairy tail. There are two main groups of genera: white-toothed shrews, (an example is *Crocidura*), of Eurasia, and red-toothed shrews, which are mainly American, though the best-known genus, *Sorex*, has a Holarctic distribution. Among the latter are the water shrews *(Neosorex)*, ranging northward into Alaska, and the pigmy shrews, including the smallest mammals *(Microsorex)*. Shrews have a venomous bite, including the common shrew of North American deciduous forest *(Blarina)*. All are active predators, the prey including insects, small fishes, and mice. The pigmy shrews, on account of the relatively large surface area with respect to mass, lose heat so fast as to require their living to an activity rhythm of about half a day or eight hours, sleeping and hunting occurring in each cycle.

Order Chiroptera

The ordinal characters of Chiroptera are given in Table 44. Following is a key to the northern families:

Snub-nosed bats of America and Eurasia	
Tail short, imbedded in body skin; common bats	Vespertilionidae
Tail long, free from body skin; free-tail bats	Molossidae
Omega-shaped fold of skin on nose; horseshoe bats of Eurasia	Rhinolophidae
Vertical skin-fold on nose; leaf-nose bats of America	Phyllostomatidae

BATS [FAMILY VESPERTILIONIDAE]. The family includes the common bats of Eurasia and North America, about 300 species, mostly in the Old World. Many western and southwestern North American species fall in the genus *Myotis*, relatively small forms. The largest common bat of North America is *Eptesicus fuscus*, ranging Mexico, the United States, and much of Canada. The reddish colored *Lasiurus borealis* is another widely distributed North American species, which migrates south in winter. *Vespertilio murinus* ranges Europe and much of Central Asia and reaches a wingspan of about 30 cm (12 inches). All members of the family are insectivores.

OTHER BATS. The other three families are also insectivorous; the main structural features and distribution data are given in the key. *Tadarida* and *Eumops* are the northern representatives of the free-tail bats in North America, confined to the southern half of the continent. The leaf-nose bats are mainly neotropical but have several representatives in the southwestern United States; an example is *Macrotus*.

Order Carnivora

The characters of the order Carnivora are indicated in Table 44. The following key distinguishes the northern families:

Jaws elongate, three molars in each lower jaw	
Tail vestigial; sole of foot on ground (plantigrade gait)	
Hand with opposable extra joint (not thumb); giant panda of montane bamboo forest, China	Ailuropodidae
Not so; omnivorous dominant predators; bears	Ursidae
Tail well-developed, bushy; walking on toes (digitigrade)	
Arboreal herbivores; pandas of Himalayas and China	Ailuridae
Terrestrial carnivores; wolves and foxes	Canidae
Jaws shortened, only 1 or 2 molars in each lower jaw	
Two molars in each upper jaw; raccoons of North America	Procyonidae
One molar only in each upper jaw (carnassial tooth)	
Foot plantigrade (bear type); otters, skunks, and so on	Mustelidae
Foot digitigrade; 1 molar in each lower jaw; cats	Felidae

PANDAS [FAMILY AILURIDAE]. Pandas are carnivores resembling a heavily built cat of a foxy aspect, of mainly herbivorous habit, but taking eggs and insects. The single genus *(Ailurus)*, with one known species *(A. fulgens)*, lives in trees in montane forest on the high mountains of Central Asia (Nepal, Assam, Burma, and Szechwan and Yunnan, China). The thick coat is glossy red above, black below. Thought to be related to raccoons and to bears.

GIANT PANDAS [FAMILY AILUROPODIDAE]. The very rare animal *Ailuropoda melanoleuca* placed here is well known from several charming specimens that have won public esteem in a few national zoos. One has the impression that a giant panda is a kind of bear, though the anatomical characters point to a relationship with the smaller tree-dwelling animal mentioned in the previous paragraph. The giant panda inhabits montane forest in eastern Tibet and also the southwest mountains of China. Food is mainly bamboo shoots, plucked by the prehensile hand that contains an additional opposable joint beside the thumb. The pandas are mainly herbivorous animals, as bears are during much of the season, so that the order Carnivora is not exclusively constituted of carnivorous forms.

RACCOONS [FAMILY PROCYONIDAE]. The family, which includes the coatis of Central America and the raccoons, is represented in the northern forests by *Procyon lotor*, the North American raccoon, which also ranges the prairies and arid lands of the Southwest. Raccoons mainly frequent areas where water is flowing; they drink by dipping the food into water before it is eaten. Chiefly nocturnal in nature, feeding on frogs, crayfishes, birds' eggs, insects, and fruits; and on the grasslands on the seeds of grasses such as corn. The latter habit has attracted it to suburban gardens, where corn is often grown, and from frequenting gardens it has learned the mysteries of garbage cans; the species presently enters city limits, such as Cambridge, Massachusetts, where the noise it is apt to make when opening garbage bins and letting the lid fall back sometimes disturbs the residents, who believe human intruders are at work.

An improvement in the attitude of many human beings toward this charming animal has led to a considerable increase in its numbers since the population minimum of eastern wild animals in the late nineteenth century. About 1900 the nearest wild raccoons to Boston were some in central New Hampshire; now the parklands of greater Boston carry stationary wild populations that seem to be flourishing in somewhat the same way as the hedgehog has learned to live with suburban man. Raccoons range back in time to the early Miocene.

The ring-tail *Bassariscus astutus* is smaller than a raccoon, with transverse black and white bands on the tail, mainly found in the arid lands of the southwest United States, but ranging northward into the Californian forests and the coastal forest of Oregon; it favors chaparral, especially near water; it is believed to control mouse populations, hence is classified as "beneficial" by state wildlife agencies in accordance with the prevailing biblical concept of wildlife as the chattels of Adam's seed.

BEARS [FAMILY URSIDAE]. Bears date from the Miocene; their characters include the humanlike plantigrade foot and the short tail almost concealed in the fur. Bears include the largest living predators, though for much of the year even the largest species are vegetable eaters. All bears are rather large animals if not very large, the smallest species being the size of a large dog, such as the pandalike Himalayan bear with contrasting furry coat of black and white, and the Malayan sun bear, which has much the aspect and disposition of a fat dog and can be kept as a house pet without the neighbors suspecting that it is in fact a bear.

The grizzly bear *(Ursus horribilis)* was made known to science in North America by the expedition of Lewis and Clark, when it was encountered in large numbers on the prairies west of the Missouri and along the course of that river. Fossils from Scotland seem to indicate that the grizzly was formerly present there, and if so then the Caledonian bear of Roman gladiatorial combats must have been *Ursus horribilis.* Most surviving grizzly bears occur in western Canada and Alaska; large specimens may reach 400 kg.

The brown bear *(Ursus arctos)* formerly ranged Europe but the species has nearly everywhere been exterminated by man. By Roman times the bear had already disappeared from the lowlands of Greece, Italy, and Gaul, so the emperors were obliged to import them from Britain (Martial, *Epi-*

grams; Plutarch *Historia*). The final solution of the ursid problem was achieved in A.D. 1057 in Britain, when Malcolm III of Scotland seized the throne from Macbeth and granted coats of arms to his retainers. The clan chief of the Gordons was awarded a shield bearing three bears' heads, tribute to his prowess in having exterminated the animal. Long afterward Edward VI, in the sixteenth century, found it needful to import bears from Europe in order to provide bear hunts along the Thames for the distraction of the Spanish ambassador. The imports were from Germany, where the brown bear had not been extinguished generally until the eighteenth century. Surviving wild remnant populations of *Ursus arctos* occur today in Lapland (one wandered into Kiruna when I was there in 1953), in the Pyrenees, Carpathians, Alps, Asia Minor, Siberia, Himalayas, Kamchatka, and Japan.

The brown bear of Alaska, or kodiak bear *(Ursus middendorfi)*, appears to be the New World representative of the brown bear of Europe, reaching 650 kg and considered by American observers to be the largest of all surviving bears. However, it would appear that eastern Siberian bears may be the same or a closely related species. It inhabits forests and also coasts in the spring, when it mainly eats seaweed. As the summer progresses the diet changes to grass and sedges; during the salmon run it frequents streams and catches the spent fish after spawning; in the fall it feeds on berries. It dens up during the winter, emerging occasionally. At all times it catches mice whenever possible and, to go from one extreme to the other, it is also addicted to eating stranded whales. This gigantic relic of the great mammals of the Pleistocene should, of course, be strictly protected for posterity but this, unhappily, is not the case. May the time never come when some American Gordon chalks up the last brown bear; but the fate of the grizzly within living memory leaves cold comfort. The American black bear, so-called, is a small variable species, formerly distributed universally in North America. The species was still common in the Boston area in 1699, when Edward Ward published his entertaining *Trip to New England* and set the Puritan fathers on their ears. Since then it has been steadily pushed back and is now restricted to wilderness areas (some near New York) where the human density is about 1 per square km; this means mountains, swamps (in the South), and the great forests of Canada. The species *(Ursus americanus)* has rather a shy though curious nature and has similar dietary preferences to those of the larger bears, but adds garbage, household garden produce, fruits, and farm honey to its menu, provoking the retaliation of the offended parties. It appears to be in no danger of eradication. When hunting falls into disrepute a more orderly method of population control will be called for, though this ethical question shows no sign of coming to the attention of public authority.

WOLVES AND FOXES [FAMILY CANIDAE]. Wolves have experienced a contraction of their range analogous to that of bears, since man invariably attacks them; they formerly ranged the entire Holarctic region but are now confined to areas where the human population density is less

Figure 130. Large forest and grassland mammals. (a) *Rangifer tarandus* (reindeer, or caribou), Cervidae, Artiodactyla; (b) *Equus caballus* (forest, or large-headed, horse), Equidae, Perissodactyla; (c) *Ursus spelaeus* (cave bear), Ursidae, Carnivora; a–c, Grotte de la Mairie; (d) *Mammuthus primigenius* (woolly mammoth), Proboscidea, Font-de-Gaume. All engraved or painted by contemporary European artists about 10,000 B.C.

than one person per square km. Again this means mountainous regions and the far northern forests and barren lands. Fossil canids range back to the Eocene.

Wolves were formerly very common in the Boston area. Governor Thomas Hutchinson, in his *History of the Colony and Province of Massachusetts Bay,* mentions a letter by a gentlewoman, dated 1637, in which the complaints of the colonists about hardship in America are dealt with:

> When I have thought again of the mean reports, and find it far better than those repourts, I have fancied the eyes of the writers were so fixed upon their old English chimney tops, that the smoke put them out. The air of the country is sharp, the rocks many, the trees innumerable, the grass little, the winter cold, the summer hot, the gnats in summer biting, the wolves at midnight howling. . . .

Wolves, in fact, saved the infant settlement from destruction, as the following passage shows, and one wonders why a wolf is not the heraldic beast of Boston, rather than the steamrollered eagle it presently sports. Hutchinson writes:

> There was a general design this year (1642) among the Indians, against the English. . . . The minds of men were filled with fear from these rumours of a generall conspiracy, and every noise in the night was alarming. A poor man in a swamp at Watertown, hearing the howling of a kennel of wolves, and expecting to be devoured by them, cried out for help, which occasioned a general alarm through all the towns near Boston. The Indians being thus prevented from surprizing the English, remained quiet. This was on September 19, 1642.

The Indian plans to which Hutchinson refers were the opening moves of King Philip's War. By the beginning of the 1800s wolves had become so rare in the eastern United States that Lewis and Clark were surprised when they encountered wolves on the prairies.

In the Middle Ages wolves ranged the whole of Europe, wherever forests existed. In the *Chronicles* of the Saxon kings of England references to their ravages are very frequent. In the Saxon calendar December was called Midwinter, but January, with its scarcity of wild game, was called Wolf Month, for the hungry packs raided the villages and ate up whatever they could get, including people. Athelstan, who reigned in East Anglia from A.D. 925 to 939, found it needful to erect refuges at intervals along the Yorkshire roads to defend travelers from the wolves "that they should not be devoured by them." Wender (1948) gives the following dates of extinction of the European wolf *(Canis lupus)*: England, about 1500; Scotland, 1743; Ireland, 1710; Denmark, 1772. In the remoter parts of Europe wolves still occur. In Lapland I was told that the Lapps still hunt them, but they have become very rare; there are wolves in the mountainous parts of Germany, France, and Italy, and in October 1953, for example, a pack of wolves killed 20 sheep in the Spoleto district of Italy. Further east, in the Balkan countries, shepherds still are obliged to watch over their flocks at night to protect them from wolves. Wolves are locally numerous on the Russian plain, in Siberia, China, and Japan. They inhabit both forest and steppe and tend to hunt in packs during the winter months when game is scarce. Recent studies indicate that the eradication of wolves in forests is a mistake, for the predation they apply is to weak, young, ailing, and aged deer, and lack of predation causes the deer populations to increase by a ratio too great for the carrying capacity of the biotope; the result is starvation for the deer and a deterioration in the health of the deer herds. So an enlightened policy will restore wolves to their appropriate habitats.

The gray wolf, or timber wolf, of North America is apparently the same species as the European wolf *(Canis lupus)*. It has recently been tamed for behavioral studies, with results highly to its credit. The wolf family is, it appears, a fundamentally happy and cooperative unit of population, the parents showing close affection for each other as also for their young. It appears that the early appearance of wolf remains with those of Paleolithic man reflect family situations in which wolf was interpolated into a human role, adopting human beings as family members. Dogs arose as the domesticated descendants, acquiring the art of barking, which is the wolf imitation of human speech, learned only under domestication. The more unsophisticated dogs of the remote peoples of the southwest Pacific are, or were, barkless varieties. Wolves tend to attack man when he is a solitary traveler and the wolves are hungry; this aspect, analogous to Athelstan's England, has been graphically described in Peter Tutein's *Larsen*, though the setting used by the Danish author is a tundra rather than a forest.

The consensus of naturalist opinion on the wolf today seems to be that control is needed since the animal is potentially dangerous and destructive, but not extermination, for natural wolf populations are an essential part of the forest fauna and promote the well-being of the deer population by preventing overpopulation and, still more important, promote the health of the forest trees, by preventing overgrazing. Here, then, is an indication of the role the fauna plays in the forest ecosystem. Indeed, some naturalists are now advocating the reintroduction of wolves into the forests of the eastern United States.

Referring, now, to the forest situation in New Zealand, the introduction of deer from the Northern Hemisphere has led to severe destruction of forest trees and consequent erosion. It is evident that predators should have been introduced if grazing animals were (it would have been far better had neither been introduced). Public opinion will under no circumstances so far permit the late introduction of predators to control the deer, so New Zealand is now stuck with a major ecosystemic unbalance that can only be restored by eradicating the deer, so far a hopeless task. The problem is exacerbated by the fact that the native New Zealand forests evolved in the absence of deer, and the thin-barked trees cannot survive winter bark-nibbling that deer carry out. What could happen if large predators were to be introduced in a further attempt to improve on nature? Probably further ecological disaster. Some hints of what it might be can be gained by considering now the smaller canids, the foxes, which are natural members of many of the northern biomes, including the forests and the grasslands.

Foxes (genus *Vulpes*), are represented by a number of species in various parts of the Holarctic region. *Vulpes fulva* is the red fox of the North American forests and more open country, and its close equivalent in Eurasia is *Vulpes vulpes*, by some authors considered identical (in which case the second name cited has priority). Foxes prey on rabbits, mice, insects,

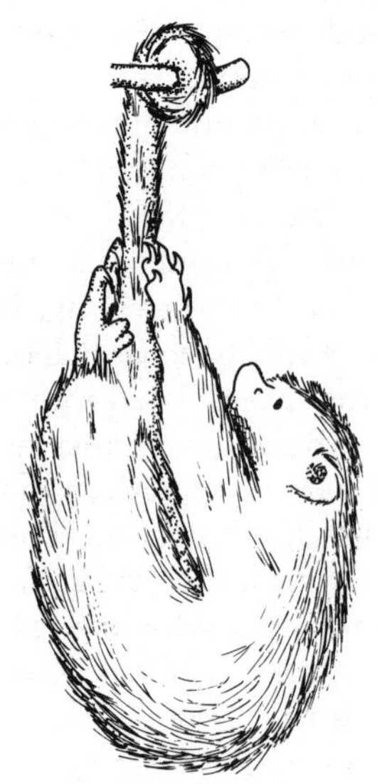

Figure 131. *Didelphys virginiana* (American opossum), young specimen, Didelphyidae, Marsupialia (23).

birds, and also eat various vegetable materials such as berries and fruit in season. Unthinking English settlers in Australia thought to recapture their lost English social milieu by introducing foxes from England for the purpose of riding to hounds in pursuit of the same. What in fact happened was that the foxes discovered a continent in which no comparable or dominant predator existed. They reproduced mightily and spread throughout the land. With no control on the fox population, soon all small species of native marsupials began to fall victim to the new predator, the hunted now assuming the commanding role of hunter as the tables were turned. The damage done to the ancient marsupial fauna of the continent appears now to be irreparable, with almost every ground-dwelling small marsupial facing the threat of extinction, or already extinguished. The painful lesson is that by apparently trivial experiments carried on by obscure and unlettered colonists, disasters of continental magnitude may strike an entire fauna before ecologists have time to learn the elements of an ecosystem. It is also apparent that millions of years of ecological evolution have gone into generating the faunal and floristic content of the northern forests and that any one element of a northern ecosystem, taken out of its natural biotope, may become an uncontrollable threat when interpolated into a different ecosystem.

The Australian and New Zealand disasters have still to disclose their real magnitude. For Australia the harm is mainly to the fauna; for New Zealand the future seems to hold the threat of almost total loss of the native vegetation biomes, with increasing erosion, and repair only possible by replacing the vulnerable native forests and mountain herblands by alien forests and heaths that alone can withstand the attacks by pigs, deer, chamois, and other importations. By these tragic mistakes we now have learned the relationship of the fauna to the flora in the forest ecosystem, but the lesson seems to have come too late.

OTTERS, SKUNKS, AND OTHERS [FAMILY MUSTELIDAE]. The family includes many different minor carnivores, well known by their individual names, but not well known by any familial group name. The earliest mustelids date from the beginning of the Oligocene period, about 40 million years ago. Mustelids tend to have long slender bodies, short rounded ears, short legs, and scent glands the quality of whose product is noted more for its repellent aspects. Insects practice gas warfare, but mustelids may be the first to employ it on a comprehensive scale.

According to lexicographers *squnck* is a Massachusetts word of proto-Algonkian derivation; the bizarre term, perhaps through its barbarous contrast with Western ideas of euphony, has given the North American animal a measure of world notoriety. This is not entirely fair, since skunks are entirely charming creatures when treated with proper respect; neither are they peculiarly American, for the Eurasian polecats have similar talents and are closely related. The commonest North American species is the striped skunk *(Mephitis mephitis)*, ranging all biomes except the Arctic. Like raccoons skunks have successfully adapted to the suburban environment and frequent gardens where the owners make them welcome, usually visiting on a rather fixed timetable at intervals through the night and taking household scraps. In nature they feed on carrion, mice, birds' eggs, insects, and berries. The contrasting black and white stripes of the coat vary widely, so that a naturalist can soon get to recognize each individual skunk operating a circuit in his area, and so determine the timetable of each animal. According to a standard handbook, one skunk per 10 acres is considered a high population density; but by that standard my own yard, a modest third-acre, is apparently supporting about 50 acres worth of skunk; rather it would seem that a growing public affection for the maligned animal is leading to higher population densities. The suburban specimens seem to be more adept at avoiding being run over than are their country cousins whose flattened bodies line the interstate highways.

Otters are widely distributed members of the Mustelidae. The best-known genus is *Lutra* which is also one of the most widely distributed genera in the world. *Lutra lutra* is the Eurasian species, also occurring in North Africa, and the North American *Lutra canadensis* is closely related, having somewhat thicker fur. Otters frequent the banks of streams and lakes, where they capture fish and frogs; fishermen dislike them as they kill more fish than they need to eat. The pelt is valued commercially, with 20,000 skins marketed annually, so otters are now becoming increasingly restricted in range. The species mentioned range the streams and lakes of all biomes of the northern and subarctic lands. Otters are sociable animals, two usually traveling together in migration to other streams or lakes, and two or more live together on the same territory. For a happier account of

Figure 132. (a) *Gulo gulo* (glutton, or wolverine), Carnivora; (b) *Lynx lynx* (Eurasian lynx), Carnivora (23).

how man can enter the social world of otters when he lays aside the role of trapper, see *Ring of Bright Water*, by Gavin Maxwell.

The fierce mustelid known in North America as wolverine, and in Scandinavia as jaerv (genus *Gulo*), is found only coincidentally in broadleaf forest, as where aspens range Alaska; essentially it is a predator of the taiga and barren lands. It is hunted very severely for its fur, and few people now have seen it alive. A zoologist in Swedish Lapland whom I questioned about the species had not seen one for many years. Badgers are more differentiated; the North American form placed in the genus *Taxidea* is more of a carnivore, eating squirrels and related animals; the Eurasian badger *(Meles)* is omnivorous. The Eurasian badger is mainly a forest animal; the American more a prairie dweller. Other mustelids are the fisher and pine marten *(Martes)*, weasel (*Mustela* spp), and mink *(Mustela vision)*, all frequenting broadleaf forests as well as taiga and open country, and all under hunting pressure. These animals are discussed in the next chapter.

CATS [FAMILY FELIDAE]. The mountain lion *(Felis concolor)* has already been referred to above as the puma of the South American fauna. The species was formerly distributed throughout the North American forests. Governor Hutchinson mentions one that was shot in Boston in 1640. The last New Hampshire specimen was killed about 1860. Uncertain records persist of a small population of mountain lion on the border of Maine and New Brunswick. Otherwise the species is now confined to the far west states of North America where, unfortunately, it is still exterminated with a price on its head. As a governor of one of these states recently remarked, "The dinosaurs died out and no one complained, why should we be concerned about wild animals now?" For a more intelligent view of the topic, see *Stalking the Mountain Lion—To Save Him* by Hornocker, in *National Geographic Magazine* 136 (5), 1969. A commonly reiterated statement in America is that "the mountain lion has never been known to attack a human being"; however, the generalization is untrue, the most recent case to the contrary being the death of a Vancouver boy in a mountain lion ambush. As in the case of wolves, what is needed is not extermination of mountain lions, but regulation, to let the species survive where it may play its proper part in the ecosystem without endangering people.

The lion *(Felis leo)* was formerly a European cat and ranged northernmost Greece in classical times, presumably in savanna country; the lion is also mentioned frequently in some of the books of the Old Testament (for example, *Job* 38), but again this must have been in savanna. Cave lions of northern Europe seem to have been a different species, exterminated in prehistoric times. Under the name nimrah, leopard *(Felis pardus)* figures in Biblical records, and some specimens survived into modern times on the mountains of Lebanon.

The tiger *(Felis tigris)* has the western extremity of its present range in the Caucasus Mountains and in northern Iran, extending eastward into the Oriental region, as already noted, and into Siberia, China, and Japan. The Persian tiger from the Caspian provinces of Iran and from the Caucasus Mountains is a relatively small form, with a shaggy coat. Eastern specimens tend to be orange, black, and white, whereas the temperate Mongolian and Siberian regions produce a race of lighter coat color, whiter in the Siberian form (which inhabits snow in winter) and yellow instead of orange in the Chinese form. In the region of Lake Baikal the tiger encounters the needleleaf taiga, and here it has a thicker coat. Tigers are typically solitary nocturnal animals, going about in pairs only during the breeding season; later the females are accompanied by the cubs.

Order Primates

Tail present; macaque monkeys of Eurasia and North Africa	Cercopithecidae
Tail absent; man; all biomes	Hominidae

MACAQUE MONKEYS [FAMILY CERCOPITHECIDAE]. Except for man, primates are absent from North American temperate forests. In Eurasia several macaques occur, including the Barbary ape (so-called; the term ape is properly restricted to the tailless anthropoids) of Gibraltar and the mountains of Morocco and Algeria *(Macaca sylvana)* seemingly the species was originally wild on Gibraltar, the only European monkey, but the existing specimens are descended from stock imported from Africa by the British in deference to a local belief that when the apes desert the rock so will the British. A Japanese species *(Macaca fuscata)* is the northernmost of all known monkeys. The Himalayan langur *(Semnopithecus entellus schistaceus)* is a subspecies of a common Indian lowland monkey that, in the mountains of Bhutan and Kashmir, ranges the montane temperate oak forests and fir forests.

16

LIFE AT HIGH LATITUDES AND HIGH ALTITUDES

To the north of the summergreen forests lies an east-west belt of cone-bearing, needle-leaf trees, collectively called the *boreal forest*, or *taiga*. This is essentially a botanical biome, characterized by the strong dominance or almost exclusive presence of gymnosperm trees that belong to about a dozen genera; nearly all of them are evergreen, with very narrow needle-shaped leaves, each leaf persisting for several seasons, so that the trees are always clothed in foliage. The most conspicuous genera are the firs *(Abies)*, the spruces *(Picea)*, and the pines *(Pinus)*. The pines range quite widely into regions outside the boreal biome, as, for example, the sandy lowlands of the southeastern United States, so they are less distinctive of the boreal forests than are the spruces and firs.

The northern limit of the boreal forests is usually near the Arctic Circle, which marks the southern limit of the treeless *tundra* lands, or Arctic barrens, as they are sometimes called. The boreal forests at their southern limits tend to form ecotones with the summergreen forests, and in such overlap regions the needle-leaf trees commonly clothe the higher slopes while the broadleaf deciduous trees occupy the sheltered valleys and the warmer southern slopes of hills.

Needle-leaf trees also occur as montane forests, above the broadleaf zone, on mountains of the Northern Hemisphere, notably the Atlas Mountains of North Africa, the Himalayas and other Eurasian highlands, and the western ranges of North America, including the Sierra Madre of Mexico. Needle-leaf trees are lacking from the Central American cordillera, where they are replaced by some podocarps, with oak occurring at the snowline. Except for a cedar, the genera do not range south of Central America. *Libocedrus* ranges south to Tierra del Fuego and also occurs in New Zealand, New Guinea, Taiwan, and China. *Cedrus* is confined to North Africa, Lebanon, and the Himalayas. Cypress *(Cupressus)* is confined to the Himalayas, China, western North America, and southern Europe. Several other genera have similarly restricted ranges, though they usually overlap somewhere with the fir and spruce. It is doubtful whether they are properly to be considered all in the same context, but for convenience the following key is extended to cover a majority of the northern gymnosperm trees. The constituent trees of the boreal forest tend to have straight unbranched trunks, so they grow close together and often form a closed canopy. In North America the term *taiga* is often applied only to boreal forest where the trees stand apart and the sky is visible between their tops. However, taiga is a Siberian word and is applied in Siberia to dense forests with a closed canopy, as, for example, the notable larch taiga of Tuva in the Altai.

The genera can be traced back to the Jurassic period, and they are thought to have constituted montane forests until late in the Tertiary cooling, with maximum extension occurring during glacial phases of the Pleistocene. One genus, *Tsuga*, or hemlock, commonly occurs as a member of predominantly broadleaf summergreen forest. The genera *Pinus* and *Juniperus* are dominant in successional phases of a subsere, such as in the regenerating forest of northern New England. The associated fauna of the boreal biome is an attenuated northern forest fauna, lacking species that favor warmer habitats but sharing almost all of the species that inhabit the northern parts of the summergreen forest. Lianes are lacking and epiphytes are restricted to inconspicuous lower plants, such as *Usnea* (lichen) and green algae that occasionally tint the bark on exposed trees. Save for the ecodominant canopy, tiers are virtually lacking, though there is a sparse field layer, sometimes of evergreen shrubs such as *Vaccinium*, or blueberry, in damp places, as well as *Ledum* and *Kalmia*. In some places mosses and lichens form a dense mat on the forest floor. Elsewhere fallen needles may form a dense mat, through which few herbs will grow. Some ground fungi are often found in pine forest.

Boreal forest yields most of the world's softwood lumber. It tends to regenerate rather rapidly after destruction but, on account of the resins produced as excretory by-products, the forests are particularly liable to sudden and fast-spreading forest fire. Fossil woods show that fire has been a feature of gymnosperm forests since the araucarians evolved in the Triassic, silicified charcoal being not uncommon, presumably the result of lightning strikes, as many fires are today. The outlook, then, is good for the survival of the boreal forests, provided chemical pollution of the air and soil can be checked. Roadside pines are presently dying from road-salt poisoning. Pines have begun to spread rapidly in the Southern Hemisphere since their introduction there by man. The rare Monterey cypress of California is becoming one of the commonest feral trees of Argentina, Australia, and New Zealand, though it is constantly uprooted by the higher wind velocities.

Trees of the Boreal Forests

The most conspicuous trees of the boreal forests belong to about a dozen genera. Most of them can be recognized by rather obvious external characters, so that it is not difficult to gain a working familiarity with their scientific names. Unfortunately local vernacular names vary considerably, and some of them are quite misleading by confusing names that in Europe have long been applied to genera other than those intended by North American usage. In the following summary names of wide international acceptance have been preferred, wherever a choice exists. The scientific names, of course, are universally accepted.

Two genera may be distinguished by the lack of cones that are otherwise so distinctive of the far northern trees; instead they have fleshy, berrylike structures at the tips of the branches, and the seed rests at the outer tip of the fleshy structure, naked as in all gymnosperms. The so-called berries occur only on female trees, the sexes being distinct in these forms, which incidentally are related rather closely to the austral podocarps. The yews (genus *Taxus*) have the leaves arranged in two ranks, one on either side of the stem. The junipers (genus *Juniperus*) have the leaves arranged in whorls around the stem, and the leaves are often very small and resemble scales. Both these genera are restricted to the Northern Hemisphere. Their fruits are valuable food for birds in the fall and early winter. *Taxus* is more temperate than boreal, and only *Juniperus communis* (a shrub) is common in boreal regions.

Among the needle-leaf conifers another genus is distinguished from all the others by the unusual fact that the leaves are shed in the fall. This is *Larix*, comprising the larches, or tamaracks. The leaves, when present, are carried in clusters on lateral spurs on the twigs, and each leaf resembles a very short pine needle. *Larix* is restricted to the Northern Hemisphere.

The remaining conifers of the boreal forests all have leaves persistent throughout the year, the evergreens. They may be subdivided into two main groups according to the form of the seed, which may either have a long terminal wing (for wind dispersal), as in the pines, the firs, the spruces, and the hemlocks, or else the seed has a narrow winglike flange around its margin, as in the redwoods, the arborvitae, the incense cedars, and the cypresses. Following are recognition points for the various genera.

The pines (genus *Pinus*) are easily recognized by their unique character of having the needle leaves arranged in clusters, enclosed at the base

Figure 133. Invading Spruce (*Picea*) developing a nearly pure stand on a formerly glaciated surface at Högtorp in Sweden (9).

by a membranous sheath; another striking feature is the massive, woody cone, which takes two or more years to form and persists afterward for several years on the branches, even after the seeds have been shed. The individual cone scales are thick, woody, and rigid. Pines are restricted to the Northern Hemisphere, but not to the boreal forests, as already noted.

The firs (genus *Abies*) have scattered leaves that are neither in clusters nor markedly needlelike. The cones mature in one season and so are never woody, the scales remaining flexible and papery, falling off at the end of the year. The cones are peculiar in standing erect on the branches instead of hanging downward as they do in the other genera with nonwoody cones. The leaves have no stalks, are rather squarish in section, and are densely crowded on the branches. Firs are restricted to the Northern Hemisphere.

The spruces (genus *Picea*) are tall, stately trees of symmetrically tapering form, the branches drooping in winter to allow the snow to fall away. Their cones are not hard and woody, though the scales on them are persistent; the cones hang downward. When the leaves fall away, they leave behind a short stumpy base that persists for years, giving the twigs a characteristic roughened appearance. If a spruce twig is allowed to dry out (as in a herbarium specimen), the leaves all drop off unless the specimen is first preserved in formaldehyde before it is dried. For the same reason they make poor Christmas trees unless grown as living specimens

Figure 134. Douglas fir (*Pseudotsuga taxifolia*) rimming the shore of Harrison Lake, British Columbia, and clothing the western foothills of the Cascade Mountains (8).

in tubs. The spruces have no stalks to their leaves, and this character distinguishes them from the otherwise similar hemlocks (genus *Tsuga*), where a short leaf stalk occurs. Spruces range the cold parts of the Northern Hemisphere, whereas hemlocks are restricted to the Himalayas, China, Japan, and North America.

Of the genera without terminal wings on the seeds, the redwoods (genus *Sequoia*) are distinguished by having the cone scales arranged in spirals, as in the conifers already mentioned; that is to say, the cone resembles what is commonly understood by that word. The two species, as well known, are restricted to the northwest coastal region of North America, where they once formed forests of unmatched grandeur. The near-extinction of these extraordinary trees was the work of the lumber industry; their natural replacement would require several thousand years, so it is doubtful if the forests will ever be regenerated. However, just as the New Zealand kauri was decimated by lumbermen, but appears to grow rapidly when planted in California, so also the Californian redwood seems to flourish in New Zealand; so possibly a thousand years hence there may be a reciprocal distribution of these two genera.

The other members of the marginal-winged seed conifers have cones of a different type from *Sequoia*. Instead of having the scales arranged in

Figure 135. Gymnosperm trees in the Lachoong valley, at an elevation of 8000 feet, in the Himalayas. To the left, two Larch trees (*Larix*); and to the right, on either side of the river, the fir (*Abies brunoniana*). See also Figure 88 for a *Pinus longifolia* forest in the Himalayas. Sketched by Sir Joseph Hooker (31).

spiral fashion, they form alternating opposite pairs (an arrangement called *decussate* by botanists; it produces curious cones, somewhat like plaited leather buttons). The cypresses (genus *Cupressus*) have globular cones of this type, rather large and conspicuous since they normally need two seasons to mature. Many seeds are concealed under each umbrella-shaped scale. Cypresses range the Northern Hemisphere and so are not restricted to cold regions, nor indeed are they particularly evident in boreal forests, the chaparral being their preferred habitat. But closely related to the true cypresses are the Japanese and North American species placed in the genus *Chamaecyparis*, where the cones are small, maturing in one season, and having only two seeds under each scale. These species are also called cypresses.

The incense cedar (genus *Libocedrus*) has oblong cones, each one with 6 cone scales. In North America this genus occurs on the west coast, but other species range the Andes and New Zealand. The arborvitae (genus *Thuja*) is related and also has oblong cones, but with up to 12 cone scales. Arborvitae is not particularly characteristic of boreal forest, for its ecology links it more with water or wet ground, wherever that may be available. Arborvitae occurs in China, Korea, Japan, and North America.

PINE MIMICS. Quite unrelated to the gymnosperms, but ranging the corresponding latitudes and altitudes in the subantarctic and

Figure 136. In Lapland *Pinus* occupies the role of *Picea*, producing slender pencil-shaped and closely grouped trees in the patches of taiga on the tundra lands. This silhouette is of *Pinus sylvestris lapponicus*, near Kiruna, where individual pines may rise to about 30 feet, the only large trees known to live within the polar regions (23).

southwest Pacific, is a remarkable assemblage of heathers that include species with needlelike foliage, in tufted sheathed groups of leaves, much like those of *Pinus longifolius* of the Himalayas, and reaching to a height of about 30 feet, with a trunk up to 4 feet in girth. These are members of the genus *Dracophyllum*, placed in the southern family Epacridaceae; the floral structure discloses a close relationship to the northern family Ericaceae, the true heathers. In New Zealand the species occur on cold and windswept exposures, ranging up to 4000 feet above sea level, the plants usually smaller and stunted, only about 3 feet high, at the upper limits of the range, above the winter snowline. In the Auckland and Campbell

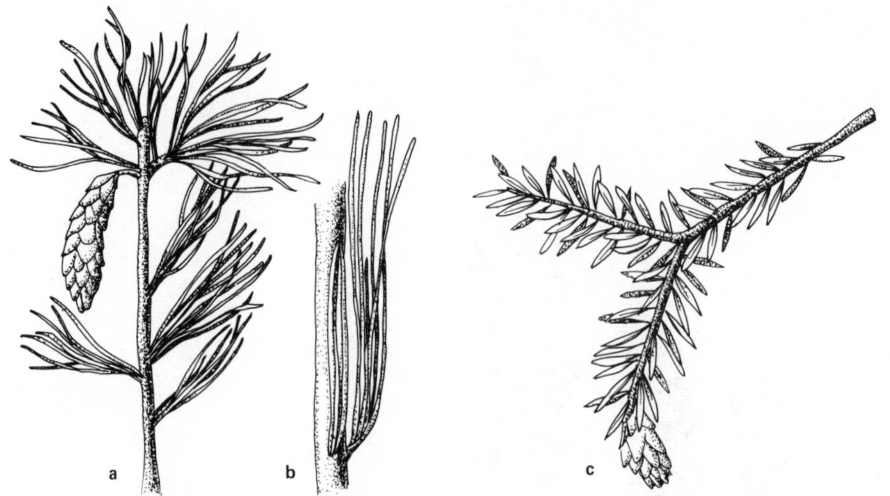

Figure 137. (a) *Pinus strobus* (Eastern white pine), North America, Pinaceae; (b) larva of moth, mimicking pine leaves on which it feeds; (c) *Tsuga canadensis* (hemlock), major forest tree of New England and adjacent states, Pinaceae. Drawn by Desmond FitzGerald.

Islands, in south latitude corresponding to about that of central Labrador or Moscow, the *Dracophyllum* trees reach their maximum development and make up a considerable part of the vegetation of those cold, windy sealands. Figure 138 illustrates inanga, one of the species *(Dracophyllum longifolius)* of these pinelike trees. The same genus also mimics grass on the subalpine plateaux of New Zealand, where they may form extensive exposed undergrowth or humpy cover to knee height or waist height; in this guise the species are often called *grass trees*; unrelated plants in Australia are known by the same name.

In New Zealand uplands *Nothofagus* (southern beech) substantially replaces the gymnosperm montane forests of the Northern Hemisphere, as already discussed (Chapter 11). Some true gymnosperms, podocarps of the genus *Dacrydium* occur, too, at high levels, often in miniaturized forms a few inches high, much resembling club mosses. At these levels the flowering shrub *Hebe* simulates the whipcord forms of cypresses, also reduced to small bushes and tufts.

Tundra

Beyond the Arctic Circle lies a treeless land where the mean annual surface temperature lies at or below 0° C and the soil is permanently frozen to a great depth (to 1000 feet or more, according to latitude). In the higher latitudes the *permafrost,* as the frozen soil is called, never thaws, and so no plant growth is possible. In the low Arctic regions, between about 65° north and 75° north latitude, the soil thaws at the surface in summer and permits annual growth, flowering, and reproduction of plants to take place. No plants are annuals, however, for many enter a dormant period during the long, cold winter and then resume growth again when the next season melts the ice. The annual range of temperatures in the northern tundra is commonly between about −50° C in winter and 10° C in summer, these figures being the averages for the warmest and coldest months. Summers are short. Within the Arctic Circle the midsummer consists of a continuous day, of variable length depending on the latitude; in the course of single summer "day" periods the temperature may rise briefly to as much as 30° C, but frost may also occur. In midwinter there are corresponding periods when the sun never rises, the length of night varying with the latitude.

Conditions of life are marginal, and so even small variations in relief or shelter may be very significant in determining what vegetation, if any, can occur. Small changes in temperature or water availability may produce quite conspicuous changes in the predominant species of plants. These variations again determine the animal life insofar as it is dependent on the plants, as in the case of reindeer. The periodic freezing and thawing of the soil causes surface irregularities in the terrain, such as polygonal honeycomblike patterns, or mounds like small rounded hills, and these, too, profoundly influence the distribution of plants. As might be expected, a life zone similar to the Arctic tundra occurs on most high mountain ranges, above the level of the needle-leaf forest in the Northern Hemisphere, or above the level of the highest austral forest in the Southern Hemisphere. The ecotones are open low savannas with spruce *(Picea)* especially participating in northern borderlands between taiga and tundra (though *Pinus sylvestris* is the main ecotone tree in Lapland); in the Southern Hemisphere *Nothofagus* (southern beech) commonly forms the ecotone forest.

Among the distinctive features of the tundra flora are:

1. Lower (nonvascular) plants are abundant. For example, in Greenland about 600 species of mosses have been reported, about 700 species of fungi, and about 300 species of lichens. These are the plants that range furthest toward the poles, and range highest on the mountain tops in regions outside the polar regions. On a high mountain the lichens are the last plants encountered on bare rock as one passes on to the permanent ice desert, approaching the summit.
2. Most other plants are herbs, particularly low grasses and sedges, generally mixed with lichens on the ground, where a shallow kind of peaty turf may form. The vascular herbs often form rosettes, or alternatively, make low, dense mats; both these life forms are in the nature of protective responses against the severity of the weather.
3. Dwarf species belonging to a few genera of trees occur. In the Northern Hemisphere these are species of willow *(Salix)* and of birch *(Betula)*. In the Southern Hemisphere mountains dwarf species of austral forest tree genera occur, such as *Metrosideros* and *Dacrydium*. These differences indicate independent evolution of the alpine tundras in the two hemispheres, though in the cases of plants with easily dispersed spores, such as the nonvascular crypto-

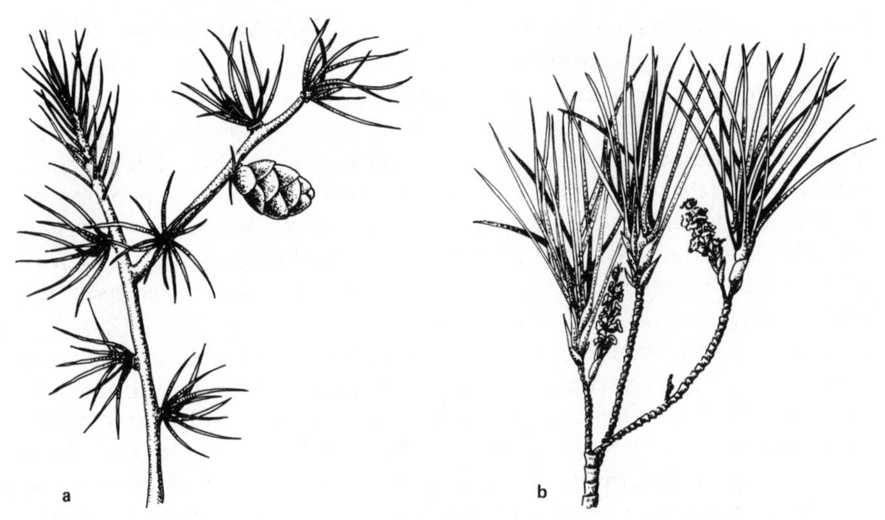

Figure 138. (a) *Larix laricina* (tamarack, or larch), North America, Pinaceae; (b) *Dracophyllum longifolius* (inanga), a Southern Hemisphere montane and subantarctic tree of the heather family Epacridaceae, occupying a similar ecological role to that of pines in northern wind-swept latitudes. Drawn by Desmond FitzGerald.

gams, there is no observable difference between the northern and southern alpine tundras.

LINNAEUS AND LAPLAND. Early exploration of the Arctic flora was carried out by Linnaeus in 1732, when he traveled from Uppsala northward to cross the polar circle at Jokkmokk, in Swedish Lapland, bringing back a large collection of dried plants and a written diary of his observations. The herb *Linnaea borealis* was one of his finds on that occasion. Afterward he and his students made similar explorations in other parts of the world. An eventual outcome of these explorations was the discovery that vegetation zones exist all over the world such that if one ascends a mountain, the plants encountered at progressively higher levels are analogous to or even identical with those which are encountered if one travels toward the polar regions.

NORTHERN PLANTS IN SOUTHERN UPLANDS. The details given in Tables 45–47 illustrate the truth of a fundamental discovery made by the successors of Linnaeus, namely, that *identical species* occur in southern uplands of the Northern Hemisphere as occur in more northern lowlands. For of the herbs noted—fire pink, buttercup, cinquefoil, mountain sorrel (or Oxalis), violet, and gentian, and also the rhododendron—every one is a species characteristic of the Canadian zone; that is to say, they are common Canadian wildflowers and not members of the southern woodland flora.

Table 45. Altitudinal zonation in the great Smokies, southern Appalachians, North America (about 35° north).

ALTITUDE (IN METERS)	CONSPICUOUS ELEMENTS OF THE VEGETATION
1900	Shrub, *Rhododendron catawbiense*
	Herbs, *Gentiana quinquefolia, Potentilla canadensis* (cinquefoil)
1800	Buttercup, *Ranunculus hispidus*
	Oxalis montana (also ranges the Alps, Atlas Mountains, Himalayas)
	Red spruce, *Picea rubens* (stenothermal species, descending to sea level in the north)
	Eastern hemlock, *Tsuga canadensis* (stenothermal, descending to sea level in the north)
1500	Striped maple, *Acer pennsylvanicum* (eurythermal, to sea level) Also following stenothermal species, confined to higher levels in south, but descending to sea level in north: mountain ash, *Sorbus americanus*; yellow birch, *Betula alleghaniensis*; canoe birch, *Betula papyrifera*; and the herb fire pink, *Silene virginica.*
1200	Canadian violet, *Viola canadensis* (stenothermal, descending to sea level in the north)
	Pitch pine, *Pinus rigida* (stenothermal, descending to sea level in the north)
900	Lower limit of *Rhododendron catawbiense*; upper limit of sugar maple, *Acer saccharum*
Below 900	Spruce, hemlock, mountain ash, birch, and pine predominate
Below 600	Southern broadleaf trees: elm, liquidambar, *Celtis, Gleditsia, Diospyros, Populus, Quercus, Betula, Carpinus, Fagus, Acer*

POLAR PLANTS LATER DISCOVERED TO BE ALPINE. The early explorations of botanists brought to light various species of herbs and shrubs discovered in Arctic Lapland, and named to signify the fact, and then later discovered to be identical with similar plants living in the European Alps and other high mountains of Eurasia. Some examples are:

Pedicularis lapponica	Lapland troll wort
Gerastium arcticum	Arctic chickweed
Arenaria norvegica	Norwegian sandwort
Linnaea borealis	A previously unnamed plant found by Linnaeus in Lapland

Comparable also are plants originally known from northern temperate lands and later discovered in alpine environments nearer the equator. For example, Sir Joseph Hooker's exploration of the Himalayas brought him one day to a hanging alpine valley at Lachoong where, at a height of 2400 m (8000 feet) he found the fields "full of such English weeds as shepherd's purse, nettles, *Solanum nigrum* and dock."

Conversely, some plants that were supposedly alpine were afterward found in far northern lands. Examples are:

Geranium pyrenaicum	Mountain crane's bill
Hypericum montanum	Mountain St. John's wort
Potentilla crantzi	Mountain cinquefoil
Cerastium alpinum	Alpine chickweed

Many others could be listed.

PLANTS SHARED BY REMOTE MOUNTAIN RANGES. Another category of geographical puzzles became known when it was discovered that plants supposedly characteristic of a particular mountain region occurred also on other mountains, though absent from the intervening regions. For example, the famed Swiss edelweiss *(Leontopodium alpinum)* was found to grow also in the Caucasus mountains and in the distant Himalayas.

Table 46. Zonation of biomes in the southeastern Himalayas (about 25° North), compiled from the diaries of Sir Joseph Hooker.

ALTITUDE (IN METERS)	BIOME	NATURE OF THE VEGETATION
5000 and above	Perpetual ice desert	Polar lichens on rock, algae in snow
4500 to 5000	Mountain tundra	*Arenaria*, woolly compositae, fleshy-leaved umbelliferous plants; lower limit of perpetual snowbanks in shady hollows
3500 to 4500	Mountain taiga	Fir *(Abies excelsa)* extends to upper limit at 4300 m; also juniper, rhododendron; upper limit of bamboo forest occurs at 3700 m, where pheasants and sandflies *(Simulium)* also occur; *Pinus* marks the lower limit of zone
2000 to 3500	Temperate broadleaf	Oak, pear, magnolia, chestnut; rhododendron grows to 12 m high; lowest part of zone is the upper limit of fig
1200 to 2000	Temperate rain forest	Tree fern, palm; upper part of zone is the upper limit of palm *(Calamus, Plectocomia)*, and of tree fern *(Alsophila)*; lower part of zone is the lower limit of oak and tree fern, also upper limit of giant bamboo *(Dendrocalamus)*; *Gordonia* and *Cedrela* commonest trees at lower level
1000	Transitional ecotone	Orange, banana; *Arum* replaces *Caladium*
600	Transitional ecotone	Upper limit of mango *(Mangifera)*
Below 600	Monsoon savanna	

The Discontinuous Distribution of Arctic and Alpine Biota

While these discoveries had been made by botanists, similar though fewer cases were found by zoologists among animals. For example, it was found that the Alpine hare is identical with the Arctic hare, and that a single species *(Lepus timidus)* ranges Arctic Lapland and Siberia, mountain moorlands of Britain and southern Scandinavia, and the peaks of the Alpine region in central Europe.

Some naturalists argued that the discontinuous distribution was the result of extinctions, perhaps caused by man, of populations in the intervening lowlands. An explanation of this type would doubtless satisfactorily explain the presence of wolves in high mountainous regions and in the Arctic but was much less convincing in the case of the Arctic hare. For the latter species is clearly adapted to life in regions where severe winter snows occur, having a white coat in winter, and a dark coat in summer. Such a habit would not suit it to life in lowland regions where severe winter snows do not occur. Some naturalists argued, too, that it is unlikely that man or any other cause could have led to extinctions of some of the plants noted above, for man ignores such herbs. Divine ordering of the universe was cited by some as the explanation, the Creator having placed suitable animals and plants in localities prone to winter snows, the Arctic hare having thus been given its appropriate distribution at the time of the creation of the world. Thus matters lay until 1847, when Jean Louis Rodolphe Agassiz announced his Ice Age theory.

Footprints of the Ice Age

Jean Louis Agassiz (1807–1873) was professor of natural history at Harvard University for the last 25 years of his life, but his earlier years were spent in Switzerland, where, among other topics, he studied the action of ice on mountains, the structure of glaciers and their effects on the mountains on which they develop. Initially he was skeptical of ideas proposed by earlier geologists that the former extent of Alpine glaciation may have been greater than it is at present; but after studying the question in the field, he changed his views and convinced himself, and then others, that such indeed had been the case. Some of the relevant geological facts are:

1. An elevated massif, when invested with a thick ice sheet, becomes subject to a special type of erosion that differs from that caused by ordinary weathering caused by wind and rain, for the effect of ice upon underlying rock is unlike that of water. Snow tends to accumulate in snow fields, or neves, on the sides of such mountains; the snow becomes deeper with increased mass from constant precipitation, and eventually the greater part of the mass under a neve consists of impacted ice. Such ice slowly flows outward, tearing away the sides and floor of the part of the mountain on which it rests and creating a concave amphitheater-shaped excavation in the underlying rock. Similar excavations on the other sides

Table 47. Variation of altitudinal zonation with latitude.

ALTITUDE (METERS ABOVE SEA LEVEL)	KILIMANJARO (3° SOUTH)	NEW GUINEA (0°–10° SOUTH)	JAVA (7°–10° SOUTH)	NEW ZEALAND (VOLCANIC PLATEAU, 36° SOUTH)	MACQUARIE ISLAND (55° SOUTH)
6000	Glaciers to 5951 m				
5000	Permanent ice from 4500 m	Permanent ice from 4500 m			
4000	Subalpine scrub and moor (tundra) from 3000 m	Alpine moor (tundra) from 3300 m; scrub ecotone to 3600 m	Scrub-forest ecotone from 3000 to 3500 m		
3000	Montane rain forest from 2000 m	Wet montane rain forest from 2500 m (cloud forest)	Montane cloud forest to 3000 m	Permanent ice from 2300 to 2823 m	
2000	Savanna	Lower montane rain forest from 900 m to 2700 m	Lower montane rain forest to 2250 m	Subalpine herb field (tundra) from 1700 m; scrub ecotone from 1400 m; montane *Nothofagus* forest from 900 m	Ice
1000		Tropical rain forest and savanna	Savanna and lowest rain forest from 600 m 1300 m; Mangrove, fig, and scrub below 600 m	Podocarp-broadleaf forest below 900 m	Subantarctic tundra
0	Sea level				

NOTE: From left to right, stratified biomes at successively more southern latitudes in the Southern Hemisphere. Permanent ice occurs at sea level at about 60° south latitude. For more detailed zonation in two northern latitudes, see Tables 45 and 46.

Figure 139. (a) *Abies balsamea* (balsam fir), northern North America, Pinaceae; (b) *Picea mariana* (black spruce), North America, Pinaceae. Drawn by Desmond FitzGerald.

of the mountain eventually meet and at the upper intersection of their contours a sharp prism-shaped peak is produced, called a *horn*. If an amelioration of climate occurs the upper part of the horn is exposed, and it is then provided with several lateral semicircular neves, called *cirques*. The upper margin of each cirque is usually demarcated by a semicircular crevasse, called a *bergschrund*, marking the fracture region between the part of the snow or ice mass that is moving downhill and the portion that adheres to the mountainside through pressure congelation. The Swiss Alps today, and most other high mountains, are presently in this phase, indicating a partial amelioration of the climate, for otherwise the entire horn would be concealed under ice.

2. Between adjacent cirques are sharp rock ridges called *arrêtes*. In the Alps and other glaciated mountains today the arrêtes are seen to extend far below the snow line. This implies that the cirques must formerly have operated at lower levels than they do today. This in turn implies that more ice must have been present and so the climate must have been more severe.

3. A *glacier* is a tongue of ice extruded from an ice mass as the ice

Figure 140. An ice-age horn still under active glaciation, Mount Cook, 12,349 feet, New Zealand. (a) Cirque with (b) bergschrund; (c) col; (d) arrête; (e) aiguille; (f) nevé. The ice from the two fields accumulates in a mile-high ice fall (below the picture) and then enters a 10-mile glacier in a U-shaped Tasman valley. The melting glacier forms a finger lake, Tekapo, in the lower part of the same valley (23).

slowly moves to lower levels. The amount of abrasive action of a glacier on its floor varies with the depth of the ice in the glacier and does not vary with the height of the glacier floor above sea level; on the other hand, a river cannot cut its valley floor below the base level of the water table, of which sea level is the lowest possible level (unless the drainage is toward an inland depression). The Alpine valleys flow outward toward the sea, yet their floors are found often to have been cut below the level of the water table. This means that if partial melting of any glacier occurs, the empty glacial valley floor is found to fill up with water and to become a long finger-shaped lake, which occupies the lower part of a continuous valley whose upper parts contain (or can be seen once to have contained) a glacier. The conditions exist in Alpine mountains. Therefore a melting of formerly greater glaciers has occurred sometime in the past.

4. Rivers cut V-shaped valleys, whereas glaciers cut U-shaped valleys, for the force of gravity, not lateral corrosion caused by momentum, is the directive force governing the ablation. Such U-shaped valleys, now empty of ice, are seen in the lower parts of large Alpine mountain regions. Hence glaciers have melted.

5. The *moraines*, or terminal and lateral mounds of rock debris brought

down to lower levels by glaciers, do not lie all at the present lower level of glaciers, but occur at much lower levels still. This indicates a former greater extension of glaciers.

6. Large boulders, called *erratics*, are commonly found perched in peculiar places where rivers could not possibly have transported them and where only ice transport is conceivable, yet no ice now lies around or near them. Such erratics often carry the long gouge marks, or *striae*, produced when a glacier that has the boulder imbedded in its mass scrapes past an adjacent valley-side or ridge.

7. At places far removed from existing glaciers—for example, the lowlands of the island of Sjaelland in Denmark—exposed sheets of bedrock are sometimes seen and observed to be deeply gouged with striae, indicating the passage of glacial ice (containing boulders that scored the striae). The direction of motion of the glacial ice can be determined and is found to have been from north to south.

From these and related data Agassiz determined that ice had formerly been much more extensive in Europe than it is today. On visiting Cambridge, Massachusetts, in 1846 to lecture at Harvard, he took the opportunity of examining the evidence in North America and satisfied himself that there had been a similar extension of the ice in both continents. The following year he published his book *The Glacial System*, thereby placing the theory of an Ice Age on a secure foundation of evidence.

THE PLEISTOCENE GLACIATIONS. Work by later students disclosed that there had in fact been a succession of extensions of the ice fields, alternating with periods when the ice retreated. At least four such fluctuations are now recognized. Table 48 gives an approximate historical sequence, as suggested by dating methods currently available.

Table 48. The Pleistocene glaciations.

YEARS AGO (BEFORE PRESENT)	NORTH AMERICAN STADE NAMES		EUROPEAN STADE NAMES
0	Present epoch	Warm	Present epoch
12,000–80,000	Wisconsin glacial	Cold	Wuerm glacial
	Sangamon interglacial	Warm	Riss-Wuerm interglacial
300,000–400,000	Illinoisan glacial	Cold	Riss glacial
	Yarmouth interglacial	Warm	Mindel-Riss interglacial
1,100,000 1,200,000	Kansan glacial	Cold	Mindel glacial
1,300,000	Aftonian interglacial	Warm	Guenz-Mindel interglacial
1,400,000	Nebraskan glacial	Cold	Guenz glacial
1,500,000		Warm	

Figure 141. Formerly glaciated terrain at Loch Ericht, Scotland. A long finger lake, or loch, occupies the former glacier bed; in the background a horn, now free from ice, a col to the right, and various corries or hanging valleys mark the site of former cirques. In the foreground erratic blocks from the moraine dropped at the ice melt, about 10,000 B.C. Red deer, or royal stags (*Cervus elaphus*), frequent this type of open terrain as well as deciduous forests, and salmon favor these highland rivers and lochs. For a comparable ecosystem in France 10,000 years ago, see Figure 54 (27).

ALPINE BIOTA AS ICE AGE RELICTS. With the realization that former ice ages have occurred, during which Arctic conditions extended over much wider regions than is the case today, an explanation was forthcoming for all the phenomena of discontinuous distribution of plants and animals. Clearly the alpine representatives are isolated relicts that retreated vertically, following their appropriate isotherms, when the temperatures began to rise after the last ice age, and the snows began to melt. Mountainous regions can be thought of as areas where the last ice age (Wisconsin or Wuerm stade) has not yet ended.

Arctic Mammals

Four orders of mammals live on the ice and treeless tundra that lies in the high Arctic latitudes within the Arctic Circle, and only 10 species are involved. These are:

Carnivora: Polar bear, ermine weasel, wolverine, wolf, Arctic fox
Rodentia: Greenland lemming, brown lemming
Lagomorpha: Arctic hare
Artiodactyla: caribou (or reindeer), musk ox

Figure 142. Arctic and subarctic moorland, or tundra, in Europe and in the Southern Hemisphere subantarctic region is commonly mountainous, in North America and Siberia more usually flat. The soil is peaty, frozen in winter, but productive of herbs in the short summer. Formerly glaciated horns of subdued outline were totally buried under ice during glacial stades. Roches moutonneeś, or ice-smoothed erratic blocks, are seen to the left of the picture (10).

All these species entered England, Denmark, and France during the ice ages, and although fossils are presently few from North America, we may take it as given that all occurred there too, probably to about 40° north latitude, where the southern limit of the continental ice generally lay.

POLAR BEAR [*Thalarctos maritimus*]. The polar bear is considered to be a late Pleistocene immigrant to North America from Eurasia. The species is now circumpolar. The polar bear ranged as far south as the North Sea in Europe, and fossils from the Ice Age have been found in Denmark. It presumably ranged the New England coast during the Wisconsin stade.

ERMINE WEASEL [*Mustela erminea*]. The weasel is found chiefly in taiga, but the range extends across the tundra to the northernmost Arctic islands, including Ellesmere, north of Greenland. It feeds on small mammals, chiefly mice, presumably small lemmings in the extreme north. It is dark brown above in summer, white in winter.

WOLVERINE [*Gulo gulo*]. It is known also by its European name of glutton; the American form is sometimes distinguished as *Gulo*

Figure 143. Ecotones in the subalpine and montane zones of a volcanic terrain near Mount Egmont, 8,500 feet, New Zealand. Alpine herbs lie buried under winter snow on the lower slopes, above which is a small permanent glacier relict. In the foreground, right, are *Dracophyllum* shrubs; the palmlike trees are agaves, (*Cordyline indivisa*). The tree to the right, and the forest in the middle distance, are *Nothofagus menziesii* (southern beech), here replacing conifers as the montane forest. The corresponding scene in the Northern Hemisphere would have spruce thinning into alpine tundra or moorland (23).

luscus. It resembles a small bear, save for the bushy tail, and inhabits high mountains, keeping near the timberline, as well as the Arctic tundra. It feeds on all available flesh, killed by itself or found as carrion. In winter it dens in sheltered places. It may grow up to 28 kg (60 lbs) exceptionally, usually less. See Figure 132, Chapter 15.

WOLVES [*Canis lupus*]. See page 332.

ARCTIC FOX [*Alopex lagopus*]. The species differs from other foxes in having short, rounded ears and furred feet, both structural adaptations to life in a cold climate, for they reduce the exposed areas from which body heat may be lost, at the same time protecting the weakly insulated pinna from frostbite and aiding snow traction in the case of the paws (compare notes on the polar bear). As befits an animal so well adapted to polar life, the Arctic fox is active all through the long polar night. During the winter the fox is reputed to follow polar bear in order to scavenge from its kills in the manner of a jackal. In the summer months it takes dead fish from the beach, eggs and young of Arctic-breeding birds, lemmings, and berries in season. It has two color phases: either (1) brownish above in summer and white in winter, or (2) brownish above in summer and bluish-gray in winter. On Pribilov Island only the second mentioned phase occurs. In the steppes of Asia, between the Caspian Sea and Mongolia, a fox is known, under the name corsac *(Alopex corsac)*, which is closely related, and may be a red phase of the polar species.

LEMMINGS. Lemmings comprise a group of three genera of molelike rodents, with thick, soft fur, short, rounded ears and a relatively very short tail (about one-sixth of the body length). They are found mainly in boreal and tundra situations, where they form the natural prey of smaller carnivores, particularly of the Arctic fox on the tundra. They are well known for the dramatic variations in their population density and the consequential forced migrations. The population density of Arctic foxes has been reported to follow the curve for lemming numbers, lagging by about one year. Two genera are found on the high Arctic tundra lands. Best-known are the Greenland lemming, *(Dicrostonyx groenlandicus)* and related species of the same genus, having a brown phase in summer (when it is mainly active by day) and a white coat in winter; they feed on vegetation, construct burrows, and operate a sanitary system in which several kilograms of small fecal pellets are stored in particular locations, presumably as some kind of communications sign. Another tundra lemming is the brown lemming *(Lemmus trimucronatus)*, which does not have a white coat in winter. This species constructs underground nests only in summer, is active by day and night, and ranges much further south, to include the alpine tundras of the coastal ranges of western Canada. *Lemmus lemmus* is a similar species occurring in Norway, one that experiences great fluctuations in number, with mass drownings occurring in the fjords in years when starvation pressure is severe. In the Siberian tundra the representative species is *Lemmus obensis*. The Norwegian species is the most brightly colored: black and yellow above, yellow below. Marmots are larger rodents with similar habits, found in alpine and polar tundras and also on prairies (genus *Marmota*).

ARCTIC HARE [*Lepus arcticus*]. The Arctic hare is one species of a group of three American hares more closely related to one another than to other hares. The three species most closely related to the present one are characterized by assuming a white coat in winter: the typical Arctic hare, the Greenland hare, and the varying hare. Some writers (for example, Burt, 1964) consider all three as one species, *L. arcticus*; the nature of the differences between them relates to the coat color, whether it is white all the year as in the Greenland hare *(L. groenlandicus)*, or only in the winter months, as in typical *L. arcticus*. Here all are considered as one variable species. Similar semantic differences concern the Eurasian representative of the group, the so-called blue hare *(Lepus timidus)*, also called *L. variabilis*; the latter includes representatives in Scandinavia, Russia, Scotland, Ireland, and the European Alps. Active through the year, it may hop kangaroo fashion or rear up on the hind legs. The food is the tundra vegetation. Like lemmings, Arctic hare are liable to great population fluctuations. The chief predators are fox, wolf, and man (Eskimo).

These closely related hares may be considered as scattered relicts of a former continuous Ice Age population.

CARIBOU, OR REINDEER [*Rangifer tarandus*]. The genus *Rangifer* includes the only deer in which both the male and female develop antlers. There is considerable difference in opinion as to the number of species, and the differences between them. European writers take the view that the reindeer *(R. tarandus)* is confined to Arctic Europe, and has only been introduced in recent years by man into Arctic North America. However, American writers, while acknowledging the (rather unsuccessful) introduction into Alaska and Canada of reindeer from Siberia, consider that the Greenland caribou is in fact the same species. Inquiries made by me in Lapland elicited the fact, acknowledged by all zoologists consulted, that Greenland caribou had been introduced into Scandinavia in an effort to improve the herds and that the results had been satisfactory, the new stock proving both stronger and larger. The view taken here, then, is that the true reindeer *(R. tarandus)* is native to both Arctic Europe and Greenland. Two other American species are commonly acknowledged, namely, the woodlands caribou *(R. caribou)* and the barren grounds caribou *(R. arcticus)*, having a more western tundra dispersion than the Greenland caribou. The two tundra species, *R. tarandus* and *R. arcticus*, differ in that the former is brown, the latter nearly white, always much paler. The two northern species migrate in the summer to the highlands to graze on the exposed lichen *(Cladonia)* and other ground plants, and in the winter to the lowlands where the snows are less severe. The Lapps who are bound in an almost commensal relationship with the reindeer are obliged to migrate with the herds.

The Lapps, whose culture seems very similar to that of European man during the Würm glaciation, depend almost wholly on the reindeer for skins (clothing, tents, or kaate), various bone or antler tools (such as knife handles, knife scabbards), ornaments, flesh for meat (either fresh or dried), mats and bed clothing, milk and cheese; they live as nomads, migrating every six months as the earth's situation in its orbit dictates—from high-

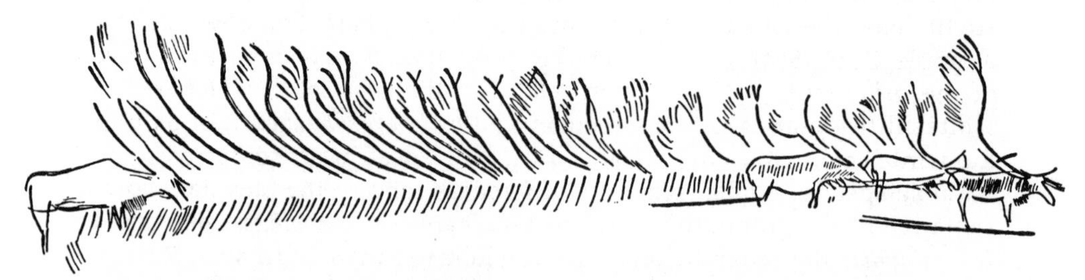

Figure 144. Herd of reindeer (*Rangifer tarandus*), as observed on the French tundra lands during the closing phases of the last glaciation and engraved on an eagle's wing bone by a Paleolithic artist of Grotte de la Mairie (1). The Magdalenian culture of the period probably resembled that of Lapps today.

Figure 145. As the ice retreated, herds of Mammoth (*Mammuthus primigenius*) roamed the tundra margins, feeding on polar willow and on spruce in winter in the valleys. The large cave bear (*Ursus spelaeus*) accompanied the herds as the major predator, in the same role as the Alaskan and Siberian bears today, taking salmon in season, berries when available, and stranded whales. Illustrations by Paleolithic artists of Font-de-Gaume and Combarelles (1).

land to lowland in the autumn, and from lowland to highland in the spring. In modern civilized conditions the Swedish government has made a praiseworthy attempt, more or less successful, to adapt the ancient nomadic regime to present-day requirements of hygiene and education, by establishing winter schools for the Lapp children, thereby teaching them modern ways, without interfering with the age-old customs of these Ice Age people, and permitting the survival into modern times of Stone Age customs of the greatest intrinsic interest to all people concerned for the past prehistoric development of society. The Swedish royal house has a long history (300 years) of personal interest with the remarkable Ice Age peoples of the farthest north. In the chaotic state of the twentieth century

Figure 146. *Ovibos moschatus* (musk ox), Bovidae, presently restricted to Greenland and Arctic America, formerly ranging Europe south to Paris in the ice ages (26).

world Lapland stands out as an exciting and inspiring example of how a people with a culture reaching back into the most remote epochs of man's involvement with nature, have somehow been able to retain their ancient character, still living in close commune with the Ice Age, yet still enjoying the greater part of the benefits of civilization.

MUSK OX [*Ovibos moschatus*]. In modern times the musk ox was only brought to human ken after the first explorers reached the northernmost confines of northern Canada and Greenland: for there are the last survivors of this remarkable Ice Age reminder of an animal our remote paleolithic ancestors knew so well, in the Thames valley near London and in the Parisian environs where the court of Versailles was later to arise. Little in the history of natural science can excel the impact made by the chance discovery of the fragments of a recognizable skeleton of a musk ox near Paris in the mid-nineteenth century. Here was proof of the former continuity of the Arctic Ice Age faunas as postulated by the new theory of Louis Agassiz. Soon after, when the paleolithic cave paintings of Ice Age animals were brought to light, the musk ox and the reindeer took their place as former denizens of a Europe before history was written. The musk ox as we know it today is a remote Arctic animal, living in Greenland and in the farthest Canadian barrens, is sometimes a solitary animal, more usually a herding species, feeding on tundra grasses, willows, and sedges and enduring the long polar night without migrating southward. The older bulls protect the herd, forming circles around the young and pregnant cows when danger threatens. The color of the coat is brownish, and it hangs in long silky threads almost to the ground. The broad, flat horns are plastered almost flat to the skull. Both sexes have horns. These animals, now so restricted as to range, once wandered widely when the glaciers spread over much of the present temperate regions of the Northern Hemisphere.

In addition to the Arctic mammals a number of peculiar mammals still survive in various high mountain areas of Eurasia and North America. Evidence from fossil deposits tells us that some, perhaps all, of these are relicts from the ice ages.

High Mountain Mammals of North America

Chief of these are the Rocky Mountain goat and the bighorn and dall sheep.

ROCKY MOUNTAIN GOAT [*Oreamnos americanus*]. These are the only American representative of the goat-antelopes—hair long and white, horns in both sexes, the horns bent backward slightly. These animals live in the Rocky Mountains in British Columbia, Montana, Idaho, Washington, and the coastal ranges northward to southern Alaska. The fur is white. There is a definite beard. They live on steep slopes, near the timberline. Mainly diurnal, they are usually seen in groups of 10 or fewer. They are believed to be most closely related to the chamois of the mountains of Europe, though little is known of the nature of the evolutionary relationship. No Asian species are known. Before the Pleistocene no American ancestors are known, so it is generally supposed that the goat-antelopes evolved in Europe and, late in the Pleistocene, migrated into America by way of the Bering Strait land bridge during a glacial phase. But this is speculation.

BIGHORN SHEEP [*Ovis canadensis*]. Nothing is known of the origin of these animals. They range western North America from Alaska and Yukon to Mexico and also occur in northeast Siberia, eastern Kamchatka in Asia. The species was first made known to science in the diaries of Lewis and Clark at the opening of the nineteenth century, where it is referred to as the "big-horned animal." The horns are relatively short but may occur in both sexes. The color is a dirty gray or brownish gray, the belly and legs whitish. In some North American forms the coat is entirely white. The species is gregarious. The sexes usually separate out in summer but rejoin in the fall; in winter they tend to move to lower elevations. It is a browser on trees and shrubs, and a grazer on grass, taking a wide variety of food. The species has been nearly exterminated by man, especially by ranchers and by so-called sportsmen, and it now survives in only a very restricted remnant of the range it had at the time Jefferson sent his explorers to report on the products of the continent. Little is known of the Siberian representatives of the species, but it may be assumed that man has treated them in a manner analogous to that in the new world.

DALL SHEEP [*Ovis dalli*]. These animals inhabit similar areas to that occupied by the preceeding species, ranging from whitish in the northern part of the range to nearly black in British Columbia, the southern extremity of the distribution. The species is mainly Alaskan, frequenting the most inaccessible terrain. European writers regard the dall sheep as merely a subspecies of the bighorn. It is supposed that this sheep reached North America from Eurasia.

High Mountain Mammals of Europe

CHAMOIS [*Rupicapra rupicapra*]. This is the only goat-antelope now occurring in Europe. Horns are present in both sexes, bent backward, hook-shaped, about the length of the head, like those of the Rocky Mountain goat. The coat is brownish, a little plare in winter. Chamois live in the Alps, Appenines, and Carpathians, with local races in the Pyrenees, Caucasus, and Asia Minor; a discontinuous distribution pointing to a former Ice Age continuity. It goes about in small flocks, chiefly in the upper part of the tree belt, but ascending in summer to the terminal moraines of the glaciers. It was introduced by man into the Southern Alps of New Zealand, with disastrous results to the native alpine herbs, which are not able to withstand grazing. See also Tahr, below.

WILD SHEEP, OR MOUFLON [*Ovis musimon*]. They are one of the ancestors of the domestic sheep *(Ovis aries)*. It is believed that in prehistoric times the urial of Asia *(Ovis vignei)* was brought into Europe by wandering tribes from Asia and crossed with the native mouflon to produce the ancestors of modern domestic sheep. The mouflon is presently found wild only in the mountains of Corsica and Sardinia, but it probably once was widespread in Europe. It is chiefly a blackish-brown, the ewes either horned or hornless.

ALPINE IBEX AND WILD GOAT [*Capra ibex*]. There are several subspecies (or ? species) of wild goat with large scimitarlike horns, of triangular cross section and with knobbly surfaces. These animals used once to inhabit the whole European Alpine region but now have become extinct, except for a few isolated occurrences such as on the island of Crete. Similar or perhaps conspecific goats occurred in north Africa and Asia, possibly relicts of a once far-ranging species. Under a variety of names the various fragments of the supposedly once-continuous population exist today: as tur (in the Pyrenees), western tur (Caucasus), sakin (Asiatic highlands), nubian ibex or beden (Sinai and Nubia east of the Nile), and wali (Ethiopia). Domestic goats are believed to have been one of the oldest domestic animals, tamed from different wild goats, divided into forms with spirally wound horns *(Capra prisca)* and forms with scimitarlike horns.

Figure 147. *Capra falconeri* (markhor), a wild goat of the central Asian Mountains, allied to the Ibex, Bovidae (26).

High Mountain Mammals of Asia

SNOW LEOPARD [*Felis uncia*]. This is a catlike animal resembling a leopard, about 2 m long, 1 m of which is tail. The fur is long and thick. It lives in the mountain ranges of Central Asia, Altai, and Thibet, also in Kashmir and at the eastern end of the Himalayas. Like all spotted cats, it has come under very severe hunting pressure in recent years owing to the extraordinary demands for its pelt made by the patrons of the fur trade. It is one of the most handsome living cats, the coat spotted with black broken rings on a white ground tinged with red. In nature it preys on wild sheep and ibex, also raiding mountain villages in the old days. Try to save this animal for posterity by exerting an influence on society to prohibit the disgraceful traffic in its pelt.

CHIRU [*Panthalops hodgsoni*]. The Tibetan antelope, or chiru, is a little-known Asiatic animal living at a height of between 3500 and 5500 m. It is related to the saiga but, unlike the latter, is not a plains-

Figure 148. *Ovis ammon* (Himalayan and Siberian sheep), Bovidae (26).

dweller. Horns occur in the male. The nose is inflated and overhangs the jaw. The coat is yellowish in summer, whitish in winter.

YAK [*Bos grunniens*]. These are central Asiatic wild cattle with long, soft hair that hangs down to the ground, blackish brown. The animal looks rather as if it had a tasseled tablecloth thrown over it. It has a shoulder height of about 1.5 m. There is a tamed form, somewhat smaller than the wild species, with a shorter head, smaller horns (or no horns). The wild species ranges higher than any other cattle, to heights in Tibet up to 6000 m (nearly 4 miles). Tamed specimens are used as milk cattle and also as beasts of burden and for riding. It ranges from Bokhara to Mongolia.

SEROWS [*Capricornis species*]. The serows are goat-antelopes of Asia; about four species range from the Himalayas to Japan. Short, straight almost parallel horns occur in both sexes.

GORAL [*Nemorhaedus spp*]. These goat-antelopes resemble the serows but are smaller. Short conical horns occur in both sexes, directed backward. The hair is long and smooth. Five species occur, in the Himalayas, western China, Korea, and the Amur.

TAKINS [*Budorcas spp*]. These are also Asiatic goat-antelopes, living in the eastern Himalayas and in Assam. It is a large, heavily built animal with relatively short hair of an unkempt harsh quality, power-

Figure 149. *Marmota marmota* (Alpine marmot), Alps and Carpathians, Rodentia (23).

ful horns in both sexes, having a gnulike aspect. They live in pairs or in small herds gathering in midsummer into large herds; in winter they migrate down to lower altitudes.

TAHR [*Hemitragus spp*]. These are similar to goats, from which they differ in having very small horns and no beard in the male. The hair is long and smooth and on the neck assumes the character of a mane, reddish-brown, legs almost black. The horns are nearly contiguous at the base, compressed, flattened on either side, bent backward. It lives on steep wooded slopes of the Himalayas, India, and Arabia. There are several species. The Himalayan species *(Hemitragus jemlahicus)* was introduced some years ago into the Southern Alps of New Zealand, where it is known as the "thar"; it has caused great destruction of native flora and is now a serious problem.

Goats and sheep similar to the species of Europe occur in Asiatic highlands.

KIANG [*Equus kiang*]. These are the only horses to occur as a high alpine species. They live on the high plateaux of Tibet and Sikkim at 4000 to 5500 m. Resembling the wild ass, the color above is reddish-brown, the belly and legs white; it has a dorsal stripe, a mane, and a tail with a terminal tuft. They are usually in small herds but sometimes are solitary.

High Mountain Mammals of Africa

The African mountains carry essentially similar fauna to

that of the mountains of southern Europe, save only that the fauna does not transgress the Sahara, as already detailed above.

High Mountain Mammals of South America

CORDILLERA FOXES [*Pseudalopes culpaeus*]. Ranging from Ecuador to Patagonia, they are the second largest of the South American wild canids (after the maned wolf); the body is 90 cm long (plus a tail of 40 cm), black and gray above, flanks brownish, belly lighter.

TUCO-TUCOS [*Ctenomys species*]. These are burrowing South American ratlike rodents with small eyes, concealed ears, short, stumpy tail and short legs with large scraping claws. The name is an imitation of its call. There are numerous species ranging the Andes to the upper limit of life. The best-known species is *Ctenomys magellanicus*, brownish-gray above, sprinkled with black, about 20 cm long. It digs like a mole and feeds on roots. Some species live on the pampas.

CHINCILLAS. These are rabbitlike South American rodents having bushy tails and are placed in a separate family Chinchillidae. There are three genera: *Chinchilla*, *Lagidium* (mountain chinchilla), and *Lagostomus* (viscacha). Some of these rodents live on the Andes, where they form colonies, the animals spending the day underground and the night above ground foraging. They range up to the limit of life and can endure severe cold. They are greatly persecuted by pelt hunters.

The South American camels, llama, guanaco, and domesticated variants are able to live at great heights on the Andes and also on the pampas lowlands. The mountain herds migrate in much the same manner as the reindeer in the Northern Hemisphere. These animals are referred to under the grasslands biome, in the next chapter.

Figure 150. *Chinchilla* sp., gregarious nocturnal rodent of the Andes, to 20,000 feet above sea level (23).

17

ARID AND SEMIARID LANDS: PRAIRIE, STEPPE, THORN SCRUB, AND DESERT

To support a forest an annual rainfall of the order of 100 cm per annum is required. Those parts of the earth's surface where dry land is exposed and where the annual precipitation is less than 100 cm are commonly referred to as semiarid or arid, depending on the annual rainfall, and the vegetation biomes produced are either grasslands or deserts, depending on the amount of precipitation. Table 49 summarizes these relationships.

Grasses (family Gramineae) make up one of the largest groups of flowering plants (angiosperms), with about 5000 different species. Most grasses are herbs and have fibrous root systems, but some tropical species—walled bamboos—grow to a height of 30 m (for example, *Dendrocalamus*). Of those grasses whose seeds are used by man, most are annuals, though a few are perennials and reproduce by spread of underground roots. The stems are mostly hollow, though they are solid in the case of maize *(Zea mais)*. The leaves are typically long and narrow, arranged along the stem in two ranks. The stem is known as the culm. Grasses of the temperate regions rarely branch, but many tropical grasses branch freely.

Grass culms, especially bamboo culms, grow rapidly, sometimes as much as 30 m in 3 months. Some can grow 1 m or more in 24 hrs. In temperate climates grasses produce a turf, the result of the intercommunication of a vast number of separate shallow roots in a surface soil. The flowers of grasses are produced in what is called spikelets enclosed in leaflike structures called glumes. There are no petals or sepals. Stamens are usually conspicuous in groups of three. In most cases grasses are pollinated by the wind.

Grasses first evolved in the Cretaceous period, about 100 million years ago. Grasslands first become recognizably important in the Oligocene period, about 30 million years ago. At that time herds of herbivorous mammals were associated with the grasslands, and the forests were dwindling in importance.

Table 49. Vegetation type and precipitation.

BIOME NAME	ANNUAL PRECIPITATION	EXAMPLES
Long-grass prairie	50–100 cm	East American prairie
Short-grass prairie	25–50 cm	West American prairie
Thorn-scrub desert	Less than 25 cm	Mesquite
Arid desert	Less than 25 cm	Sahara

LONG-GRASS PRAIRIE, OR TALL GRASSLAND. On continents this biome lies immediately adjacent to the broadleaf summergreen biome and extends to meet the shortgrass prairie, or steppe, where the rainfall is lower. The summer is dry and the winter relatively cold, often with snow lying for months at a time. The rainfall lies between 100 and 50 cm per annum. The predominant grasses include species of the genera *Andropogon, Sporobolus, Stipa, Agropyron, Koeleria, Muhlenbergia,* and *Panicum,* and also there are numerous species of herbs associated with these grasses. Tall trees are sometimes present along the margins of rivers and creeks, particularly willows. The rainfall is mainly concentrated over the early part of the summer, and moisture is otherwise mainly available during the spring snow melt.

SHORT-GRASS PRAIRIE, OR STEPPE. This occupies a region of lesser rainfall between the long-grass prairie on the wetter side and the thorn scrub or desert on the drier side. This biome is also continental and occupies regions where the rainfall is less than about 50 cm per annum and more than about 25 cm per annum. The summers tend to be long and dry and the winters long and cold, with a great deal of wind. The grass species present include representatives of the genera *Buchloe, Bouteloua,* and *Muhlenbergia*; also sedges such as *Carex.* Somewhat taller grasses that may occur include *Hilaria, Agropyron, Sporobolus, Stipa, Festuca,* and *Elymus.* Various broadleaf herbs may be common, for example, species of various

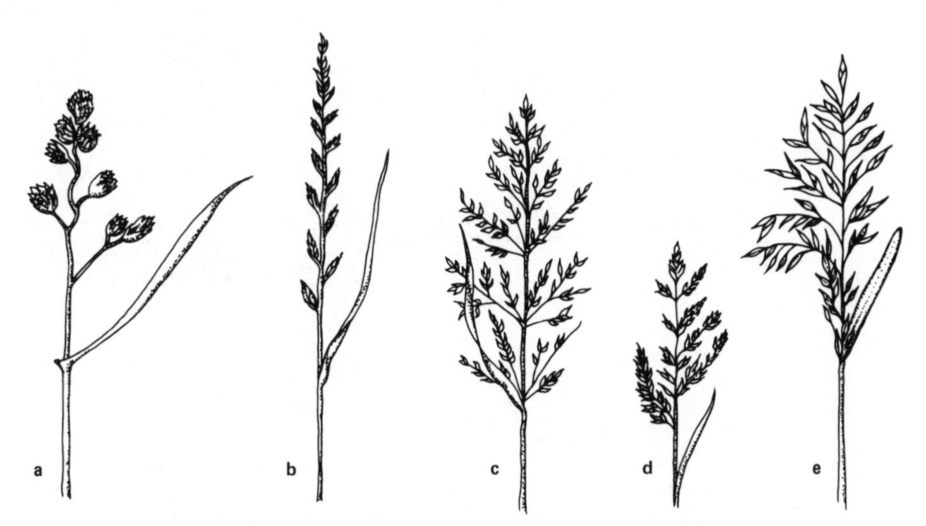

Figure 151. (a) *Dactylis glomerata* (cocksfoot); (b) *Lolium perenne* (perennial rye grass); (c) *Poa trivialis* (rough-stalked meadow grass); (d) *Festuca pratensis* (meadow fescue); (e) *Arrhenatherum arenaceum* (tall oat grass). Drawn by Irene Fell.

genera of Compositae, such as *Aster, Artemisia*, and *Senecio*, and Leguminosae, such as *Lupinus*. This biome, in temperate regions, roughly corresponds to the savanna of the tropics. Extensive areas of short-grass prairie, or steppe, occur in North America, between the long-grass prairie and the western ranges; also in Eurasia between latitudes 30° and 50° north from the Middle East to Mongolia, and in the Southern Hemisphere in South America and Australia.

THORN SCRUB. An open biome dominated by spiny, thorny, or oily scrubs, such as the mesquite *(Prosopis)* and similar plants, wait-a-bit thorns *(Acacia)*, and creosote bush *(Larrea)*. Rain is of irregular occurrence and during the dry periods the leaves of the nonoily or succulent species are shed. The root may either penetrate to considerable depths in the soil to deeper water-bearing layers (phreatophytes) or else spread over the surface, adapted to catch surface water when it is available, with devices such as thick superficial layers of tissue to retain moisture in the plant body during dry periods, as in cacti (Americas) and desert milkweeds, or Euphorbiaceae (Africa). After rain a temporary layer of herbs, including annual grasses, appears over the soil. This arid biome occurs in mainly subtropical and tropical regions and commonly lies between steppe and dry desert.

DRY DESERT. This biome occupies regions where precipitation is between zero and about 25 cm per annum, the occurrence of rain being quite erratic. Dew occurs at night and humidity during the day is very low. Temperature may vary widely, sometimes very high during the

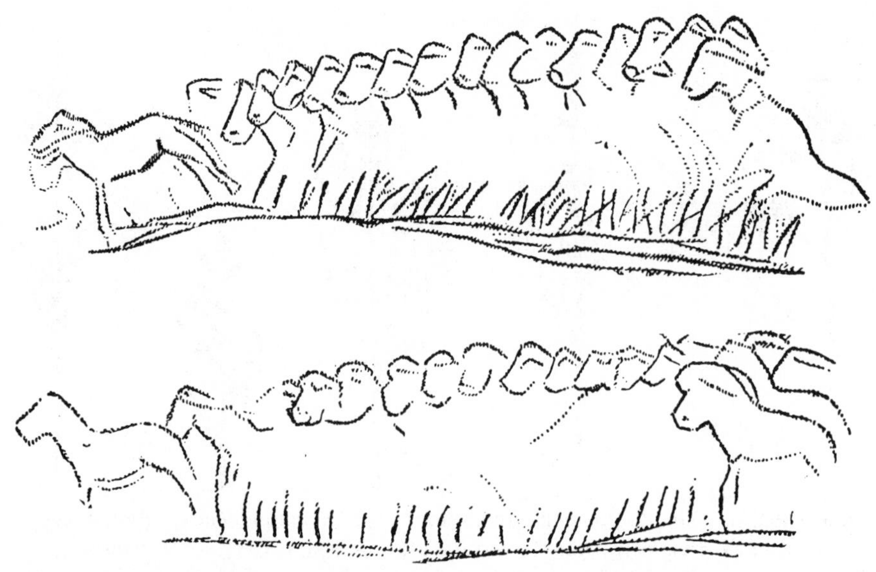

Figure 152. Herds of horses (*Equus caballus*) seen on the postglacial steppe of France by Paleolithic men who engraved these scenes 10,000 years ago at Vienne (1).

day, with possibility of frost at night on account of radiation cooling under the clear cloudless skies. Wind may be considerable and, owing to the lack of herbaceous cover, is apt to strip the soil from exposed areas, leaving a stony *desert pavement*, with the fine soil particles transferred to regions of accumulation where it forms *sand dunes*. The dunes wander, under the influence of the wind, with gentle slopes on the windward side and steep slopes on the lee side. These conditions are found in the interiors of the continents, particularly Africa, Asia, America, and Australia.

An American naturalist with experience in these matters amplifies the foregoing paragraph with the following comment: "All dune fields in arid North America are related to major rivers or to Pleistocene lakes, or (in the case of Nebraska) to glacial outwash. The richest source of sand for the Arizona dunes is the erosion of ancient sandstone, climatic control not being of primary significance here."

The desert areas form belts to the north and south of the equator, roughly corresponding to the zones of high atmospheric pressure on the poleward margin of the trade winds, where air is descending, after losing its moisture at high altitudes. The Sahara (an Arabic word meaning sand) Desert extends across northern Africa, through Arabia into southern Asia, then northward into eastern central Asia, where it is known as the Gobi Desert. To the south, across the Himalayas, is the Desert of Sind. In North America there are desert tracts across the continent east and south of the Sierra Nevada mountains. In the Southern Hemisphere, the Atacama Desert lies on the west coast of America, the Kalahari Desert in southwest Africa, and the Victoria and Sandy deserts occupy central Australia. All these regions occupy areas where the moisture content of the planetary winds has been dissipated by passage across large land areas. The barren

Figure 153. Lioness (*Felis leo*) and horses (*Equus caballus*) depicted by Paleolithic mural artists of the Font-de-Gaume (1).

dryness of some deserts is intensified when there are mountain barriers between them and the direction of the prevailing winds, for moisture is always stripped from the air when it rises to heights where the low temperature produces precipitation. Deserts ringed by mountains are therefore dry basins, with the Tarim Basin of western China and the central desert region of North America as examples. If such basins occur outside of the tropics, they may be intensely cold in winter.

All deserts tend to heat up rapidly during the day, as the soil is warmed by the sun, and during the night the heat is equally rapidly lost. During the day the air overlying the surface of the desert becomes a shimmering haze with layers of unequal density and resultant varying refractive index; the resultant bent pathways followed by light rays passing over deserts lead to the frequent occurrence of mirages. When light from the sky is bent so as to curve upward to reach the eye of the viewer, the resultant appearance simulates pools of reflecting water on the ground (the commonest type of mirage). When light from the ground is bent downward so as to reach the viewer's eye seemingly from above, images of surface features are seen in the sky, above the horizon; this also is a not infrequent type of mirage. Rarer effects include the magnification of distant objects, including oases below the horizon, which appear near the viewer. Similar effects occasionally occur over a sea surface and, for example, the coast of France is occasionally seen from the south coast of England on a larger scale and with greater clarity than would be normally expected.

During the night the desert air becomes cold and clear and the stars shine very brightly. When crossing a desert, therefore, best progress is made by traveling at night rather than by day. The term *desert devil* is sometimes applied to the swirls of airborne sand particles that form and are carried about the terrain when locally induced thermal air currents develop as a result of ascending air masses caused by differential heating by the sun's rays. Severer winds may occur in which the moving air is at a temperature of 40° C or more, laden with blinding and choking clouds of dust and sand, and visibility is reduced to little more than a meter; the term *simoon* is applied to such storms. The winds, as noted above, also

Figure 154. *Cynomys* sp. (prairie dog), burrowing colonial rodent, Montana to Mexico (23).

cause a gradual drift of the great dunes. The sand particles are blown up the gentle slope and fall down the steep lee slope; the motion of the dunes is always to the direction that the steep side faces. Thus the topography of a sandy desert changes constantly and there are few or no fixed landmarks. Wandering dunes of this type overwhelmed the Roman cities of North Africa during the Dark Ages, when land misuse or overly used caravan routes cut into the vegetation veneer and let the wind extract the sand of previously fixed dunes. By a converse process, wandering dunes can be stabilized, and a fixed topography developed, if suitable drought-resistant *(xerophytic)* vegetation is able to gain a foothold on the surface of the dunes. Examples of dune-fixing plants are marram grass *(Spinifex),* often successfully introduced by man along sandy foreshores to prevent the overwhelming of arable land by sand.

Occasional rains fall in desert regions, producing temporary streams, such as the *wadi* or *billabongs.* After rainfall plants that may have lain dormant for years spring into growth and flowering. Wadis eventually flow under the surface of the ground and disappear from view; the underground water, however, persists as a water table or may emerge elsewhere at a lower level as springs, producing local pools of water. If the soil is laden with salts, such pools are highly saline or sometimes toxic, depending on the dissolved substances. Examples are the bitter lakes of the Sahara, the Dead Sea of Asia Minor, and the Great Salt Lake in Utah. On the other hand, the soil may be relatively free from salts, in which case the pools or lakes produced by emergent springs are potable *sweetwater lakes* and the resultant vegetation and associated animal life constitutes a desert *oasis.*

Because of the mode of origin of oases, just outlined, it follows that oases commonly occur in linear chains (corresponding to the course of the underground rivers), and they tend also to occur near mountains, for the later induce precipitation of rain and the streams that flow down the sides of such ranges are the sources of the most reliable underground waters, where the sands have swallowed up the surface of the water. Thus desert settlements tend to form linear series, and the main caravan routes run from one oasis to the next. Among the vegetation elements that occur in oases are date palm *(Phoenix)*, maize *(Zea)*, cotton *(Gossypium)*, various citrus trees and *Acacia*, melons, wheat, millet, various desert grasses, the latter often providing grazing for herds of ruminants. Of various literary works that have treated these environmental features as a substantial background for interesting narrative, notable examples are the books of Antoine de Saint-Exupéry.

Plants of Arid Regions

Succulent plants are those which contain a large proportion of fluid (sap) together with various adaptations to enable the fluid to be retained for long dry periods without excessive loss by evaporation, often also with protective structures such as thorns to discourage excessive grazing or browsing by herbivorous animals. The plant body often assumes curious or bizarre forms and this, coupled with their ability to endure long periods without water, makes succulent plants especially suitable for artificial cultivation as house plants in dry centrally heated rooms, often with forgetful owners who deluge them with water on rare occasions separated by intervals of neglect. So it happens that in city environments the one-time little-known cacti, euphorbias, and the like have now become some of the best known of all plants.

A few types of desert plant develop very long and deeply penetrating

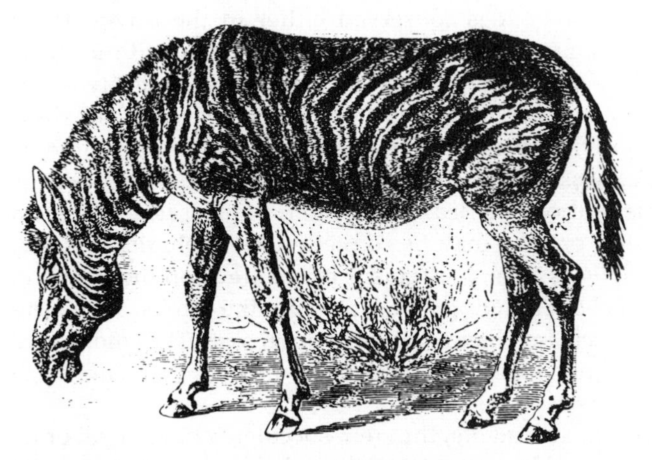

Figure 155. *Equus quagga burchelli* (burchell's zebra), South African veldt, nearing extinction (18).

Figure 156. *Muntiacus munrjak* (muntjak), southeast Asia, Cervidae (26).

root systems, able to seek out moisture at deeper levels in the soil. *Welwitschia* is a gymnosperm with this habit. Plants of this type, called *phreatophytes,* are unsuitable, of course, for growing indoors. In some arid regions phreatophytes extract so much moisture from the soil as to make it difficult for other, more useful, plants to grow; so agricultural practices at present tend to concentrate on destroying the phreatophytes in regions where man plans an arid economy based on desert grasses or herbs more useful as fodder.

Most desert plants are not phreatophytes. Instead, they have very shallow and far-spreading root systems, adapted to the rapid absorption of dew at night, as well as of rain on the occasions when it falls. Once water is taken into the plant body it is conserved either in the leaves or the stems, or both; so these are usually very fleshy and juicy, with up to 95 percent of the weight being water. Examples of plants with fleshy leaves are the agaves, aloes, and fig marigolds *(Mesembryanthemum)*: in these there is a thin superficial layer of green assimilating (photosynthetic) tissue surrounding a massive, colorless watery inner flesh. To reduce the evaporation by transpiration, the leaves are often adpressed to the stem (for example, *Crassula*); or they may be converted into spherical masses, as in the pebblelike *Lithops*. Many succulent plants develop the leaves in dense rosettes, each leaf overlapping and protecting the others, as in *Sempervivum*. Nearly all succulents have a leathery outer skin. The chlorophyll may be protected against excessive irradiation by red pigment, as in *Aeonium*.

In contrast to the leaf succulents, the stem succulents have small or no leaves, and the functions of transpiration and photosynthesis are taken over by layers of green assimilative tissue on the surface of the stems. The

Figure 157. *Lama glama* (wild llama), plains of Patagonia and Andes in northern part of range (to 16,000 feet in Ecuador), Camelidae. Drawn by Desmond FitzGerald.

stems in these cases are usually very fleshy, with many species of *Euphorbia* as examples. The leaves may form and then be shed soon afterward, or the plants may be leafless at all times. In *Euphorbia,* where leaves may appear to be absent, they may in fact be present but converted into spines. The latter, by discouraging browsing animals, must be regarded as water-conservation adaptations. The loss of water through transpiration varies with the temperature and dryness of the environment and the surface area of the exposed tissue. So in conditions of extreme aridity the surface area of the plant is always reduced to a minimum. Extreme cases of such adaptation are seen in the columnar and spherical cacti; for example, *Echinocactus,* which looks like a sea urchin, and in the spherical euphorbias, for example, *Euphorbia obesa.* Another adaptation involves protec-

Figure 158. *Macropus gigantaeus* (great grey kangaroo), Australian grasslands, Macropodidae (23).

tive camouflage, in which the plant resembles the environment in color or shape or both. *Lithops* so closely resembles the pebbles of the desert pavements on which it lives as to make the plant quite hard to find. These are also water-conserving adaptations, for the protective simulation is directed against detection by grazing animals.

The effectiveness of these various adaptations is so great that many

Figure 159. *Dromaius novaehollandiae* (emu), Australian grasslands, Casuariiformes (23).

Figure 160. Succulent desert vegetation. (a) *Lobivia binghamiana*, Argentina, Cactaceae; (b) *Haworthia cymbiformis*, South Africa, Liliaceae; (c) *Gibbaeum*, South Africa karroo, Aizoaceae; (d) *Euphorbia balsamifera*, North Africa and Canary Islands, Euphorbiaceae. Drawn by Irene Fell.

different and quite unrelated families of plants have adopted the same life forms. The cacti of the American arid regions closely resemble the euphorbias of the African arid lands, and it is only when the flowers are produced that the differences are obvious—for the short-lived reproductive stages retain their family characters, unaffected by the environment. In southern lands the family Aizoaceae has produced many succulents.

Some desert regions experience regular annual rains at certain seasons, followed by 6 or 9 months of drought. These cyclic arid climates occur, for example, on the high interior of Asia where a desert steppe experiences regular spring rains and floods, followed by prolonged summer, autumn, and winter drought, the latter with snow and ice. Plants typical of such regions develop succulent underground leaf bases (bulbs) or enlarged underground stems (corms). Well-known examples are the tulips

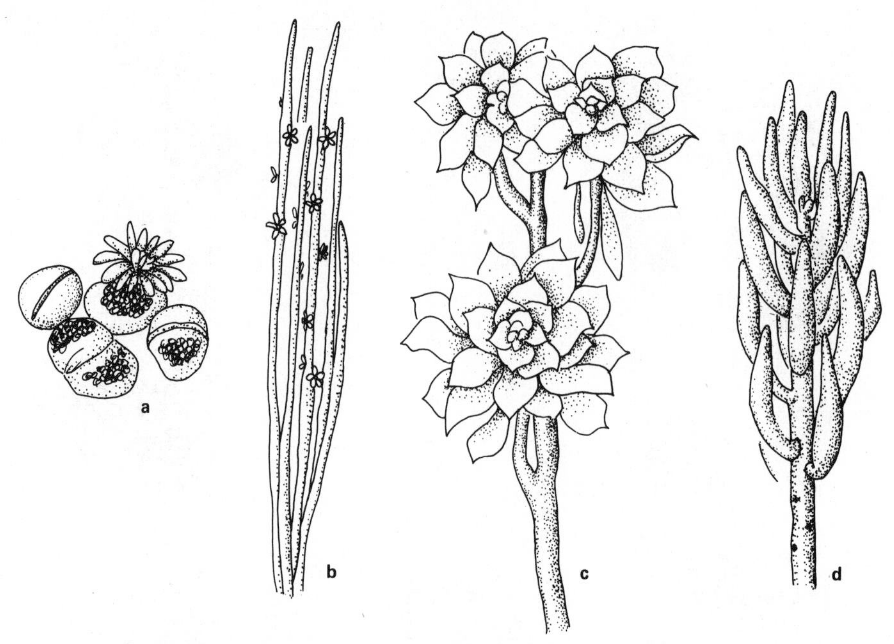

Figure 161. Succulent desert vegetation (continued). (a) *Lithops* sp., South Africa, Aizoaceae; (b) *Euphorbia antisyphilitica*, Mexican desert, Euphorbiaceae; (c) *Aeonium arboreum*, southern Mediterranean and Sahara, grows to 3 feet high, Crassulaceae; (d) *Kleinia tomentosa*, Cape Province, South Africa, Compositae. Drawn by Irene Fell.

and onions. The brilliantly colored horticultural varieties of these plants are derived from various wild species of the Pamirs, Himalayas, and Turkestan arid steppes. These plants are hardy in northern and southern temperate regions, but they do best if the summer drought of their natural environment is simulated by lifting the plants after flowering and storing the succulent underground parts for 6 months on dry racks until the late fall, when they can be replanted to spend the winter in the soil, in accordance with their natural state.

Adaptations of Desert Animals

Since food and water supplies are available only at intervals of space or time, desert animals are obliged to develop special adaptations to overcome these environmental obstacles to life. Many desert animals are able to store water within their bodies. They may drink deeply when water is encountered and may then be obliged to live without water for days or weeks. Camels have water-storing cellular compartments in the lining of a part of the stomach. The two surviving desert camels are the dromedary, or Arabian camel *(Camelus dromedarius)*, with one hump, of North Africa and western Asia, and the Bactrian camel *(C. bactrianus)*, with two humps, of central Asia. The humps comprise fatty storage tissue not related to the water storing property of the stomach. It is possible that

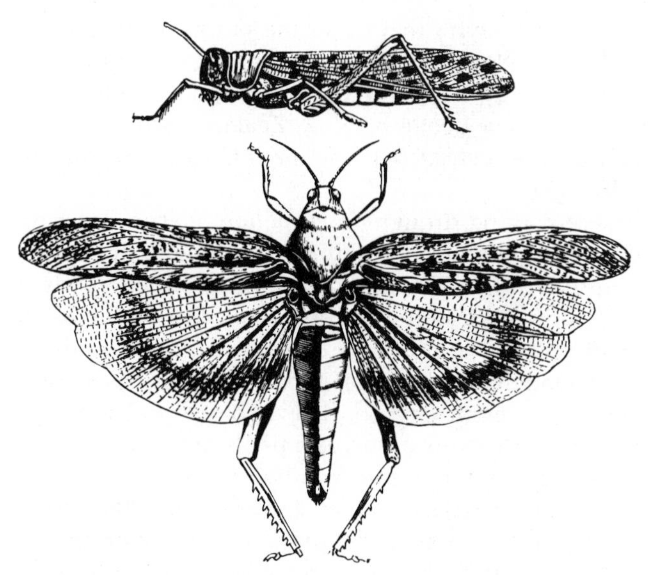

Figure 162. *Locusta migratoria* (migratory locust), tropical and temperate savannas and grasslands of the world, crossing oceans on the wind, occasionally reaching New Zealand, Orthoptera (34).

all surviving camels may be descended from domesticated stock distributed by man in ancient times.

Reptiles are generally tolerant of desert or semidesert conditions, and many of these animals store water in internal organs or cavities of the body. Desert peoples are familiar with this fact and obtain potable body fluids from desert tortoises and some lizards. Amphibia have poorly developed lungs as a general rule and therefore depend on skin respiration; the latter requires a semipermeable damp membrane, so the skin has to be kept moist by the constant secretion of fluids on to the skin by dermal glands. Thus the amphibians are generally absent from deserts, for their water requirements cannot be met. There are some exceptions. An Australian desert frog *(Chiroleptes)* constructs underground burrows whose walls it keeps damp by continuous secretions supplied by large internal body reservoirs of fluid; this animal, with its body swollen when water is replenished in rainy periods, is able to maintain a damp micro-climate in the burrow and there survive even when the surrounding soil is baked hard. A second class of desert adaptations involves a reduction in the rate of water loss. For example, many desert animals become nocturnal in habit, reducing the exposure to sun and dry atmosphere, emerging only when cool air occurs, frequently drinking the night dews also. Among the animals that follow this routine may be mentioned lions, hyenas, desert foxes, coyotes, desert owls, and most of the insects, spiders, and scorpions. Animals with thickened skins or scales frequently ignore this pattern of behavior and emerge in the hottest sun—for example, desert beetles, apparently protected by the thick elytra that cover the wings. In temperate climates, incidentally, tiger beetles are to be seen running about on sun-baked soil, hunting their prey by day; and another diurnal beetle is the brilliantly metallic green manuka beetle of New Zealand, which occurs in great numbers by day on the *Leptospermum* scrub of the savanna during the seasonal summer drought.

A third method of overcoming drought is *estivation,* that is, entering on a period of relative or complete inactivity during prolonged dry spells, with revival and resumption of feeding, growth, and reproduction when the rains occur. Desert snails are examples. Certain species, collected in deserts, and subsequently placed in museum cabinets, have been found to be alive years later. Aquatic crustaceans of the Australian desert produce resting eggs when their temporary ponds evaporate, and the eggs lie dormant for long periods until rains occur, and pools form, whereupon full growth to a length of 6 cm may occur within two weeks.

A fourth adaptation to life in semiarid and arid conditions is the evolution of locomotor organs adapted to swift running. The limbs of horses, antelopes, pronghorns, emus, ostriches, rheas, and kangaroos permit very rapid traverse of unfavorable territory and frequent migration from one region to another whenever drought becomes intensified. Such adaptations may be highly efficacious in arid and semiarid environments but, from their almost irreversible character, may have the effect of permanently fixing the range of the animal to such arid, open, hard-ground (prairie, pavement), or sandy terrains. Thus, horses are extremely successful on open grasslands and open savannas, high semidesert steppes, and even arctic barrens (for example, Przewalski's horse in Ice Age Eurasia), yet they play little or no part in the adjacent forest biomes, cannot enter swamplands, and are liable to extinction if a new predator enters the region.

Professor Paul S. Martin, who has paid particular attention to problems related to Pleistocene extinctions, argues persuasively that the American native horses were extinguished by Amerindians, and few who have heard him lay out the evidence would be inclined to debate the matter. But he himself cautions against dogmatism here, for much of the evidence is circumstantial. We know from archeological excavations that horse bones appear in kitchen-midden material of early Amerindians, that horses lived in North America for millions of years before man appeared, that no native horses remained alive on the continent when the conquistadores explored it, and that the wild mustang of today descends from Arab stock that crossed the Atlantic on board Spanish galleons. A cautious jury would not care to convict on such evidence alone, but each juryman would probably entertain dark suspicions! Much of paleontological and archeological evidence is necessarily of this character, and nowhere in American sites do we find tablets engraved with statements such as British history can produce; there is no American equivalent to the Earl of Gordon, who could claim with pride "I destroyed the last bear in Britain in the year 1057." However, relentless naturalists are probably going to identify before long certain modern hunters as having destroyed the last known living example of this and that species, and some ghastly roll of dishonor is somewhere now in the making, for the edification of a posterity that will never know the beasts who shared their forefathers' world.

The quality of vegetation on arid lands, including the semiarid prairies and steppes, is such that grass predominates or other hard-textured vegetation, and here again evolution has led to the development of dental patterns suited to grinding such harsh forage, particularly true in the horses. In deer, sheep, cattle, and goats the dentition is less specialized, and the ridges on the molars are such as permit soft herbage, such as leaves of forest trees, to be eaten; these animals are not restricted to grasslands and, as already noted, can be found in a number of forest biomes. Some birds have developed running progression at the expense of the powers of flight, ostriches and other large ratite birds being examples; less drastic terrestrial adaptation in birds is seen in the road runners, which are ground cuckoos having a rapid gait; desert owls, which burrow or occupy the burrows of rodents; and ground larks. Adaptation to sandy soil is seen in the spreading toes of camels; some skinks (lizards) of sandy deserts are so adapted to moving over sand as to give the appearance of swimming through the sand.

Burrowing for protection against the hot sun also serves as protection against predators. Various groups of rodents, such as the prairie dogs, have been so successful in this regard as to be able to establish veritable cities underground, sometimes extending over hundreds of square miles.

Figure 163. (a) *Polemaetus bellicosus* (martial eagle), large predator of savannas and semidesert terrain, frequenting the top of an acacia, it preys on antelopes and some smaller game, in the savannas taking also monkeys, Falconiformes; (b) *Gazella dorcas* (gazelle), plains and deserts of North Africa, height at shoulder only 2 feet, one of the smallest antelopes (23).

Desert rodents have acute sight and hearing and habitually feed in an erect kangaroolike position, with the head above the level of the grass or other ground cover. Jerboas are examples. Some of these rodents are so well adapted to desert life that the occurrence of rainy weather disturbs their normal life pattern, or perhaps even their physiology, and they pass into a torpid resting state during rains, a situation analogous but opposite to estivation.

Desert animals frequently adopt a body color or coat color of a brownish or sandy tint, matching the surroundings. This is true of the predators, such as the desert cats, as well as of the herbivores. Valuable in their natural environment, such coat color becomes a dangerous advertisement should the animal move from the desert into a green vegetation biome; thus, like the locomotor adaptations, this feature tends to prohibit desert animals from leaving the arid environment and tends to stabilize and perpetuate the desert fauna.

Disturbances in the rhythm of desert communities occasionally lead to ecological disasters, of which the best-known are the periodic plagues of migratory locusts. The sequence appears to operate in the following order: (1) an abnormally wet year occurs sporadically in a desert region; (2) this produces an unusually rich growth of vegetation from the dormant seeds of herbs; (3) this produces an unusually large population of herbivorous animals with one-season life spans, notably grasshoppers, including the flying grasshoppers called locusts; (4) the latter rapidly exhaust the available food supplies when the climate reverts to its normal arid state; (5) long-range migration of enormous flocks of locusts follows, with widespread devastation of all edible vegetation in adjacent biomes; (6) mass mortality of herbivorous animals ensues.

Locust plagues are uncommon in North America at this epoch in geological history, but they may have been frequent in the past and may occur in the future, should the relative proportions of the appropriate biomes undergo a change. After a wet spring in Arizona large numbers of flying grasshoppers frequent the mercury vapor lights of Tucson, a sign that would in Africa spread fear of worse to come. And visitors to Utah to this day are reminded of the plague of locusts in 1855, soon after the first Mormon disturbance of the biome, and the divine miracle that brought great flocks of gulls to save the saints of the desert. In 1874 the high plains west of the Missouri River were devastated by a plague of locusts. "Within a week grain fields, shrubs, vines, gardens had been eaten down to the ground or to the bark. Nothing could be done. You sat by and saw everything go." Many settlers gave up their holdings and left. Not so Stuart Henry, who penned the passage just cited. He said his prayers, and the following year brought good crops. These were folk who read their Bibles, but only the strong in faith dared follow the example of Henry, who read that milk and honey flowed again after the years which the locust ate.

Predators, such as the carnivorous mammals, and the raptorial birds, such as eagles, are of their nature swift and also capable of obtaining much of their fluid needs from the bodies of their prey. This combination of characters permits them to be eurytopic, ranging from one biome to an-

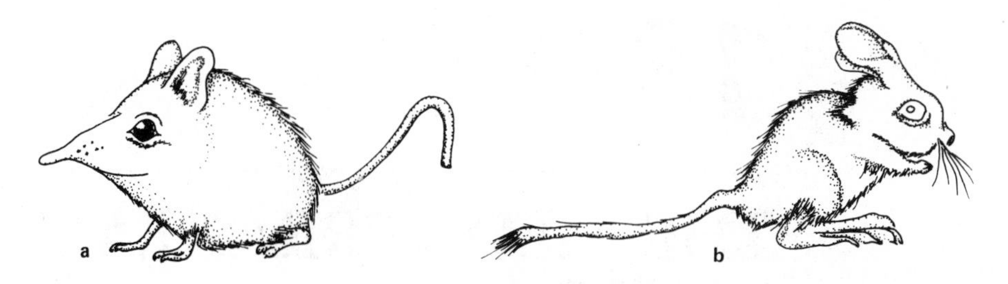

Figure 164. (a) *Petrodromus* sp. (rock elephant shrew), burrowing diurnal desert and savanna insectivore of East Africa, Macroscelidae; (b) *Allactaga* sp. (desert jerboa), burrowing mainly nocturnal rodent of north Africa and Middle East arid lands, Rodentia (23).

other. Consequently these animals show much less specific adaptation to desert environments, though they freely enter them and freely quit them, as need or chance dictates.

Chaparral Biomes

In these vegetation regions aridity is seasonal, in summer, but the winters are wet and often rather cold, with snow on the higher ground. Wind is nearly constant. The chaparral areas lie in the temperate zones, on the western sides of continents. The chaparral oaks have already been mentioned (Chapter 12) and the chapter that follows deals in particular with the type example, the Mediterranean lands themselves, but it should be stressed that this type of climate and the associated vegetation is widely distributed in the locations indicated. In the New Zealand region, which is largely chaparral in character in the drier districts, the trees tend to develop leathery leaves and a fur, or tomentum, as a wind-resistant covering on the lower side. Also shrubs and even trees may arise as species in genera that elsewhere are merely herbs. Thus *Senecio*, a daisy-like plant such as groundsel in northern biomes, becomes a tree up to 30 feet high. Figure 166 illustrates a twig of such a tree, *Senecio rotundifolius*, ranging the South Island of New Zealand at a height of 3500 feet, down to sea level.

18

THE MEDITERRANEAN LANDS

For ecologists the world over the Mediterranean lands have a special meaning—for it is here, and here alone, that we can find a continuous record of man's impact on his environment, written at large in the picturesque thought patterns of our ancestors, inscribed in the ever-changing hieroglyphs, syllabaries, cuneiform and alphabetic scripts that mark the course of evolution of our common literary and scientific heritage. The record here, as in other ancient contexts, was set down by poets, in this case the poet-kings of Israel. To their witness we can add the silent testimony of fossils, less eloquent it is true, but able to carry our inquiry back another ten millennia to the closing phases of the last great advance of the northern ice fields.

NORTH AFRICA AFTER THE ICE AGE. From carbon-dating we learn that about 9000 B.C. the lands of North Africa presently overwhelmed by the sands of the Sahara were still covered over by a veneer of vegetation. Palynology tells us that the climate must have been relatively cool and humid, for the North African forest of this period comprised the gymnosperms Aleppo pine, Atlantic cedar, juniper, and cypress, together with the broadleaf angiosperm trees, oak, olive, alder, and linden. Linden *(Tilia)*, an index of cold temperate conditions, has no modern species in Africa. Forests of this type apparently extended right across northern Africa, for traces of them have been recognized among the archeological remains of Egypt.

The fauna of North Africa in late prehistoric times has been eloquently depicted for us in the cave paintings and rock engravings of the Paleolithic, Mesolithic, and Neolithic peoples of that region. They set before us clear evidence that the African savanna animals ranged far north of their present limits. Vivid vignettes of charging black rhinoceroses, of giraffes, and of other beasts that cannot survive in the desert tell us that there was in fact no desert and that the savanna lands to the north of the equatorial rainforest must have spread out in one uninterrupted sweep to the shorelines of the Mediterranean, varied only by the intensified forest biomes investing the mountains of Morocco, Algeria, and Tunisia.

THE FATE OF THE NORTH AFRICAN FOREST BIOME. Only traces now remain of this once rich forest. The tree species survive in part, restricted to the highlands whose snows are glimpsed along the southern horizon by the modern seafarer on the Mediterranean, or seen on the Atlas Mountains in the course of the Gibraltar passage. The fauna has been decimated; there survive today only those few beasts that can withstand drought or that are at home in the mountains, such as the deer, bear, mountain sheep, and others noted in the foregoing pages; the zebra, buffalo, giraffe, and rhino have gone, and the forest peoples who hunted or were coexistent with them have also gone, leaving a trail of flaked stone weapons and a gallery of pictures of life as they knew it. What was the fate that overtook the North African forest community? We can only infer its probable character by turning to much later events and extrapolating backward in time to guess at the missing sequences.

THE MEDITERRANEAN WORLD OF 1000 B.C. To place yourself in the context of Mediterranean life 3000 years ago, take out a map and begin by altering the names to match those used by ancient writers. Egypt, under the pharaoh Shishak I, was heir to its own and the civilized world's culture, occupying no overwhelming role in the affairs of men. It was a land of fertile cornfields, and had been so at least since the twelfth dynasty (1991–1786 B.C.), when the nameless author of a papyrus text known as "The Admonitions" so described it. To the west lay Libya, Tunisia, and Algeria, to appear in subsequent classical history under the name Numidia. On the Atlantic coast of north Africa is Mauretania, and that is its ancient name. North of Mauretania lies Spain, called Hispania by the classical world, known to the peoples of Asia Minor by the names of

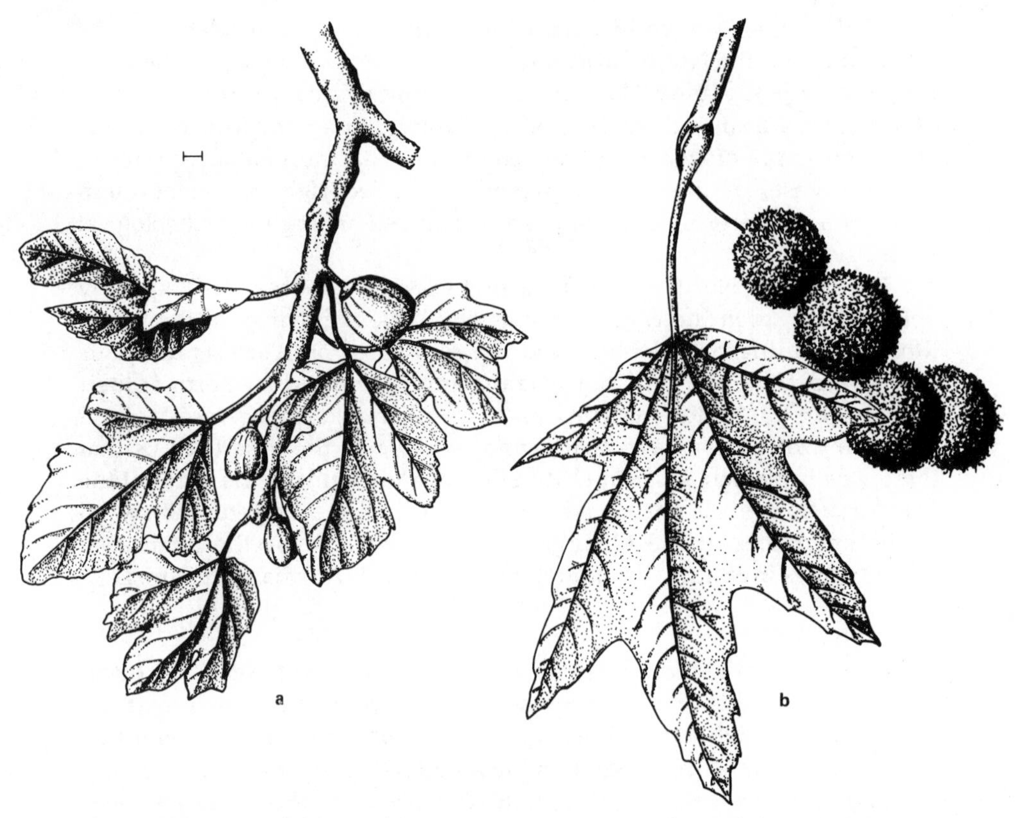

Figure 165. Trees of chaparral biomes. (a) *Ficus carica* (classical fig), Mediterranean arid lands, Moraceae; (b) *Platanus* sp. (plane), Platanaceae. Drawn by Desmond FitzGerald.

its first Semitic colonies, Gades (Cadiz) and Tarshish (Tartessus). Northeast of Spain was the unknown barbaric land of France, called Gallia in later times, already occupied by Celts speaking a southern dialect of their race. Germany was a wild forestland, inhabited by Teutons with a lively Bronze Age culture, who traded with the south but were regarded as savages by the Mediterranean traders who sought their amber. Italy was the home of many races of whom only the Etruscans were civilized; we cannot even read their inscriptions, let alone guess at their origins. Of the Romans archeology tells us little but supports their own traditions, which set the founding of the village on the Tiber in 753 B.C., much later than the period we now are considering. To the east of Italia lay the savage Balkan lands, with Hellas to the southeast. Greece was in eclipse—the Mycenean civilization had crumpled under invasions by unknown assailants, and no voice was heard from Greek writers until Homer and Hesiod recreated the past in poems that have come down to us.

East of Hellas lay Pontus, the modern Turkey, largely semibarbaric though with the germ of the Greek Ionian settlements perhaps developing on the Mediterranean coast. South of Pontus lay Syria, where now live peoples who maintain the ancient name of the land of the Assyrian kings.

East of Syria the Euphrates River flowed into the Persian Gulf, and the country it drains was known at this epoch as Babylon, called Iraq on modern maps. Palestine lay to the south of Syria, its northern part called Lebanon; the southern parts Philistia, near Egypt, adjacent to the Mediterranean; the rest Israel. East of Israel, across the Jordan, stretched the forest country of Bashan and Gilead, and beyond that the land of Arabia and extending southward to Sheba, where Aden now stands, tied by trading routes to east and west, then as now. Tyre guarded the coast of Lebanon and at the epoch we are considering was ruled by a king called Hiram, the second by that name. Israel, after 500 years of rule by so-called judges, had recently adopted a monarchy and in the period we are examining was ruled by two kings in succession; first David, then his son Solomon. According to tradition, both men would seem to have been keen observers of nature and also poets of the first order. However, the relentless researches of biblical scholars and of archeologists tell us that probably most of the writings attributed to David and Solomon were in fact the work of later composers. In the paragraphs that follow, I use their names in the traditional way, but they must be understood rather as referring to literary schools. These writers, whoever they really were, had a profound feeling for the environment.

From these surprising manifestations it might be expected that man's sympathy with his fellow creatures might promote a balanced ecology. But sympathy was not enough; in the light of after-knowledge it seems that no inductive reasoning was applied toward developing a stable environment. Invasion, captivity, and evident neglect of the land doubtless accentuated the subsequent loss of productivity.

A LAND OF MILK AND HONEY. The traditions of these men recorded that their forebears had been a wandering tribe of nomads who, about 1250 B.C., successfully occupied by military force the territory of Canaan, straddling the River Jordan. At the time of its seizure from the prior occupants, called in history Philistines, it was reported to be "a good land, a land of brooks of water, of fountains and depths that spring out of valleys and hills. A land of wheat and barley and vines and fig trees and pomegranates, a land of oil olive and honey." So wrote one of the anonymous authors of Deuteronomy, perhaps about 400 B.C., when the events recorded were remote in the past. Though we cannot now accept the traditional attribution of the earliest books of the Hebrew canon to Moses, there are no grounds for rejecting the belief that they contain sound records derived from the times with which they deal.

From Genesis we learn that the year was divided into four seasons, named seedtime, summer, harvest, and winter. This, with casual references to snow, hailstones, winter rain, and heat in summer tells us that Palestine about 1000 B.C. must have had much the same kind of climate as it has today—what ecologists call the Mediterranean climate, to which reference has already been made. The evergreen broadleaf forest that characterizes the Mediterranean, or chaparral, climate is adapted to endure a summer drought, when growth is suspended; the leaves tend to be simple, often leathery or hairy, wind resistant. Some trees are deciduous, others are not, but there is no pronounced seasonal fall. The autumn is a time of ripe fruit and grain rather than of falling leaves, and the forests are never totally bare of leaves.

The Song of Songs, thought to date from about 250 B.C., tells of spring in Israel: "The winter is past, the rain is over and gone, the flowers appear on the earth, the time of the singing birds is come, and the voice of the turtle is heard in our land. The fig tree puts forth her green leaves, and the vines with the tender grape give a good smell." Elsewhere we learn that among his many other possessions and activities, "Solomon had a vineyard at Baalhamon; he let out the vineyard unto keepers." Another entry: "I went down into the garden of nuts to see the fruits of the valley and to see whether the vines flourished, and the pomegranates budded." Apparently they did, for ". . . I am come into my garden, I have gathered the myrrh with my spices; I have eaten my honeycomb with my honey; I have drunk my wine with my milk; eat O friends!" Shades of Virgil or Horace, there is no mistaking the pride of a farmer in this king's diaries.

The histories of Solomon's reign tell of the multitude of horses he purchased from Shishak of Egypt, to equip the Israeli army with cavalry, yet in his own journal Solomon says simply, in the third person he sometimes affected, "King Solomon made himself a chariot out of the wood of

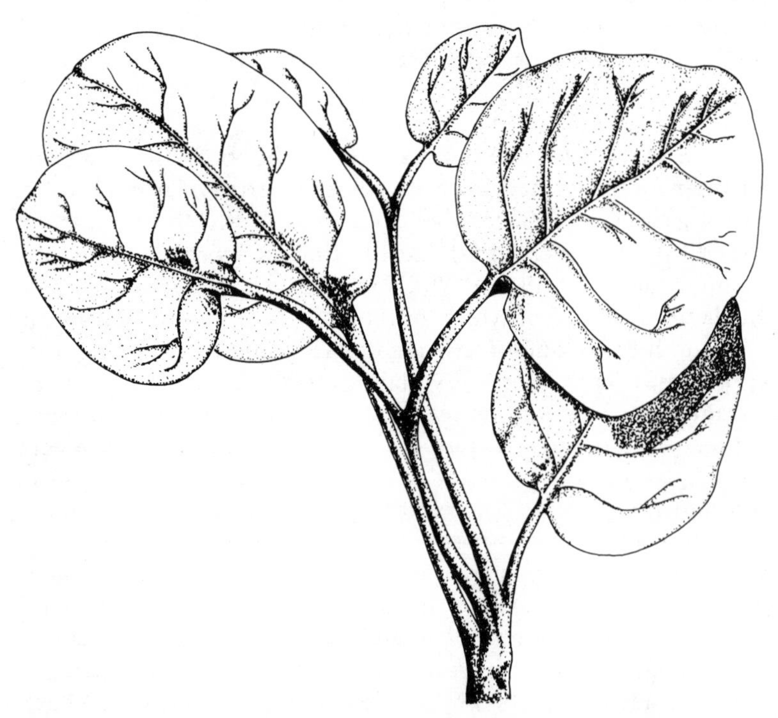

Figure 166. *Senecio rotundifolius* (New Zealand tree daisy), Compositae, chaparral exposure to 3500 feet. Drawn by Desmond FitzGerald.

Lebanon." This is the same man who made a treaty with Hiram of Tyre as a result of which ten thousand hewers were sent to cut the cedars of Lebanon; the lumber was shipped by the Phoenicians by rafts from Tyre to Joppa and used in the decoration of Jerusalem and in the construction of the Red Sea fleet.

Elsewhere Solomon boasts that he made cedar as common in Israel as sycomore. By this we can interpret the nature of the Israeli forests as lacking in cedar (also lacking fir, as other passages imply), but rich in *Ficus sycomorus*, a deciduous softwood that produces fruit but was, and is, regarded as worthless as lumber. The sycomore, or mulberry fig, still grows in Israel in scattered copses. (The tree called sycamore in North America is the plane. See p. 304). The cedar still grows in Lebanon, in very rare isolated stands; so does fir. It begins to look, then, as if what is in Israel and neighboring lands today is a remnant of what once was there 3000 years ago; the change has been in quantity not specific content at least so far as the vegetation is concerned. In fact, flicking over the leaves of those Biblical books written by men who observed nature (not all did), and compiling a short list of the flora most frequently mentioned, we can make a list that will look somewhat like this (checking the translations and identifications with recent scholarship):

CONSPICUOUS FLORA IN PALESTINE ABOUT 1000 B.C.

Gymnosperms	Cedar *(Cedrus libani)*, fir *(Abies)*, cypress *(Cupressus)*
Broadleaf trees	Willow *(Salix)*, olive *(Olea)*, myrtle *(Myrtus)*, acacia *(Acacia)*; oak *(Quercus)*, fig *(Ficus)*, sycomore fig *(Ficus sycomorus)*, chestnut *(Castanea)*, poplar *(Populus)*, apple *(Malus)*, almond *(Prunus)* thyine *(Callitris)*, pomegranate *(Punica)*, bay *(Laurus)*
Shrubs	Broom *(Cytisus)*
Herbs	Aloe, mustard, wheat, grass, reeds
Lianes	Vine

The foregoing is a typical Mediterranean flora. One could make up much the same catalog by a careful examination of the paintings of medieval Italy, from the backdrops of Botticelli, or more simply by noting the trees along the road from Naples to Salerno.

The Book of Psalms (ca. 400–200 B.C.) may derive in part from the time of David, to whom the psalms are nominally attributed. David likes to remind his hearers (for the poems were sung to the harp) that he began life as a shepherd and watched over the flocks by night. We can glimpse the wild hills in his references to the animals he learned to know, for they are mostly creatures that avoid man, or fear him, and David often speaks feelingly of their terror; he seems to view the world through the eyes of the hunted rather than that of the hunter; of hunting itself he knew much, and from him we learn that in 1000 B.C. men used snares, nets, bows and arrows, darts, spears, hooks with barbs, and in fact pretty well the whole armory before gunpowder.

He watched the habits of animals, noted that the facial muscles express the emotions, as Darwin later was to point out; that lions sometimes miss their kill and go hungry; he could feel for the loneliness of the ostrich in the wilderness (wrongly translated in the Authorized Version as owl). All his writings betray a respect for the natural environment and the creatures of it, including man: "I am fearfully and wonderfully made, and in thy book were all my members written." He cared for the solitary life, and in his last poetic flight of fancy he spoke of taking the wings of the morning to find refuge in the uttermost parts of the sea. King James's translators were kind to his poetry but knew little zoology. In the Authorized Version, David's urox becomes a unicorn; his jackals are foxes; his ostrich becomes an owl, and his porcupine becomes a bittern. Adding to David's notes those of the other Biblical books where animals are mentioned, we can develop a short list of vertebrates, which looks like this:

CONSPICUOUS VERTEBRATES OF PALESTINE ABOUT 1000 B.C.

Mammals	
Carnivora	Lion, leopard, brown bear, jackal, wolf, weasel
Artiodactyla	Urox, ox, sheep, goat, camel, fallow deer
Hyracoidea	Coney
Perissodactyla	Wild ass, horse (some, or all, imported from Egypt)
Rodentia	Porcupine, jerboa
Lagomorpha	Hare
Chiroptera	Bat
Primates	Man (apes mentioned, but imported from Asia by Solomon)
Birds	
Passeriformes	Sparrow, raven, swallow
Galliformes	Quail (peacock mentioned, but imported from Asia by Solomon)
Falconiformes	Eagle, hawk, gier eagle (vulture), osprey
Strigiformes	Owl (but confused with ostrich in translation)
Struthioniformes	Ostrich
Caprimulgiformes	Nightjar
Columbiformes	Rock pigeon, turtle dove
Others	
	Turtle, serpent, frog

DISTRIBUTION. Other passages in the Hebrew chronicles tell us something of the distribution of the biota. Solomon associates the lions with the high mountains of the northeast of Palestine (Mount Hermon, 9400 feet) and with the mountains of Lebanon; he also speaks in the same passage of the mountains of the leopards. Bear and leopard were still living in Lebanon within memory; the lion disappeared in medieval times. Except for jackals, or possibly foxes, little is said of any animals apart from those in domestication by the writers of the New Testament books. The disciples of Jesus had many adventures in rough, stony places, on the Sea of Galilee, on all sorts of mountains and high places; they were struck by storms, nearly drowned by waves, and encountered a fig tree that bore no fruit. But somehow we never hear of them being chased by bears, or jumped on by leopards, or lost in a great forest, nor do they ever seem to have so much as glimpsed a large number of trees in one place.

This does not at all agree with the environmental picture we gain from the Old Testament books so far cited. From those latter we learn of the great forests of the Lebanon whence the Phoenicians cut the cedar, cypress, and fir they needed for their ships. We hear of a great oak forest east of Jordan, the woods of Ephraim, whence the Phoenicians obtained the timbers they needed for their oars. Ezekiel tells us that firs grew on Hermon. One of Solomon's brothers met his end in an oak forest that no longer exists. To the west, along the seaboard, were the rich plains of Sharon, famed for their flowers and honey. So it is clear that great changes occurred in Palestine between the times of David and Solomon and the time of Christ. When did the changes occur, and what were their causes?

THE DESOLATION OF PALESTINE. The approximate date of the changes is not difficult to determine. There are vague hints of impending disaster in the closing verses of a poem attributed to the unknown preacher, or Ecclesiastes, who claims to be a king of Israel and a son of David: scholars have taken this to be poetic license, that the author was some later writer attributing his philosophy to Solomon the Wise. Whereas the earlier writings stress the "good land, a land of brooks of water," Ecclesiastes warns vaguely of a pitcher broken at the fountain, the wheel broken at the cistern, then shall the dust return to the earth as it was. By about 350 B.C. a new writer appeared on the scene, Joel, the author of a book by that name. In the first chapter he reports a scene of desolation that has come over the country. "That which the palmerworm hath left hath the locust eaten" and he goes on to catalog a whole food chain, the details of which can be omitted here. The net result is, he says: "My vine is wasted, my fig tree barked, the field is wasted, the land mourneth, the corn, and the vine, are dried up and wasted, so also the pomegranate tree, and the palm also, and the apple tree, in fact, all the trees of the field are withered. The herds of cattle are perplexed for they have no pasture. The rivers of waters are dried up and fire hath devoured the pasture." Here we can recognize the outraged farmer, not unknown in the evil days of living memory.

Hebrew writers attributed the cause of the desolation to divine intervention. Long before Joel, one of the authors of Isaiah (760–700 B.C.) had reported: "The earth is defiled under the inhabitants thereof. They that dwell therein are desolate and few men are left. The vine languisheth. The earth is utterly broken down. Lebanon is ashamed and hewn down. Sharon is like a wilderness and Bashan and Carmel shake off their fruits.

The highways lie waste. The wayfaring man ceaseth." Their theory seems to have been that men were full of sin (noted in considerable detail) and the Lord therefore punished the whole earth. The men of Tyre and Sidon were bad enough but, being heathens, could perhaps at least be understood; but the wickedness of Israel was most reprehensible; the drunkenness of the men of Ephraim was particularly obnoxious, so their oak forest had to go.

Shortly after the time of Christ, Roman engineers surveyed the situation and concluded apparently that a proper water conservation program, coupled with judicious use of the land and avoidance of overgrazing, would restore the situation. Such a program was put into effect from Numidia to Palmyra in Arabia. It succeeded.

THE RESTORATION OF FERTILITY IN ROMAN TIMES. It was not until about the year 1920 that the tremendous influence of Roman engineering on the fertility of the North African and eastern Mediterranean lands began to be realized. Archeological exploration disclosed that numerous districts that are now desert were once irrigated by canals, aqueducts, and reservoirs. Cities with amphitheaters, public baths, and temples have now been uncovered, and it is obvious that they must have been surrounded by fertile lands yielding wheat and olives and fruits. The Roman written records themselves also disclose the nature and extent of the trade carried on with these settlements. Mauretania, on the Atlantic coast, yielded vines, and sheep were pastured on the hills; and ebony and citrus wood was exported to Spain for furniture making. During the time of the empire Numidia was developed and proved to be much more fertile than Mauretania. Further east wheat, vines, olives, fruit trees, almonds, horses, asses, cattle, and sheep were all raised. The annual export to Rome of millions of bushels of wheat was handled by a series of ports that were established, such as Hippo Regius, Hadrumetum, Tacape, and Oea. These cities, and others that grew up inland from original legionary camps, were connected to the east with Egypt by a highway that ran through Cyrene in Libya.

In Asia Minor, after Trajan annexed Arabia Petraea in A.D. 106, the city of Damascus and the Sinai Peninsula became part of Syria; it had, until then, been the home of nomadic tribes living under arid semidesert conditions. For the next four centuries this region flourished. Paved roads were constructed on the eastern side of the River Jordan, and these connected to the east with Palmyra, to the northeast with Babylonia, and to the south with the cities of Bostra and Petra in the Sinai area. Out of the desert sands that now overflow most of these places there still rise the remains of palaces, public baths, temples, aqueducts, reservoirs, and all the material signs of a former civilization that must have demanded a rural setting for its support.

Interpreting these data leads one to conclude that the forest and savanna lands of the Middle East and of north Africa were, during the Stone Age, gradually entering a delicate state of balance in which natural replacement and maintenance would be possible so long as no extraneous factor entered to cause an unbalance. The apparent overgrazing and the fires to which Joel makes reference remind us of similar situations that developed in modern times in places such as California, temperate Australia, and New Zealand, where similar climates also occur, with dry summer. If the forest is felled, and a delicate veneer of grass established in the soil on the initial basis of the minerals liberated in the potash after the fire, such pasture can only be maintained so long as rigorous control is exerted over the density of the grazing animals. If wolves and foxes are exterminated, and rabbits gain entry, then immediate overgrazing occurs, with destruction of the turf, desiccation in the summer, and disastrous loss of topsoil, leaving only infertile subsoil, from which a desert pavement may easily develop. Winter rains drain off in sudden floods, insufficient moisture is retained in the soil (such little as there is) and an agricultural and ecological disaster is the outcome.

These modern experiences, and the modern remedies adopted, so strongly parallel what seems to be implicit in the record of north Africa and Palestine that it is easy to believe that both are aspects of the same ecosystem, one in which a delicate water balance exists. The maquis of southern France, and the similar districts called guarigue, are now thought by ecologists to represent the arid sere following human interference with an original Mediterranean oak forest (*Quercus ilex* and *Q. suber*). Similarly, in western North America it has been inferred that the chaparral vegetation dominated by *Rhamnus* and *Arctostaphylos* is a degradation phase in a subsere following the destruction by man or his client-beasts of an original oak-madrone *(Quercus-Arbutus)* forest cover. The annual bush fires of eastern Australia and California, the winter floods and mudslides and transfer of topsoil as sediment in overflowing rivers, and decline in productivity appear to be the repetition of what may have led to the desolation of the Biblical records.

THE FALL OF ROME AND THE ADVANCE OF THE DESERT. After the fall of Roman civilization written records vanish from most of Europe, save Ireland, and we are left to guess at what happened next, or to fathom it out by archeological or palynological research.

From Britain comes one telling document, written after the events that followed the barbarian invasions, but sufficiently near the period of the collapse to show what England was like in the Dark Ages. This precious manuscript forms a fragment of an Anglo-Saxon poem, sometimes known as *The Ruined City* (its original title is lost). The fragment was bound into a book known as the Codex Exoniensis, the history of which is rather precisely known, for it has lain in its original library for 900 years, ever since its last owner, Leofric, tenth bishop of Crediton, placed it there in the year 1050 or thereabouts. On that occasion he celebrated the construction of his new cathedral at Exeter by presenting to it a library of 60 books, one of which was "I mycel englisc boc be gehwilcum thingum on leoth wisan gewohrt . . ." (One large English book about various things in lay wise wrought). *The Ruined City* tells us of how a Saxon minstrel found him-

Figure 167. (a, b) *Lilium candidum* (lilies of Babylonia), as seen by an Assyrian artist of the seventh century B.C., Nineveh (3); (c) *Panicum miliaceum* (millet), as drawn by an anonymous scribe of Dioscorides, A.D. 512 (7).

self in a lonely forest and came upon a great city built by giants who wrought in stone, not in wattle and daub, and who somehow inconceivably constructed huge palaces of wondrously fashioned masonry. Works of wonder are described, stone carved in human form lies tumbled, towers in ruins. The singer goes on to speculate on what happened to the giants who built it—Fate intervened, death swept off the inhabitants, earth holds the mighty workmen, only the weed-grown, lichen-spotted walls remain. The city has been recognized from internal evidence in the poem describing wondrous hot pools, as Akemanceaster, destroyed by the Saxons in 577. So soon was this forgotten that a near-descendant of the destroyers a few generations later could not believe that men could have built such a wonder.

We may well infer that further southward, as the eastern irruption spread across north Africa, the untutored forefathers of the later splendid Moorish civilization inflicted on Roman Africa the same blow that our Teutonic forebears dealt to Britannia. Only Africa and Palestine, unlike rain-sodden Britain, could not withstand the gross disruption of the water cycle. Inevitable desiccation followed.

RESTORATION IN MODERN TIMES. The remedy is known. Israeli and Arab scientists and agriculturists are already making the desert bloom again in north Africa and Negev. There can be little question that a climatic change occurred after the retreat of the ice, and the north African and Palestinian lands became liable to desiccation; but it also appears plain that man and his beasts and his wars destroyed the balance of nature that would probably have held the sands in check. Some writers believe that Arab caravans broke the veneer of vegetation and let the ancient wind-blown sands erupt. Deserts are very old features of the earth, and their fossil remains go back for at least half a billion years. What is new is the destructive effects of man and, latterly, his alarming rate of increase in population density.

Through long experience in the conservation of their sea-beleaguered land the people of the Netherlands have discovered how to turn an advancing dune of sea sand to man's advantage. Carefully sowing phreatophytes and wind-resistant plants on the sand hills they have contrived not only to anchor the dune, but even to make it carry a forest or other standing crop. Grazing animals it may seldom carry, for in time of dry weather the beasts eat the grass down to its roots, and their habit of walking along the same tracks day after day cuts open wounds into the soil, and so the sand erupts. Not even men may walk across such reclaimed dunes, and public warnings at the toe of every anchored sea dune tell the traveler to keep to the marked trails. From the experience of the Dutch folk others have learned, and now in many parts of the world imported marram grasses and lupins from Texas hold the sands in check and make the forelands bloom. What has been done on relatively minor modules where the sea casts its sands ashore can also be done on a far greater scale on and around the vaster arid lands of the northern and southern desert belts of our planet. The Greeks have discovered in modern times how to fill a huge plastic bag with a thousand tons of fresh water, where a river debouches into the sea, and how to tow such a floating cache of fresh water across the Mediterranean to the parched Aegean Islands, where it brings fertility and new life to desolate communities. Here, too, we may expect to see pleasant oases and later even forests arise where once they also flourished before the sons of Adam set out to subdue the earth. Given peace to pursue these endeavors, and the wisdom to limit family size to the capacity of the soil, given the will to develop attractive foods from plant rather than animal proteins and carbohydrates, the flowering earth can yet awhile support fairly large populations. But in the long run the world's resources are being exhausted and the days of the apocalypse must in the end come upon us.

When this happens, the last planet of the solar system will have yielded up its free water and apparently a Martian phase will follow; the sands will ultimately spread everywhere, and no organic binder will be able to hold them in check. Lichens, apparently the first creatures to colonize the dry land, will also be the last survivors. After that only fossils will remain to record the pageant of life on the earth. Had man never evolved, this desiccation would doubtless have required several billion years to complete. The destructive effects of human technology are so great that the process has been accelerated by about seven or eight orders of magnitude, perhaps. Taken together with the various kinds of pollution—thermal, gaseous, and chemical—the life expectation of the biosphere as the abode of life is now quite limited. The one crucial question presently unanswered

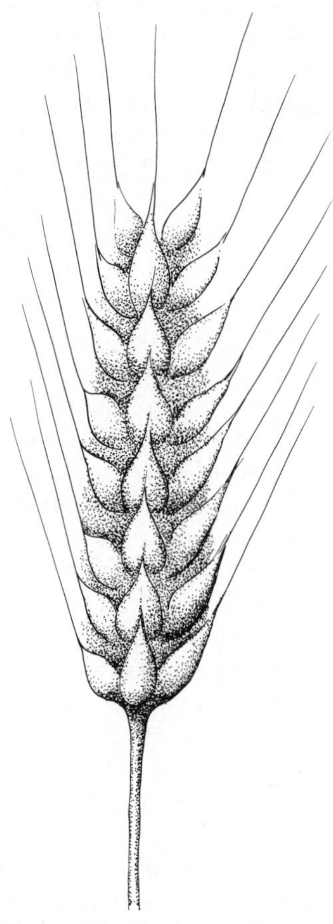

Figure 168. *Hordeum vulgare* (barley), as developed by the sixth century B.C. and engraved by one Pythagoras of Croton, son of Mnesarchos (4).

is whether man himself, already recognized by some biologists as an endangered species, will in fact succeed in extinguishing himself. If that should happen soon enough, the biosphere could probably recover from the grievous injuries that now afflict it.

A Reversed Equation?

At various times in the past, and in various civilizations, thinking men perceived the dualistic aspect of nature. The annual Nile floods that evoked the scientific sense of men in Egypt in the flourishing years of the fourth dynasty were seen to be destructive yet creative, for the silt they laid upon the fields of lower Egypt was the source of its fertility; those were the days when Egypt was the granary of the world, and the corn sprang from soil that now has been blown into the depths of the Mediterranean, its coarser particles remaining behind as the dunes of the Sahara. In India similar events were observed, and dualistic deities were conceived. Shiva, the creator and destroyer, is manifest in human guise, multibrachiate as befits one skilled in many arts. Shiva is surely man himself. If that is so, man can be thought of as only one more of a succession of evolutionary developments that in the end became self-destructive; it has happened so many times before on a lesser scale in the organic world that perhaps the present human phase is merely a larger manifestation, hopefully as transient as the others proved to be. Like some giant carnivore his very mastery may prove to be his own death warrant.

Those who take such a view—and they are now many, and they may well be correct—overlook, nonetheless, man's extraordinary adaptability. Man occupies every known biome and has even contemplated the real estate on the moon with measured glance and detailed surveys. No other organism has achieved this feat before. It is obvious that so adaptable a species can make yet one more adaptation, that of controlling the reproduction rate to match the natural resources, and thus reversing the equation that has always operated in past evolutionary history. A reversed equation could lead to totally unforeseen and unprecedented courses in evolution, and these might well hold attractive prospects. It will be a difficult task to persuade a largely illiterate world population to accept the idea that one-child or childless couples are deserving of honor as benefactors of mankind and of nature; but truth has a habit of winning men's hearts and minds, so this new ethic may well prevail. Thus the prospect for survival still seems hopeful.

The full impact of atmospheric pollution was felt in Britain in the 1950s, and after one frightening London experience that nation turned resolutely to the task of eliminating particulate contaminants from the air. Now, after twenty-five years, the lovely old buildings of Christopher Wren and the stately cathedral he created have begun to emerge from the black grime of smoke deposits, and the Portland stone gleams bright against a clearer sky; only the ugliness of twentieth-century skyscrapers mars the once charming skyline. American cities fall in a different category; desolate canyons often fifty floors down below the rooftops are called streets—streets where the sun penetrates for barely one hour a day and few trees could survive even if there were room for them. But it would be a mistake to think of these man-made deserts as typical, or to judge the urban ecology by such examples. Here it is, I think, the visitor to the land who sees most clearly the achievement of American town planners. For the true urban America is not the concentrated skyscraper metropolis, but the host of beautiful towns that are scattered halo-wise around each of the business centers.

American town planners have, during the past two centuries, rediscovered the lost art of the Bronze Age peoples—how to live in a forest, without destroying the trees. More accurately, Americans have shown

Figure 169. *Cupressus sempervirens* (Mediterranean cypress), which has endured the devastation of man and still graces ancient Greek temples and Italian palaces. Silhouette in Florence by Sandro Botticelli, *ca.* 1470.

modern man how to create a forest within and around the places where one lives, does much business, where banks and universities and shopping centers lie, none striving to rise many floors above the ground level, all intent on the planting and maintaining of what seem to be endless avenues of fine trees along both sides of every street. Where else in the world does one find towns that regularly maintain a separate municipal tree department, whose sole responsibility is to plant and care for trees at the taxpayers' expense! An aerial view of the average northeastern city or town discloses more trees than houses, indeed most houses are invisible from the air save in winter, for the dwellings lie under a canopy of foliage. Whereas in most countries constant warfare is waged between electricity suppliers and tree lovers, the former forbidding the latter to bring offending branches anywhere near the sacred power line—in America, on the other hand, I am told by a tree department friend, "we consult all the time with the power corporations as to how to coordinate our services, and we have good cooperation."

In the charming New England town that has been my family's home for ten years the power lines pass from tree to tree, looping under the boughs and, far from being hideous, constitute major cross-town migration ways for the countless squirrels who share the woodlands with the human beings whose houses are concealed beneath the oaks, maples, and ashes. Added to these woodland amenities is the sheer beauty of the eighteenth-century style of classical architecture so favored in eastern America, and so much admired by visitors from abroad. The woodlands are so extensive that one may travel by train for the whole length of Megalopolis—that is from Boston to Washington, some five hundred miles—and save for about five metropolitan wastelands of only ten miles or so maximum diameter, the whole of the journey appears to be conducted through charming woodlands (second growth, of course) and peaceful countryside.

No, for the visitor—whether he be from Europe or from the Antipodes—North America is not to be compared with the deadly industrial horrorscapes of the other side of the Atlantic—not yet, nor probably ever. Here trees are supreme, and a town may be torn apart by factional strife over when to spray or whether to spray, and the insect-borne viruses that afflict elms and chestnuts often arouse more community concern than a threat of some other kind. Some industrial development within the past five years has at last begun to follow the shining precepts set by the trading and dwelling sections of the community. For example, a substantial part of the advanced electronics industry, whose technology made the space program a reality, is now located in low-roofed, inconspicuous buildings, each surrounded by groves of planted trees, or by copse left over from the former second-growth woods. Along one pleasing highway where, I am told, some 160,000 engineers have their homes, schools, laboratories, and factories, one may travel in near oblivion of the surrounding and interpenetrating human activity. In such surroundings, unspoiled by noise or ugliness, people can live and work in dignity, and animals from the remoter woodlands not infrequently wander by without molestation. Russian correspondents tell me that something similar has been undertaken near Lake Baikal. These are surely hopeful signs that many thinking people are now insisting that natural beauty be a part of normal existence, just as did the ancient Egyptians, who were perhaps the most skillful workers in stone and bronze that the world has ever known.

I began this book by speaking of the achievements of Roman engineers as friends of the environment under the aegis of the noblest of the Roman emperors. Let me end by these references to an ethos that may well capture the minds of young engineers of our own day—how best to develop that art which conceals art, and how to work with the natural environment without marring its beauty or harming its denizens. For of such is the genius of our age.

Biographic Index

Subject Index

74 75 76 77 9 8 7 6 5 4 3 2 1